CAMBRIDGE LIBRARY COLLECTION

Books of enduring scholarly value

Physical Sciences

From ancient times, humans have tried to understand the workings of the world around them. The roots of modern physical science go back to the very earliest mechanical devices such as levers and rollers, the mixing of paints and dyes, and the importance of the heavenly bodies in early religious observance and navigation. The physical sciences as we know them today began to emerge as independent academic subjects during the early modern period, in the work of Newton and other 'natural philosophers', and numerous sub-disciplines developed during the centuries that followed. This part of the Cambridge Library Collection is devoted to landmark publications in this area which will be of interest to historians of science concerned with individual scientists, particular discoveries, and advances in scientific method, or with the establishment and development of scientific institutions around the world.

A Popular History of Astronomy During the Nineteenth Century

Agnes Mary Clerke (1842-1907) first published *A Popular History of Astronomy* in 1885. The work was received with widespread acclaim and brought Clerke an international reputation as a science writer. The *History* surveys the progress made in the field of astronomy during the nineteenth century. It is split chronologically into two parts, dealing with the first and the second half of the century. Part 1 focuses on the career of the astronomer William Herschel (1738–1822) and the development of sidereal astronomy; part 2 deals with the discovery of spectrum analysis and the progress of knowledge about sun spots and the magnetic disturbances which cause them. Clerke's work, a classic example of Victorian popular scientific literature, stands alongside Grant's earlier *History of Astronomy* in its success in popularising the subject. The work is important today for scholars researching the history of the discipline and its place in educated Victorian society.

A Popular History of
Astronomy During the
Nineteenth Century

AGNES MARY CLERKE

CAMBRIDGE
UNIVERSITY PRESS

CAMBRIDGE UNIVERSITY PRESS

Cambridge, New York, Melbourne, Madrid, Cape Town, Singapore,
São Paolo, Delhi, Dubai, Tokyo

Published in the United States of America by Cambridge University Press, New York

www.cambridge.org
Information on this title: www.cambridge.org/9781108014328

This edition first published 1885
This digitally printed version 2010

ISBN 978-1-108-01432-8 Paperback

HISTORY OF ASTRONOMY.

A POPULAR

HISTORY OF ASTRONOMY

DURING

THE NINETEENTH CENTURY

BY

AGNES M. CLERKE

EDINBURGH: ADAM & CHARLES BLACK

MDCCCLXXXV.

PREFACE.

———◆———

THE progress of astronomy during the last hundred years has
been rapid and extraordinary. In its distinctive features,
moreover, the nature of that progress has been such as to lend
itself with facility to untechnical treatment. To this circumstance the present volume owes its origin. It embodies an
attempt to enable the ordinary reader to follow, with intelligent interest, the course of modern astronomical inquiries,
and to realise (so far as it can at present be realised) the full
effect of the comprehensive change in the whole aspect, purposes, and methods of celestial science introduced by the
momentous discovery of spectrum analysis.

Since Professor Grant's invaluable work on the *History of
Physical Astronomy* was published, a third of a century has
elapsed. During the interval, a so-called "new astronomy"
has grown up by the side of the old. One effect of its advent
has been to render the science of the heavenly bodies more
popular, both in its needs and in its nature, than formerly.
More popular in its needs, since its progress now primarily
depends upon the interest in, and consequent efforts towards
its advancement of the general public; more popular in its
nature, because the kind of knowledge it now chiefly tends to
accumulate is more easily intelligible—less remote from ordinary experience—than that evolved by the aid of the calculus
from materials collected by the use of the transit-instrument
and chronograph.

It has thus become practicable to describe in simple language
the most essential parts of recent astronomical discoveries ;
and being practicable, it could not be otherwise than desirable
to do so. The service to astronomy itself would be not inconsiderable of enlisting wider sympathies on its behalf; while
to help one single mind towards a fuller understanding of the
manifold works which have, in all ages, irresistibly spoken to
man of the glory of God, might well be an object of no ignoble
ambition.

The present volume does not profess to be a complete or
exhaustive History of Astronomy during the period covered
by it. Its design is to present a view of the progress of
celestial science, on its most characteristic side, since the time
of Herschel. Abstruse mathematical theories, unless in some
of their more striking results, are excluded from consideration.
These, during the eighteenth century, constituted the sum and
substance of astronomy; and their fundamental importance
can never be diminished, and should never be ignored. But,
as the outcome of the enormous development given to the
powers of the telescope in recent times, together with the
swift advances of physical science, and the inclusion, by means
of the spectroscope, of the heavenly bodies within the domain
of its inquiries, much knowledge has been acquired regarding
the nature and condition of those bodies, forming, it might be
said, a science apart, and disembarrassed from immediate dependence upon intricate, and, except to the initiated, unin-
telligible formulæ. This kind of knowledge forms the main
subject of the book now offered to the public.

There are many reasons for preferring a history to a formal
treatise on astronomy. In a treatise, *what* we know is set
forth. A history tells us, in addition, *how* we came to know
it. It thus places facts before us in the natural order of their
ascertainment, and narrates instead of enumerating. The story

to be told leaves the marvels of imagination far behind, and requires no embellishment from literary art or high-flown phrases. Its best ornament is unvarnished truthfulness, and this at least may confidently be claimed to be bestowed upon it in the ensuing pages.

In them unity of treatment is sought to be combined with a due regard to chronological sequence by grouping in separate chapters the various events relating to the several departments of descriptive astronomy. The whole is divided into two parts, the line between which is roughly drawn at the middle of the present century. Herschel's inquiries into the construction of the heavens strike the keynote of the first part ; the discovery of sun-spot and magnetic periodicity and of spectrum analysis, determine the character of the second. Where the nature of the subject required it, however, this arrangement has been disregarded. Clearness and consistency should obviously take precedence of method. Thus, in treating of the telescopic scrutiny of the various planets, the whole of the related facts have been collected into an uninterrupted narrative. A division, elsewhere natural and helpful, would here have been purely artificial, and therefore confusing.

The interests of students have been consulted by a full and authentic system of references to the sources of information relied upon. Materials have been derived, as a rule with very few exceptions, from the original authorities. The system adopted has been to take as little as possible at second-hand. Much pains have been taken to trace the origin of ideas, often obscurely enunciated long before they came to resound through the scientific world, and to give to each individual discoverer, strictly and impartially, his due. Prominence has also been assigned to the biographical element, as underlying and determining the whole course of human endeavour. The advance of knowledge may be called a vital process. The lives of men are absorbed into and assimilated by it. Inquiries into the kind and mode of the surrender in each separate case must always possess a strong interest, whether for study or for example.

The acknowledgments of the writer are due to Professor Edward S. Holden, director of the Washburn Observatory, Illinois, and to Dr. Copeland, chief astronomer of Lord Crawford's Observatory at Dunecht, for many valuable communications.

CONTENTS.

INTRODUCTION.

Part I.

PROGRESS OF ASTRONOMY DURING THE FIRST HALF OF THE NINETEENTH CENTURY.

CHAPTER I.

FOUNDATION OF SIDEREAL ASTRONOMY.

CHAPTER II.

PROGRESS OF SIDEREAL ASTRONOMY.

CHAPTER III.

PROGRESS OF KNOWLEDGE REGARDING THE SUN.

CHAPTER IV.

PLANETARY DISCOVERIES.

CHAPTER V.

COMETS.

CHAPTER VI.

INSTRUMENTAL ADVANCES.

𝔓art 𝔌𝔌.

RECENT PROGRESS OF ASTRONOMY.

CHAPTER I.

FOUNDATION OF ASTRONOMICAL PHYSICS.

CHAPTER II.

SOLAR OBSERVATIONS AND THEORIES.

CHAPTER III.

RECENT SOLAR ECLIPSES.

CHAPTER IV.

SPECTROSCOPIC WORK ON THE SUN.

CHAPTER IX.

THEORIES OF PLANETARY EVOLUTION.

CHAPTER X.

RECENT COMETS.

CHAPTER XI.

RECENT COMETS (*continued*).

CHAPTER XII.

STARS AND NEBULÆ.

CONTENTS.

CHAPTER XIII.

METHODS OF RESEARCH.

ERRATA.

Page 4, line 3 from bottom, *for* 1748, *read* 1758.

Page 183, line 10 from bottom, *for* "the sciences," *read* "the physical sciences."

HISTORY OF ASTRONOMY

DURING THE NINETEENTH CENTURY.

INTRODUCTION.

WE can distinguish three kinds of astronomy, each with a different origin and history, but all mutually dependent, and composing, in their fundamental unity, one science. First in order of time came the art of observing the returns and measuring the places of the heavenly bodies. This was the sole astronomy of the Chinese and Chaldeans; but to it the vigorous Greek mind added a highly complex geometrical plan of their movements, for which Copernicus substituted a more harmonious system, without as yet any idea of a compelling cause. The planets revolved in circles because it was their nature to do so, just as laudanum sets to sleep because it possesses a *virtus dormitiva*. This first and oldest branch is known as "observational," or "practical astronomy." Its business is to note facts as accurately as possible; and it is essentially unconcerned with schemes for connecting those facts in a manner satisfactory to the reason.

The second kind of astronomy was founded by Newton. Its nature is best indicated by the term "gravitational;" but it is also called "theoretical astronomy."[1] It is based on the idea of cause; and the whole of its elaborate structure is reared

---

[1] The denomination "physical astronomy," first used by Kepler, and long appropriated to this branch of the science, has of late been otherwise applied.

according to the dictates of a single law, simple in itself, but the tangled web of whose consequences can be unravelled only by the subtle agency of an elaborate calculus.

The third and last division of celestial science may properly be termed "physical and descriptive astronomy." It seeks to know what the heavenly bodies are in themselves, leaving the How? and the Wherefore? of their movements to be otherwise answered. Now such inquiries became possible only with the invention of the telescope, so that Galileo was, in point of fact, their originator. But Herschel was the first to give them a prominence which the whole progress of science during the nineteenth century has served to confirm and render more exclusive. Inquisitions begun with the telescope have been extended and made effective in unhoped-for directions by the aid of the spectroscope and photographic camera; and a large part of our attention in the present volume will be occupied with the brilliant results thus achieved.

The unexpected development of this new physical-celestial science is the leading fact in recent astronomical history. It was out of the regular course of events. In the degree in which it has actually occurred it could certainly not have been foreseen. It was a seizing of the prize by a competitor who had hardly been thought qualified to enter the lists. Orthodox astronomers of the old school looked with a certain contempt upon observers who spent their nights in scrutinising the faces of the moon and planets, rather than in timing their transits, or devoted daylight energies, not to reductions and computations, but to counting and measuring spots on the sun. They were regarded as irregular practitioners, to be tolerated perhaps, but certainly not encouraged.

The advance of astronomy in the eighteenth century ran in general an even and logical course. The age succeeding Newton's had for its special task to demonstrate the universal validity, and trace the complex results of the law of gravitation. Its accomplishment occupied just one hundred years. It was virtually brought to a close when Laplace explained to the

French Academy, November 19, 1787, the cause of the moon's accelerated motion. As a mere machine, the solar system, so far as it was then known, was found to be complete and intelligible in all its parts; and in the *Mécanique Céleste* its mechanical perfections were displayed under a form of majestic unity which fitly commemorated the successive triumphs of analytical genius over problems amongst the most arduous ever dealt with by the mind of man.

Theory, however, demands a practical test. All its data are derived from observation; and their insecurity becomes less tolerable as it advances nearer to perfection. Observation, on the other hand, is the pitiless critic of theory; it detects weak points, and provokes reforms which may be the beginnings of discovery. Thus, theory and observation mutually act and react, each alternately taking the lead in the endless race of improvement.

Now, while in France Lagrange and Laplace were bringing the gravitational theory of the solar system to completion, work of a very different kind, yet not less indispensable to the future welfare of astronomy, was being done in England. The Royal Observatory at Greenwich is one of the few useful institutions which date their origin from the reign of Charles II. The leading position which it still occupies in the science of celestial observation was, for near a century and a half after its foundation, an exclusive one. It was absolutely without a rival. Systematic observations of sun, moon, stars, and planets were, during the whole of the eighteenth century, made only at Greenwich. Here materials were accumulated for the secure correction of theory, and here refinements were introduced by which the exquisite accuracy of modern practice in astronomy was eventually attained.

The chief promoter of these improvements was James Bradley. Few men have possessed in an equal degree with him the power of seeing accurately, and reasoning on what they see. He let nothing pass. The slightest inconsistency between what appeared and what was to be expected, roused his keenest attention; and he never relaxed his mental grip of a subject until it had yielded to his persistent inquisition. It was to these qualities that he owed his discoveries of the aberration of light and the nutation of the earth's axis. The first was announced in 1729. It means that, owing to the circumstance of light not being instantaneously transmitted, the heavenly bodies appear shifted from their true places by an amount depending upon the ratio which the velocity of light bears to the speed of the earth in its orbit. Because light travels with enormous rapidity, the shifting is very slight; and each star returns to its original position at the end of a year.

Bradley's second great discovery was finally ascertained in 1748. Nutation is a real "nodding" of the terrestrial axis produced by the dragging of the moon at the terrestrial equatorial protuberance. From it results an *apparent* displacement of the stars, each of them describing a little ellipse about its true, or "mean" position, in a period of eighteen years and about seven months.

Now an acquaintance with the fact and the laws of each of these minute irregularities is vital to the progress of observational astronomy; for without it the places of the heavenly bodies could never be accurately known or compared. So that Bradley, by their detection, at once raised the science to a higher grade of precision. Nor was this the whole of his work. Appointed Astronomer-Royal in 1742, he executed during the years 1750–62 a series of observations which formed the real beginning of exact astronomy. Part of their superiority must, indeed, be attributed to the co-operation of John Bird, who provided Bradley in 1750 with a measuring instrument of till then unequalled excellence. For not only was the art of observing in the eighteenth century a peculiarly English art, but the means of observing were furnished almost exclusively by British artists. John Dollond, the son of a Spitalfields weaver, invented the achromatic lens in 1748, removing thereby the chief obstacle to the development of the powers of refracting telescopes; James Short, of Edinburgh, was without a rival in the construction of reflectors; the sectors, quadrants, and circles of Graham, Bird, Ramsden, and Cary were inimitable by Continental workmanship.

Thus practical and theoretical astronomy advanced on parallel lines in England and France respectively, the improvement of their several tools—the telescope and the quadrant on the one side, and the calculus on the other—keeping pace. The whole future of the science seemed to be theirs. The cessation of interest through a too speedy attainment of the perfection towards which each spurred the other, appeared to be the only danger it held in store for them. When, all at once, a rival stood by their side—not, indeed, menacing their progress, but threatening to absorb their popularity.

The rise of Herschel was the one conspicuous anomaly in the astronomical history of the eighteenth century. It proved decisive of the course of events in the nineteenth. It was unexplained by anything that had gone before; yet all that came after hinged upon it. It gave a new direction to effort; it lent a fresh impulse to thought. It opened a channel for the widespread public interest which was gathering towards astronomical subjects to flow in.

Much of this interest was due to the occurrence of events calculated to arrest the attention and excite the wonder of the uninitiated. The predicted return of Halley's comet in 1759 verified, after an unprecedented fashion, the computations of astronomers. It deprived such bodies for ever of their portentous character; it ranked them as denizens of the solar system. Again, the transits of Venus in 1761 and 1769 were the first occurrences of the kind since the awakening of science to their consequence. Imposing preparations, journeys to remote and hardly accessible regions, official expeditions, international communications, all for the purpose of observing them to the best advantage, brought their high significance vividly to the public consciousness; a result aided by the facile pen of Lalande, in rendering intelligible the means by which these elaborate arrangements were to issue in an accurate knowledge of the sun's distance. Lastly, Herschel's discovery of Uranus, March 13, 1781, had the surprising effect of utter novelty. Since the human race had become acquainted with the company of the planets, no addition had been made to their number. The event thus broke with immemorial traditions, and seemed to show astronony as still young, and full of unlooked-for possibilities.

Further popularity accrued to the science from the sequel of a career so strikingly opened. Herschel's huge telescopes, his detection by their means of two Saturnian and as many Uranian moons, his piercing scrutiny of the sun, picturesque theory of its constitution, and sagacious indication of the route pursued by it through space; his discovery of stellar revolving systems, his bold soundings of the universe, his grandiose ideas, and the elevated yet simple language in which they were con-veyed—formed a combination powerfully effective to those least susceptible of new impressions. Nor was the evoked enthusiasm limited to the British Isles. In Germany, Schröter followed—*longo intervallo*—in Herschel's track. Von Zach set on foot from Gotha that general communication of ideas which gives life to a forward movement. Bode wrote much and well for unlearned readers. Lalande, by his popular lec-tures and treatises, helped to form an audience which Laplace himself did not disdain to address in the *Exposition du Système du Monde.*

This great accession of popularity gave the impulse to the extraordinarily rapid progress of astronomy in the nineteenth century. Official patronage combined with individual zeal sufficed for the elder branches of the science. A few well-endowed institutions could accumulate the materials needed by a few isolated thinkers for the construction of theories of wonderful beauty and elaboration, yet precluded, by their abstract nature, from winning general applause. But the new physical astronomy depends for its prosperity upon the favour of the multitude whom its striking results are well fitted to attract. It is, in a special manner, the science of amateurs. It welcomes the most unpretending co-operation. There is no one "with a true eye and a faithful hand" but can do good work in watching the heavens. And not unfrequently prizes of discovery which the most perfect appliances failed to grasp have fallen to the share of ignorant or ill-provided assiduity.

Observers, accordingly, have multiplied; observatories have been founded in all parts of the world; associations have been constituted for mutual help and counsel. A formal astronomical congress met in 1798 at Gotha—then, under Duke Ernest II. and Von Zach, the focus of German astronomy—and instituted a combined search for the planet suspected to revolve undiscovered between the orbits of Mars and Jupiter. The Astronomical Society of London was established in 1820, and the similar German institution in 1863. Both have been highly influential in promoting the interests, local and general, of the science they are devoted to forward; while functions corresponding to theirs have been discharged elsewhere by older or less specially constituted bodies, and new ones are springing up on all sides.

Modern facilities of communication have helped to impress more deeply upon modern astronomy its associative character. The electric telegraph gives a certain ubiquity which is invaluable to an observer of the skies. With the help of a wire, a battery, and a code of signals, he sees whatever is visible from any portion of our globe, depending, however, upon other eyes than his own, and so entering as a unit into a widespread organisation of intelligence. The press, again, has been a potent agent of co-operation. It has mainly contributed to unite astronomers all over the world into a body animated by the single aim of collecting "particulars" in their special branch for what Bacon termed a History of Nature, eventually to be interpreted according to the sagacious insight of some one among them gifted above his fellows. The first really effective astronomical periodical was the *Monatliche Correspondenz*, started by Von Zach in the year 1800. It was followed in 1822 by the *Astronomische Nachrichten,* later by the *Memoirs* and *Monthly Notices* of the Astronomical Society, and by the host of varied publications which now, in every civilised country, communicate the discoveries made in astronomy to divers classes of readers, and so incalculably quicken the current of its onward flow.

Public favour brings in its train material resources. It is represented by individual enterprise, and finds expression in an ample liberality. The first regular observatory in the southern hemisphere was founded at Paramatta by Sir Thomas Makdougall Brisbane in 1821. The Royal Observatory at the Cape of Good Hope was completed in 1829. Similar establishments were set to work by the East India Company at Madras, Bombay, and St. Helena, during the first third of the nineteenth century. The organisation of astronomy in the United States of America was due to a strong wave of popular enthusiasm. In 1825 John Quincy Adams vainly urged upon Congress the foundation of a National Observatory; but in 1843 the lectures of Ormsby MacKnight Mitchel on celestial phenomena stirred an impressionable audience to the pitch of providing him with the means of erecting at Cincinnati the first astronomical establishment worthy the name in that great country. On the 1st of January 1882 no less than one hundred and forty-four were active within its boundaries.

The apparition of the great comet of 1843 gave an additional fillip to the movement. To the excitement caused by it the Cambridge Observatory—called the "American Pulkowa"—directly owed its origin; and the example was not ineffective elsewhere. Corporations, universities, municipalities, vied with each other in the creation of similar institutions; private subscriptions poured in; emissaries were sent to Europe to purchase instruments and procure instruction in their use. In a few years the young Republic was, in point of astronomical efficiency, at least on a level with countries where the science had been fostered since the dawn of civilisation.

A vast widening of the scope of astronomy has accompanied,

and in part occasioned, the great extension of its area of cultivation which our age has witnessed. In the last century its purview was a comparatively narrow one. Problems lying beyond the range of the solar system were almost unheeded, because they seemed inscrutable. Herschel first showed the sidereal universe as accessible to investigation, and thereby offered to science new worlds—majestic, manifold, "infinitely infinite" to our apprehension in number, variety, and extent—for future conquest. Their gradual appropriation has absorbed, and will long continue to absorb, the powers which it has served to develop.

But this is not the only direction in which astronomy has enlarged, or rather has levelled, its boundaries. The unification of the physical sciences is perhaps the greatest intellectual feat of recent times. The process has included astronomy; so that, like Bacon, she may now be said to have "taken all knowledge" (of that kind) "for· her province." In return, she proffers potent aid for its increase. Every comet that approaches the sun is the scene of experiments in the electrical illumination of rarefied matter, performed on a huge scale for our benefit. The sun, stars, and nebulæ form so many celestial laboratories, where the nature and mutual relations of the chemical "elements" may be tried by more stringent tests than sublunary conditions afford. The laws of terrestrial magnetism can be completely investigated only with the aid of a concurrent study of the face of the sun. The positions of the planets will perhaps one day tell us something of impending droughts, famines, and cyclones.

Astronomy generalises the results of the other sciences. She exhibits the laws of Nature working over a wider area, and under more varied conditions, than ordinary experience presents. Ordinary experience, on the other hand, has become indispensable to her progress. She takes in at one view the indefinitely great and the indefinitely little. The mutual revolutions of the stellar multitude during tracts of time which seem to lengthen out to eternity as the mind attempts to traverse them, she does not admit to be beyond her ken; nor is she indifferent to the constitution of the minutest atom of matter that thrills the ether into light. How she entered upon this vastly expanded inheritance, and how, so far, she has dealt with it, is attempted to be set forth in the ensuing chapters.

PART I.

PROGRESS OF ASTRONOMY DURING THE FIRST HALF
OF THE NINETEENTH CENTURY.

——◆——

CHAPTER I.

FOUNDATION OF SIDEREAL ASTRONOMY.

UNTIL nearly a hundred years ago the stars were regarded by
practical astronomers mainly as a number of convenient fixed
points by which the motions of the various members of the
solar system could be determined and compared. Their
recognised function, in fact, was that of milestones on the
great celestial highway traversed by the planets, as well as
on the byeways of space occasionally pursued by comets.
Not that curiosity as to their nature, and even conjecture as
to their origin, were at any period absent. Both were from
time to time powerfully stimulated by the appearance of
startling novelties in a region described by philosophers as
"incorruptible," or exempt from change. The catalogue of
Hipparchus probably, and certainly that of Tycho Brahe,
some seventeen centuries later, owed each its origin to the
temporary blaze of a new star. The general aspect of the
skies was thus (however imperfectly) recorded from age to
age, and with improved appliances the enumeration was
rendered more and more accurate and complete; but the
secrets of the stellar sphere remained inviolate.

In a qualified, though very real sense, Sir William Herschel may be called the Founder of Sidereal Astronomy. Before his time some curious facts had been noted, and some ingenious speculations hazarded, regarding the condition of the stars, but not even the rudiments of systematic knowledge had been acquired. The facts ascertained can be summed up in a very few sentences.

Giordano Bruno was the first to set the suns of space in motion, but in imagination only. His daring surmise was, however, confirmed in 1718, when Halley announced [1] that Sirius, Aldebaran, Betelgeux, and Arcturus had unmistakably shifted their quarters in the sky since Ptolemy assigned their places in his catalogue. A similar conclusion was reached by J. Cassini in 1738, from a comparison of his own observations with those made at Cayenne by Richer in 1672 ; and Tobias Mayer drew up in 1756 a list showing the direction and amount of about fifty-seven proper motions,[2] founded on star-places determined by Olaus Römer fifty years previously. Thus the stars were no longer regarded as "fixed," but the question remained whether the movements perceived were real or only apparent ; and this it was not yet found possible to answer. Already, in the previous century, the ingenious Robert Hooke had suggested an "alteration of the very system of the sun," [3] to account for certain suspected changes in stellar positions ; Bradley in 1748, and Lambert in 1761, pointed out that such apparent displacements (by that time well ascertained) were in all probability a combined effect of motions, both of sun and stars ; and Mayer actually attempted the analysis, but without result.

On the 13th of August 1596, David Fabricius, an unprofessional astronomer in East Friesland, saw in the neck of the Whale a star of the third magnitude, which by October had

[1] *Phil. Trans.*, vol. xxx. p. 737.

[2] Out of eighty stars compared, fifty-seven were found to have changed their places by more than 10″. Lesser discrepancies were at that time regarded as falling within the limits of observational error. *Tobiæ Mayeri, Op. Inedita*, t. i. pp. 80–81, and Herschel in *Phil. Trans.*, vol. lxxiii. pp. 275–278.   [3] *Posthumous Works*, p. 506.

disappeared. It was, however, visible in 1603, when Bayer marked it in his catalogue with the Greek letter o, and was watched through its phases of brightening and apparent extinction by a German professor named Holwarda in 1638–39.[1] From Hevelius this first known periodical star received the name of "Mira," or the Wonderful, and Boulliaud, in 1667, fixed the length of its cycle of change at 334 days. It was not a solitary instance. A star in the Swan was perceived by Janson in 1600 to show fluctuations of light, and Montanari found in 1669 that Algol in Perseus shared the same peculiarity to a marked degree. Altogether the class embraced in 1782 half a dozen members. When it is added that a few star-couples had been noted in singularly, but it was supposed accidentally, close juxtaposition, and that the failure of repeated attempts to find an annual parallax pointed to distances *at least* 400,000 times that of the earth from the sun,[2] the picture of sidereal science, when the last quarter of the eighteenth century began, is practically complete. It included three items of information—that the stars have motions, real or apparent; that they are immeasurably remote; and that a few shine with a periodically variable light. Nor were the facts thus scantily collected ordered into any promise of further development. They lay at once isolated and confused before the inquirer. They needed to be both multiplied and marshalled, and it seemed as if centuries of patient toil must elapse before any reliable conclusions could be derived from them. The sidereal world was thus the recognised domain of far-reaching speculations, which remained wholly uncramped by systematic research until Herschel entered upon his career as an observer of the heavens.

The greatest of modern astronomers was born at Hanover, November 15, 1738. He was the fourth child of Isaac

---

[1] Arago in *Annuaire du Bureau des Longitudes*, 1842, p. 313.

[2] Bradley to Halley, *Phil. Trans.*, vol. xxxv. (1728), p. 660. His observations were directly applicable to only two stars, γ Draconis and η Ursæ Majoris, but some lesser ones were included in the same result.

Herschel, a hautboy-player in the band of the Hanoverian
Guard, and was early trained to follow his father's profession.
On the termination, however, of the disastrous campaign of
1757, his parents removed him from the regiment, and he
went to England to seek his fortune. He was then nearly
nineteen, his military service having lasted four years. Of the
life of struggle and privation which ensued little is known beyond
the circumstances that in 1760 he was engaged in training the
regimental band of the Durham Militia, and that in 1765 he
was appointed organist at Halifax. This post he exchanged
a year later for the more distinguished one of organist at the
Octagon Chapel in Bath. The tide of prosperity now began
to flow for him. The most brilliant and modish society in
England was at that time to be met at Bath, and the young
Hanoverian quickly found himself a favourite and the fashion
in it. Engagements multiplied upon him. He became
director of the public concerts; he conducted oratorios, engaged singers, organised rehearsals, composed anthems, chants,
choral services, besides undertaking private tuitions, at times
amounting to thirty-five or even thirty-eight lessons a week.
He in fact personified the musical activity of a place then
eminently and energetically musical.

But these multifarious avocations did not take up the whole
of his thoughts. His education, notwithstanding the poverty
of his family, had not been neglected, and he had always
greedily assimilated every kind of knowledge that came in his
way. Now that he was a busy and a prosperous man, it might
have been expected that he would run on in the deep professional groove laid down for him. On the contrary, his
passion for learning seemed to increase with the diminution of
the time available for its gratification. He studied Italian,
Greek, mathematics; Maclaurin's Fluxions served to "unbend
his mind;" Smith's Harmonics and Optics and Ferguson's
Astronomy were the nightly companions of his pillow. What
he read stimulated without satisfying his intellect. He desired
not only to know, but to discover. In 1773 he hired a

small telescope, and through it caught a preliminary glimpse of the rich and varied fields in which, for so many years, he was to expatiate. Henceforward the purpose of his life was fixed. It was to obtain "a knowledge of the construction of the heavens;"[1] and to this sublime ambition he remained true until the end.

A more powerful instrument was the first desideratum; and here his mechanical genius came to his aid. Having purchased the apparatus of a Quaker optician, he set about the manufacture of specula with a zeal which seemed to anticipate the wonders they were to disclose to him. It was not until fifteen years later that his grinding and polishing machines were invented, so the work had at that time to be entirely done by hand. During this tedious and laborious process (which could not be interrupted without injury, and lasted on one occasion sixteen hours), his strength was supported by morsels of food put into his mouth by his sister,[2] and his mind amused by her reading aloud to him the Arabian Nights, Don Quixote, or other light works. At length, after repeated failures, he found himself provided with a reflecting telescope—a five-foot Newtonian—of his own construction. A copy of his first observation with it, on the great Nebula in Orion—an object of continual amazement and assiduous inquiry to him—is preserved by the Royal Society. It bears the date March 4, 1774.[3]

In the following year he executed his first "review of the heavens," memorable chiefly as an evidence of the grand and novel conceptions which already inspired him, and of the enthusiasm with which he delivered himself up to their guidance. Overwhelmed with professional engagements, he still contrived to snatch some moments for the stars; and between

---

[1] *Phil. Trans.*, vol. ci. p. 269.

[2] Caroline Lucretia Herschel, born at Hanover, March 16, 1750, died in the same place, January 9, 1848. She came to England in 1772, and was her brother's devoted assistant, first in his musical undertakings, and afterwards, down to the end of his life, in his astronomical labours.

[3] Holden, *Sir William Herschel, his Life and Works*, p. 39.

the acts at the theatre was often seen running from the harpsi-
chord to his telescope, no doubt with that "uncommon precipi-
tancy which accompanied all his actions."[1]   He now rapidly
increased the power and perfection of his telescopes.   Mirrors
of seven, ten, even twenty feet focal length were successively
completed, and unprecedented magnifying powers employed.
His energy was unceasing, his perseverance indomitable.   In
the course of twenty-one years no less than 430 parabolic
specula left his hands.   He had entered upon his forty-second
year when he sent his first paper to the *Philosophical Transactions ;* yet during the ensuing thirty-nine years his contributions—many of them elaborate treatises—numbered sixty-nine,
forming a series of extraordinary importance to the history of
astronomy.   As a mere explorer of the heavens his labours
were prodigious.   He discovered 2500 nebulæ, 806 double stars,
passed the whole firmament in review four several times, counted
the stars in 3400 "gauge-fields," and executed a photometric
classification of the principal stars, founded on an elaborate
(and the first systematically conducted) investigation of their
relative brightness.   He was as careful and patient as he was
rapid ; spared no time and omitted no precaution to secure
accuracy in his observations ; yet in one night he would
examine, singly and carefully, up to 400 separate objects.

    The discovery of Uranus was a mere incident of the scheme
he had marked out for himself—a fruit gathered, as it were,
by the way.   It formed, nevertheless, the turning-point in his
career.   From a star-gazing musician he was at once transformed into an eminent astronomer ; he was relieved from
the drudgery of a toilsome profession, and installed as royal
astronomer, with a modest salary of £200 a year ; funds were
provided for the construction of the forty-foot reflector, from
the great space-penetrating power of which he expected as yet
unheard-of revelations ; in fine, his future work was not only
rendered possible, but it was stamped as authoritative.[2]   On

---

[1] *Memoir of Caroline Herschel*, p. 37.
[2] See Holden's *Sir William Herschel*, p. 54.

Whit-Sunday 1782, William and Caroline Herschel played and sang in public for the last time in St. Margaret's Chapel, Bath ; in August of the same year the household was moved to Datchet, near Windsor, and on April 3, 1786, to Slough. Here happiness and honours crowded on the fortunate discoverer. In 1788 he married Mary, only child of James Baldwin, a merchant of the city of London, and widow of John Pitt, Esq.,—a lady endowed not only with all the domestic virtues, but with a large share of more substantial, though less precious goods. The fruit of their union was one son, of whose work—the worthy sequel of his father's— we shall have to speak further on. Herschel was created a Knight of the Hanoverian Guelphic Order in 1816, and in 1821 he became the first President of the Royal Astronomical Society, his son being its first Foreign Secretary. But his health had now for some years been failing, and on August 25, 1822, he died at Slough, in the eighty-fourth year of his age, and was buried in Upton churchyard.

His epitaph claims for him the lofty praise of having "burst the barriers of heaven." Let us see in what sense this is true.

The first to form any definite idea as to the constitution of the stellar system was Thomas Wright, the son of a carpenter living at Byer's Green, near Durham. With him originated what has been called the "Grindstone Theory" of the universe, which regarded the Milky Way as the projection on the sphere of a stratum or disc of stars (our sun occupying a position near the centre), similar in magnitude and distribution to the lucid orbs of the constellations.[1] He was followed by Kant,[2] who transcended the views of his predecessor by assigning to nebulæ the position they long continued to occupy, rather on imaginative than on scientific grounds, of "island universes,"

---

[1] *An Original Theory or New Hypothesis of the Universe*, London, 1750. See also De Morgan's summary of his views in *Philosophical Magazine*, April 1848.

[2] *Allgemeine Naturgeschichte und Theorie des Himmels*, 1755.

B

external to, and co-equal with the Galaxy. Johann Heinrich Lambert,[1] the tailor's apprentice of Mühlhausen, followed, but independently. The conceptions of this remarkable man were grandiose, his intuitions bold, his views on some points a singular anticipation of subsequent discoveries. The sidereal world presented itself to him as a hierarchy of systems, starting from the planetary scheme, rising to throngs of suns within the circuit of the Milky Way—the "ecliptic of the stars," as he phrased it—expanding to include groups of many Milky Ways ; these again combining to form the unit of a higher order of assemblage, and so onwards and upwards until the mind reels and sinks before the immensity of the contemplated creations.

"Thus everything revolves—the earth round the sun ; the sun round the centre of his system ; this system round a centre in common to it with other systems ; this group, this assemblage of systems, round a centre which is common to it with other groups of the same kind ; and where shall we have done ?"[2]

The stupendous problem thus speculatively attempted, Herschel undertook to grapple with experimentally. The upshot of this memorable inquiry was the inclusion for the first time within the sphere of human knowledge, of a connected body of facts and inferences from facts regarding the sidereal universe— in other words, the foundation of what may properly be called a science of the stars.

Tobias Mayer had illustrated the perspective effects which must ensue in the stellar sphere from a translation of the solar system, by comparing them to the separating in front and closing up behind of trees in a forest to the eye of an advanc-

---

[1] *Cosmologische Briefe*, Augsburg, 1761.

[2] *The System of the World*, p. 125, London, 1800 (a translation of the above). Lambert regarded nebulæ as composed of stars crowded together, but *not* as external universes. In the case of the Orion nebula, indeed, he throws out such a conjecture, but afterwards suggests that it may form a centre for that one of the subordinate systems composing the Milky Way to which our sun belongs.

ing spectator;[1] but the appearances which he thus correctly
described he was unable to detect. By a more searching
analysis of a smaller collection of proper motions, Herschel
succeeded in rendering apparent the very consequences foreseen
by Mayer. He showed, for example, that Arcturus and Vega
did, in fact, appear to recede from, and Sirius and Aldebaran
to approach, each other by very minute amounts ; and, with a
striking effort of divinatory genius, placed the "apex," or point
of direction of the sun's motion, close to the star λ in the constellation Hercules,[2] within a few degrees of the spot indicated
by the latest and most refined methods of research. The
validity of this conclusion was long doubted ; but it has been
triumphantly confirmed, and scarcely corrected. The question
as to the "secular parallax" of the fixed stars was in effect
answered.

With their *annual* parallax, however, the case was very
different. The search for it had already led Bradley to the
important discoveries of the aberration of light and the nutation of the earth's axis; it was now about to lead Herschel
to a discovery of a different, but even more elevated character. Yet in neither case was the object primarily sought
attained.

From the very first promulgation of the Copernican theory
the seeming immobility of the stars had been urged as an
argument against its truth ; for if the earth really travelled
in a vast orbit round the sun, objects in surrounding space
should appear to change their positions, unless their distances
were on a scale which, to the narrow ideas of the universe then

---

[1] *Op. In.*, t. i. p. 79.

[2] *Phil. Trans.*, vol. lxxiii. (1783), p. 273. He resumed the subject in
1805 (*Phil. Trans.*, vols. xcv. and xcvi.), but, though employing a more
rigorous method, was scarcely so happy in his result. It is worthy of
remark that Prévost, almost simultaneously with Herschel, executed an
investigation similar to his with very considerable success. Klugel confirmed Herschel's result by an analytical inquiry in 1789.

prevailing, seemed altogether extravagant.[1]   The existence of such apparent, or "parallactic" displacements was accordingly regarded as the touchstone of the new views, and their detection became an object of earnest desire to those interested in maintaining them.   Copernicus himself made the attempt; but with his "Triquetrum," a jointed wooden rule with the divisions marked in ink, constructed by himself, he was hardly able to measure angles of ten minutes, far less fractions of a second.   Galileo, a more impassioned defender of the system, strained his ears, as it were, from Arcetri, in his blind and sorrowful old age, for news of a discovery which two more centuries had still to wait for.   Hooke believed he had found a parallax for the bright star in the Head of the Dragon, but was deceived.   Bradley convinced himself that such effects were too minute for his instruments to measure.   Herschel made a fresh attempt by a practically untried method.

It is a matter of daily experience that two objects situated at different distances, seem to a beholder in motion to move relatively to each other.   This principle Galileo, in the third of his Dialogues on the Systems of the World,[2] proposed to employ for the determination of stellar parallax; for two stars, lying apparently close together, but in reality separated by a great gulf of space, must shift their mutual positions when observed from opposite points of the earth's orbit; or rather, the remoter forms a virtually fixed point, to which the movements of the other can be conveniently referred.   By this means complications were abolished more numerous and perplexing than Galileo himself was aware of, and the problem was reduced to one of simple micrometrical measurement. The "double-star method" was also suggested by James Gregory in 1675, and again by Wallis in 1693;[3] Huygens first,

[1] "Ingens bolus devorandus est," Kepler admits to Herwart in May 1603.

[2] *Opere*, t. i. p. 415.   [3] *Phil. Trans.*, vol. xvii. p. 848.

and afterwards Dr. Long of Cambridge (about 1750), made futile experiments with it; and it eventually led, in the hands of Bessel, to the successful determination of the parallax of 61 Cygni.

Its advantages were not lost upon Herschel. His attempt to assign definite distances to the nearest stars was no isolated effort, but part of the settled plan upon which his observations were conducted. He proposed to sound the heavens, and the first requisite was a knowledge of the length of his sounding-line. Thus it came about that his special attention was early directed to double stars.

"I resolved," he writes,[1] "to examine every star in the heavens with the utmost attention and a very high power, that I might collect such materials for this research as would enable me to fix my observations upon those that would best answer my end. The subject has already proved so extensive, and still promises so rich a harvest to those who are inclined to be diligent in the pursuit, that I cannot help inviting every lover of astronomy to join with me in observations that must inevitably lead to new discoveries."

The first result of these inquiries was a classed catalogue of 269 double stars, presented to the Royal Society in 1782, followed, after three years, by an additional list of 434. In both these collections the distances separating the individuals of each pair were carefully measured, and (with a few exceptions) the directions with reference to an invariable line, of the lines joining their centres (technically called "angles of position"), were determined with the aid of a "revolving-wire micrometer," specially devised for the purpose. Moreover, an important novelty was introduced by the observation of the various colours visible in the star-couples, the singular and vivid contrasts of which were now for the first time described.

Double stars were at that time supposed to be a purely optical phenomenon. Their components, it was thought, while in reality indefinitely remote from each other, were

[1] *Phil. Trans.*, vol. lxxii. p. 97.

brought into fortuitous contiguity by the chance of lying nearly in the same line of sight from the earth. This view, however, was not universal. The Rev. John Mitchell, arguing by the doctrine of probabilities, came to a different conclusion.

"It is highly probable in particular," he wrote in 1767,[1] "and next to a certainty in general, that such double stars as appear to consist of two or more stars placed very near together, do really consist of stars placed near together, and under the influence of some general law." And in 1784:[2] "It is not improbable that a few years may inform us that some of the great number of double, triple stars, &c., which have been observed by Mr. Herschel, are systems of bodies revolving about each other."

This remarkable speculative anticipation had a practical counterpart in Germany. Father Christian Mayer, a Jesuit astronomer at Mannheim, set himself, in January 1776, to collect examples of stellar pairs, and shortly after published the supposed discovery of "satellites" to many of the principal stars.[3] His observations, however, were neither exact nor prolonged enough to lead to useful results in such an inquiry. His disclosures were derided; his planet-stars treated as results of hallucination. *On n'a point cru à des choses aussi extraordinaires*, wrote Lalande[4] within one year of a better-grounded announcement to the same effect.

Herschel at first shared the general opinion as to the merely optical connection of double stars. Of this the purpose for which he made his collection is in itself sufficient evidence, since what may be called the *differential* method of parallaxes depends, as we have seen, for its efficacy upon disparity of distance. It was "much too soon," he declared in 1782,[5] "to form any theories of small stars revolving round large ones;"

---

[1] *Phil. Trans.*, vol. lvii. p. 249.   [2] *Ibid.*, vol. lxxiv. p. 56.

[3] *Gründliche Vertheidigung neuer Beobachtungen von Fixsterntrabanten,* 1778, and *De Novis in Cælo Sidereo Phænomenis,* 1779.

[4] *Bibliographie,* p. 569.   [5] *Phil. Trans.,* vol. lxxii. p. 162.

while in the year following,[1] he remarked that the identical proper motion of the two stars forming, to the naked eye, the single bright orb of Castor could only be explained as both equally due to the "systematic parallax" caused by the sun's movement in space. Plainly showing that the notion of a physical tie compelling the two bodies to travel together, had not as yet entered into his speculations. But he was eminently open to conviction, and had, moreover, by observations unparalleled in amount as well as in kind, prepared ample materials for convincing himself and others. In 1802 he was able to announce the fact of his discovery, and in the two ensuing years to lay in detail before the Royal Society, proofs, gathered from the labours of a quarter of a century, of orbital revolution in the case of as many as fifty double stars, henceforth, he declared, to be held as real binary combinations, "intimately held together by the bond of mutual attraction."[2] The fortunate preservation in Dr. Maskelyne's notebook of a remark made by Bradley about 1759, to the effect that the line joining the two stars of Castor was precisely coincident with that joining Castor with Pollux, added eighteen years to the time during which the pair were under scrutiny, and confirmed the evidence of change afforded by more recent observations. Approximate periods were fixed for many of the revolving suns—for Castor, 342 years ; for γ Leonis, 1200, δ Serpentis, 375, ε Bootis, 1681 years ; ε Lyræ was noted as a "double-double star," a change of situation having been detected in each of the two pairs composing the group ; and the occultation of one star by another in the course of their mutual revolutions, of which curious phenomenon two examples (in δ Cygni and ζ Herculis) occurred in 1802, was described.

Thus, by the sagacity and perseverance of a single observer, a firm basis was at last provided upon which to raise the edifice of sidereal science. The analogy long presumed to exist between the mighty star of our system and the bright points of

---

[1] *Phil. Trans.*, vol. lxxiii. p. 272.   [2] *Ibid.*, vol. xciii. p. 340.

light spangling the firmament, was shown to be no fiction of
the imagination, but a physical reality; the fundamental quality
of attractive power was proved to be common to matter so far
as the telescope was capable of exploring, and law, subordina-
tion, and regularity to give testimony of supreme and intelligent
design no less in those limitless regions of space, than in our
narrow terrestrial home. The discovery was emphatically (in
Arago's phrase) "one with a future," since it introduced the
element of precise knowledge where more or less probable
conjecture had previously held almost undivided sway; and
precise knowledge tends to propagate itself and advance from
point to point.

We have now to speak of Herschel's pioneering work in the
skies. To explore with line and plummet the shining zone of
the Milky Way, to delineate its form, measure its dimensions,
and search out the intricacies of its construction, was the
primary task of his life, which he never lost sight of, and to
which all his other investigations were subordinate. He was
absolutely alone in this bold endeavour. Unaided he had to
devise methods, accumulate materials, and sift out results. Yet
it may safely be asserted that all the knowledge we possess on
this sublime subject was prepared, and the greater part of it
anticipated, by him.

The ingenious method of "star-gauging," and its issue in
the delineation of the sidereal system as an irregular stratum
of evenly scattered suns, is the best known part of his work.
But it was, in truth, only a first rude approximation, the principle of which maintained its credit in the literature of astronomy a full half-century after its abandonment by its author.
This principle was the general equality of star distribution. If
equal portions of space really held equal numbers of stars,
it is obvious that the number of stars visible in any particular
direction would be strictly proportional to the range of the
system in that direction, apparent accumulation being produced by real extent. The process of "gauging the heavens,"
accordingly, consisted in counting the stars in successive

telescopic fields, and calculating thence the depths of space necessary to contain them. The result of 3400 such operations was the plan of the Galaxy familiar to every reader of an astronomical text-book. Widely varying evidence was, as might have been expected, derived from an examination of different portions of the sky. Some fields of view were almost blank, while others (in or near the Milky Way) blazed with the radiance of many hundred stars compressed into an area about one-fourth that of the full moon. In the most crowded parts 116,000 were stated to have been passed in review within a quarter of an hour. Here the "length of his sounding-line" was estimated by Herschel at about 497 times the distance of Sirius—in other words, the bounding orb, or farthest sun of the system in that direction, so far as was revealed by the 20-foot reflector, was thus inconceivably remote. But since the distance of Sirius, no less than of every other fixed star, was as yet an unknown quantity, the dimensions inferred for the Galaxy were of course purely relative; a knowledge of its form and structure might (admitting the truth of the fundamental hypothesis) be obtained, but its real or absolute size remained altogether undetermined.

Even as early as 1785, however, Herschel perceived traces of a tendency which completely invalidated the supposition of any approach to an average uniformity of distribution. This was the action of what he called a "clustering power" in the Milky Way. "Many gathering clusters"[1] were already discernible to him even while he endeavoured to obtain "a true *mean* result" on the assumption that each star in space was separated from its neighbours as widely as the sun from Sirius. "It appears," he wrote in 1789, "that the heavens consist of regions where suns are gathered into separate systems;" and in certain assemblages he was able to trace "a course or tide of stars setting towards a centre," denoting, not doubtfully, the presence of attractive forces.[2] Thirteen years later, he described our sun and his constellated companions as sur-

[1] *Phil. Trans.*, vol. lxxv. p. 255.   [2] *Ibid.*, vol. lxxix. pp. 214, 222.

rounded by "a magnificent collection of innumerable stars, called the Milky Way, which must occasion a very powerful balance of opposite attractions to hold the intermediate stars at rest.   For though our sun, and all the stars we see, may truly be said to be in the plane of the Milky Way, yet I am now convinced, by a long inspection and continued examination of it, that the Milky Way itself consists of stars very differently scattered from those which are immediately about us."   "This immense aggregation," he added, "is by no means uniform. Its component stars show evident signs of clustering together into many separate allotments."[1]

The following sentences, written in 1811, contain a definite retractation of the view frequently attributed to him :—

" I must freely confess," he says, "that by continuing my sweeps of the heavens my opinion of the arrangement of the stars and their magnitudes, and of some other particulars, has undergone a gradual change ; and indeed, when the novelty of the subject is considered, we cannot be surprised that many things formerly taken for granted should on examination prove to be different from what they were generally but incautiously supposed to be.   For instance, an equal scattering of the stars may be admitted in certain calculations ; but when we examine the Milky Way, or the closely compressed clusters of stars of which my catalogues have recorded so many instances, this supposed equality of scattering must be given up."[2]

Another assumption, the fallacy of which he had not the means of detecting since become available, was retained by him to the end of his life.   It was that the brightness of a star afforded an approximate measure of its distance.   Upon this principle he founded in 1817 his method of "limiting apertures,"[3] by which two stars, brought into view in two precisely similar telescopes, were "equalised" by covering a certain portion of the object-glass collecting the more brilliant rays.   The distances

[1] *Phil. Trans.*, vol. xcii. pp. 479, 495.
[2] *Ibid.*, vol. ci. p. 269.   [3] *Ibid.*, vol. cvii. p. 311.

of the orbs compared were then taken to be in the ratio of the reduced to the original apertures of the instruments with which they were examined. If indeed the absolute lustre of each were the same, the result might be accepted with confidence; but since we have no warrant for assuming a "standard star" to facilitate our computations, but much reason to suppose an indefinite range, not only of size but of intrinsic brilliancy, in the suns of our firmament, conclusions drawn from such a comparison are entirely worthless.

In another branch of sidereal science besides that of stellar aggregation, Herschel may justly be styled a pioneer. He was the first to bestow serious study on the enigmatical objects known as "nebulæ." The history of the acquaintance of our race with them is comparatively short. The only one recognised before the invention of the telescope was that in the girdle of Andromeda, certainly familiar in the middle of the tenth century to the Persian astronomer Abdurrahman Al-Sûfi; and its place was marked with dots on an old Dutch chart of the constellation, presumably about 1500 A.D.[1] Yet so little was it noticed, that it might practically be said—as far as Europe is concerned—to have been discovered in 1612 by Simon Marius (Mayer of Genzenhausen), who aptly described its appearance as that of a "candle shining through horn." The first mention of the great Orion nebula is by a Swiss Jesuit named Cysatus, who succeeded Father Scheiner in the chair of mathematics at Ingolstadt. He used it, apparently without any suspicion of its novelty, as a term of comparison for the comet of December 1618.[2] A novelty, nevertheless, to astronomers it still remained in 1656, when Huygens discerned "as it were, an hiatus in the sky, affording a glimpse of a more luminous region beyond."[3] Halley in 1714 knew of six nebulæ, which he believed to be composed of a "lucid

[1] Bullialdus, *De Nebulosâ Stellâ in Cingulo Andromedæ* (1667); see also G. P. Bond, *Mem. Am. Ac.*, vol. iii. p. 75, and Holden's Monograph on the Orion Nebula, *Washington Observations*, vol. xxv. 1878 (pub. 1882).
[2] *Mathemata Astronomica*, p. 75.   [3] *Systema Saturnium*, p. 9.

medium" diffused through the ether of space.[1] He appears, however, to have been unacquainted with some previously noticed by Hevelius. Lacaille brought back with him from the Cape a list of forty-two—the first-fruits of observation in Southern skies—arranged in three numerically equal classes ; [2] and Messier (nicknamed by Louis XV. the "ferret of comets "[3]), finding such objects a source of extreme perplexity in the pursuit of his chosen game, attempted to eliminate by methodising them, and drew up a catalogue comprising, in 1781, 103 entries.[4]

These preliminary attempts shrank into insignificance when Herschel began to "sweep the heavens" with his giant telescopes. In 1786 he presented to the Royal Society a descriptive catalogue of 1000 nebulæ and clusters, followed, three years later, by a second of an equal number; to which he added in 1802 a further gleaning of 500. On the subject of their nature his views underwent a remarkable change. Finding that his potent instruments resolved into stars many nebulous patches in which no signs of such a structure had previously been discernible, he naturally concluded that "resolvability " was merely a question of distance and telescopic power. He was (as he said himself) led on by almost imperceptible degrees from evident clusters, such as the Pleiades, to spots without a trace of stellar formation, the gradations being so well connected as to leave no doubt that all these phenomena were equally stellar. The singular variety of their appearance was thus described by him :—

"I have seen," he says, "double and treble nebulæ variously arranged ; large ones with small, seeming attendants; narrow, but much extended lucid nebulæ or bright dashes; some of the shape of a fan, resembling an electric brush, issuing from a lucid point; others of the cometic shape, with a seeming

[1] *Phil. Trans.*, vol. xxix. p. 390.   [2] *Mém. Ac. des Sciences*, 1755.
[3] Wolf, *Gesch. d. Astr.*, p. 709.
[4] *Conn. des Temps*, 1784 (pub. 1781), p. 227. A previous list of forty-five had appeared in *Mém. Ac. d. Sc.*, 1771.

nucleus in the centre, or like cloudy stars surrounded with a nebulous atmosphere; a different sort, again, contain a nebulosity of the milky kind, like that wonderful, inexplicable phenomenon about θ Orionis; while others shine with a fainter, mottled kind of light, which denotes their being resolvable into stars." [1]

"These curious objects" he considered to be "no less than whole sidereal systems," [2] some of which might "well outvie our Milky Way in grandeur." He admitted, however, a wide diversity in condition as well as compass. The system to which our sun belongs he described as "a very extensive branching congeries of many millions of stars, which most probably owes its origin to many remarkably large, as well as pretty closely scattered small stars, that may have drawn together the rest." [3] But the continued action of this same "clustering power" would, he supposed, eventually lead to the breaking up of the original majestic Galaxy into two or three hundred separate groups, already visibly gathering. Such minor nebulæ, due to the "decay" of other "branching nebulæ" similar to our own, he recognised by the score, lying, as it were, stratified in certain quarters of the sky. "One of these nebulous beds," he informs us, "is so rich, that in passing through a section of it, in the time of only thirty-six minutes, I detected no less than thirty-one nebulæ, all distinctly visible upon a fine blue sky." The stratum of Coma Berenices he judged to be the nearest to our system of such layers; nor did the marked aggregation of nebulæ towards both poles of the circle of the Milky Way, escape his notice.

By a continuation of the same process of reasoning, he was enabled (as he thought) to trace the life-history of nebulæ from a primitive loose and extended formation, through clusters of gradually increasing compression, down to the kind named by him "Planetary" because of the defined and uniform discs which they present. These he regarded

[1] *Phil. Trans.*, vol. lxxiv. p. 442.   [2] *Ibid.*, vol. lxxix. p. 213.
[3] *Ibid.*, vol. lxxv. p. 254.

as "very aged. and drawing on towards a period of change or dissolution."[1]

"This method of viewing the heavens," he concluded, "seems to throw them into a new kind of light. They now are seen to resemble a luxuriant garden, which contains the greatest variety of productions in different flourishing beds; and one advantage we may at least reap from it is, that we can, as it were, extend the range of our experience to an immense duration. For, to continue the simile which I have borrowed from the vegetable kingdom, is it not almost the same thing whether we live successively to witness the germination, blooming, foliage, fecundity, fading, withering, and corruption of a plant, or whether a vast number of specimens, selected from every stage through which the plant passes in the course of its existence, be brought at once to our view?"[2]

But already this supposed continuity was broken. After mature deliberation on the phenomena presented by nebulous stars, Herschel was induced, in 1791, to modify essentially his original opinion.

"When I pursued these researches," he says, "I was in the situation of a natural philosopher who follows the various species of animals and insects from the height of their perfection down to the lowest ebb of life; when, arriving at the vegetable kingdom, he can scarcely point out to us the precise boundary where the animal ceases and the plant begins; and may even go so far as to suspect them not to be essentially different. But, recollecting himself, he compares, for instance, one of the human species to a tree, and all doubt upon the subject vanishes before him. In the same manner we pass through gentle steps from a coarse cluster of stars, such as the Pleiades . . . till we find ourselves brought to an object such as the nebula in Orion, where we are still inclined to remain in the once adopted idea of stars exceedingly remote, and inconceivably crowded, as being the occasion of that remarkable appearance.

[1] *Phil. Trans.*, vol. lxxix. p. 225.   [2] *Ibid.*, vol. lxxix. p. 226.

It seems, therefore, to require a more dissimilar object to set us right again. A glance like that of the naturalist, who casts his eye from the perfect animal to the perfect vegetable, is wanting to remove the veil from the mind of the astronomer. The object I have mentioned above is the phenomenon that was wanting for this purpose. View, for instance, the 19th cluster of my 6th class, and afterwards cast your eye on this cloudy star, and the result will be no less decisive than that of the naturalist we have alluded to. Our judgment, I may venture to say, will be that *the nebulosity about the star is not of a starry nature.*" [1]

The conviction thus arrived at of the existence in space of a widely diffused "shining fluid" (a conviction long afterwards fully justified by the spectroscope), led him into a field of endless speculation. What was its nature? Should it "be compared to the coruscation of the electric fluid in the aurora borealis? or to the more magnificent cone of the zodiacal light?" Above all, what was its function in the cosmos? And on this point he already gave a hint of the direction in which his mind was moving by the remark that this self-luminous matter seemed "more fit to produce a star by its condensation, than to depend on the star for its existence." [2]

This was not a novel idea. Tycho Brahe had tried to explain the blaze of the star of 1572 as due to a sudden concentration of nebulous material in the Milky Way, even pointing out the space left dark and void by the withdrawal of the luminous stuff; and Kepler, theorising on a similar stellar apparition in 1604, followed nearly in the same track. But under Herschel's treatment the nebular origin of stars first acquired the consistency of a formal theory. He meditated upon it long and earnestly, and in two elaborate treatises, published respectively in 1811 and 1814, he at length set forth the arguments in its favour. These rested entirely upon the "principle of continuity." Between the successive

[1] *Phil. Trans.*, vol. lxxxi. p. 72.   [2] *Ibid.*, vol. lxxxi. p. 85.

classes of his progressive assortment of objects, there was, as
he said, " perhaps not so much difference as would be in an
annual description of the human figure, were it given from the
birth of a child till he comes to be a man in his prime." [1]
From diffused nebulosity, barely visible in the most powerful
light-gathering instruments, but which he estimated to cover
nearly 152 square degrees of the heavens,[2] to planetary
nebulæ, supposed to be already centrally solid, instances
were alleged by him of every stage and phase of condensation.   The validity of his reasoning, however, was evidently
impaired by his confessed inability to distinguish between the
dim rays of remote clusters and the milky light of true
gaseous nebulæ.

It may be said that such speculations are futile in themselves, and necessarily barren of results.   But they gratify an
inherent tendency of the human mind, and, if pursued in a
becoming spirit, should be neither reproved nor disdained.
Herschel's theory still exercises men's thoughts, and not
unworthily, although the testimony of recent discoveries with
regard to it is, at the best, hesitating and inconclusive.   It
should be added, that it seems to have been propounded in
complete independence of Laplace's nebular hypothesis as to
the origin of the solar system.   Indeed, it dated, as we have
seen, in its first inception, from 1791, while the French geo-
metrician's view was not advanced until 1796.

We may now briefly sum up the chief results of Herschel's
long years of "watching the heavens."   The apparent motions
of the stars had been disentangled; one portion being clearly
shown to be due to a translation towards a point in the constellation Hercules of the sun and his attendant planets;
while a large balance of displacement was left to be accounted
for by real movements, various in extent and direction, of the
stars themselves.   By the action of a central force similar to,
if not identical with, gravity, suns of every degree of size and
splendour, and sometimes brilliantly contrasted in colour, were

[1] *Phil. Trans.*, vol. ci. p. 271.   [2] *Ibid.*, vol. ci. p. 277.

seen to be held together in systems, consisting of two, three, four, even six members, whose revolutions exhibited a wide range of variety both in period and in orbital form. A new department of physical astronomy was thus created,[1] and rigid calculation for the first time made possible within the astral region. The vast problem of the arrangement and relations of the millions of stars forming the Milky Way was shown to be capable of experimental treatment, and of at least partial solution, notwithstanding the inexhaustible variety and boundless complexity seen to prevail, to an extent previously undreamt of, in the adjustments of that majestic system. The existence of a luminous fluid, diffused through enormous tracts of space, and intimately associated with stellar bodies, was virtually demonstrated, and its place and use in creation attempted to be divined by a bold, but plausible conjecture. Change on a stupendous scale was observed to be everywhere in progress. One star—55 Herculis—vanished, it might be said, under the very eye of the astronomer, and other disappearances were more than surmised; progressive ebbings or flowings of light were indicated as probable in many stars under no formal suspicion of variability; forces were everywhere perceived to be at work, by which the very structure of the heavens themselves must be slowly but fundamentally modified. In all directions groups were seen to be formed or forming; tides and streams of suns to be setting towards powerful centres of attraction; new systems to be in process of formation, while effete ones hastened to decay or regeneration when the course appointed for them by Infinite Wisdom was run. And thus, to quote the words of the observer who "had looked farther into space than ever human being did before him,"[2] "the state into which the incessant action of the clustering power has brought the Milky Way at present, is a kind of chronometer that may be used to measure the time

[1] Sir J. Herschel, *Phil. Trans.*, vol. cxiv. part iii. p. 1.
[2] His own words to the poet Campbell, cited by Holden, *Life and Works*, p. 109.

C

of its past and future existence; and although we do not know
the rate of going of this mysterious chronometer, it is nevertheless certain that since the breaking up of the parts of the
Milky Way affords a proof that it cannot last for ever, it
equally bears witness that its past duration cannot be admitted
to be infinite."[1]

[1] *Phil. Trans.*, vol. civ. p. 283.

CHAPTER II.

PROGRESS OF SIDEREAL ASTRONOMY.

WE have now to consider labours of a totally different character from those of Sir William Herschel. Exploration and discovery do not constitute the whole business of astronomy; the less adventurous, though not less arduous, task of gaining a more and more complete mastery over the problems immemorially presented to her, may, on the contrary, be said to form her primary duty. A knowledge of the movements of the heavenly bodies has, from the earliest times, been demanded by the urgent needs of mankind; and science finds prosperity, as in many cases it has taken its origin, in condescension to practical claims. Indeed, to bring such knowledge as near as possible to absolute precision has been defined by no mean authority [1] as the true end of astronomy.

Several causes concurred about the beginning of the present century to give a fresh and powerful impulse to investigations having this end in view. The rapid progress of theory almost compelled a corresponding advance in observation; instrumental improvements rendered such an advance possible; Herschel's discoveries quickened public interest in celestial inquiries; royal, imperial, and grand-ducal patronage widened the scope of individual effort. The heart of the new movement was in Germany. Hitherto the observatory of Flamsteed and Bradley had been the acknowledged centre of practical astronomy; Greenwich observations were the standard of

---

[1] Bessel, *Populäre Vorlesungen*, pp. 6, 408.

reference all over Europe ; and the art of observing prospered
in direct proportion to the fidelity with which Greenwich
methods were imitated.   Dr. Maskelyne, who held the post of
Astronomer Royal during forty-six years (from 1765 to 1811),
was no unworthy successor to the eminent men who had gone
before him.   His foundation of the *Nautical Almanac* (in
1767) alone constitutes a valid title to fame ; he introduced at
the Observatory the important innovation of the systematic
publication of results ; and the careful and prolonged series of
observations executed by him formed the basis of the improved
theories, and corrected tables of the celestial movements, which
were rapidly being brought to completion abroad.   But he had
in him no stirrings of the future.   He was fitted rather to
continue a tradition than to found a school.   The old ways
were dear to him ; and, indefatigable as he was, a definite
purpose was wanting to compel him, by its exigencies, along
the path of progress.   Thus, for almost fifty years after
Bradley's death, the acquisition of a small achromatic[1] was
the only notable change made in the instrumental equipment
of the Observatory.   The transit, the zenith sector, and the
mural quadrant, with which Bradley had done his incomparable
work, retained their places long after they had become de-
teriorated by time, and obsolete by the progress of invention ;
and it was not until the very close of his career that Maskelyne,
compelled by Pond's detection of serious errors, ordered a
Troughton's circle, which he did not live to employ.

Meanwhile, the heavy national disasters with which Germany
was overwhelmed in the early part of the present century,
seemed to stimulate, rather than impede the intellectual revival already for some years in progress there.   Astronomy was
amongst the first of the sciences to feel the new impulse.   By
the efforts of Bode, Olbers, Schröter, and Von Zach, just and
elevated ideas on the subject were propagated, intelligence was
diffused, and a firm ground prepared for common action in
mutual sympathy and disinterested zeal.   They were powerfully

[1] Fitted to the old transit instrument, July 11, 1772.

seconded by the foundation, in 1804, by a young artillery officer named Von Reichenbach, of an Optical and Mechanical Institute at Munich.   Here the work of English instrumental artists was for the first time rivalled, and that of English opticians—when Fraunhofer entered the new establishment—far surpassed. The development given to the refracting telescope by this extraordinary man was indispensable to the progress of that fundamental part of astronomy which consists in the exact determination of the places of the heavenly bodies.   Reflectors are brilliant engines of discovery, but they lend themselves with difficulty to the prosaic work of measuring right ascensions and polar distances.   A signal improvement in the art of making and working flint-glass thus most opportunely coincided with the rise of a German school of scientific mechanicians, to furnish the instrumental means needed for the reform which was at hand.   Of the leader of that reform it is now time to speak.

Friedrich Wilhelm Bessel was born at Minden in Westphalia, July 22, 1784.   A certain taste for figures, coupled with a still stronger distaste for the Latin accidence, directed his inclination and his father's choice towards a mercantile career. In his fifteenth year, accordingly, he entered the house of Kuhlenkamp & Sons, in Bremen, as an apprenticed clerk. He was now thrown completely upon his own resources. From his father, a struggling Government official, heavily weighted with a large family, he was well aware that he had nothing to expect; his dormant faculties were roused by the necessity for self-dependence, and he set himself to push manfully forward along the path that lay before him.   The post of supercargo on one of the trading expeditions sent out from the Hanseatic towns to China and the East Indies, was the aim of his boyish ambition, for the attainment of which he sought to qualify himself by the industrious acquisition of suitable and useful knowledge.   He learned English in two or three months; picked up Spanish with the casual aid of a gunsmith's apprentice; studied the geography of the distant lands which he hoped to visit; collected information as to

their climates, inhabitants, products, and the courses of trade.
He desired to add some acquaintance with the art (then much
neglected) of taking observations at sea ; and thus, led on from
navigation to astronomy, and from astronomy to mathematics,
he groped his way into a new world.

It was characteristic of him that the practical problems of
science should have attracted him before his mind was as yet
sufficiently matured to feel the charm of its abstract beauties.
His first attempt at observation was made with a sextant, rudely
constructed under his own directions, and a common clock.
Its object was the determination of the longitude of Bremen,
and its success, he tells us himself,[1] filled him with a rapture
of delight, which, by confirming his tastes, decided his destiny.
He now eagerly studied Bode's *Jahrbuch* and Von Zach's
Monatliche Correspondenz, overcoming each difficulty as it
arose with the aid of Lalande's *Traité d'Astronomie*, and
supplying, with amazing rapidity, his early deficiency in mathematical training.   In two years he was able to attack a
problem which would have tasked the patience, if not the
skill, of the most experienced astronomer.   Amongst the Earl
of Egremont's papers, Von Zach had discovered Harriot's
observations on Halley's comet at its appearance in 1607, and
published them as a supplement to Bode's Annual.   With an
elaborate care inspired by his youthful ardour, though hardly
merited by their loose nature, Bessel deduced from them an
orbit for that celebrated body, and presented the work to
Olbers, whose reputation in cometary researches gave a special
fitness to the proffered homage.   The benevolent physician-
astronomer of Bremen welcomed with surprised delight such a
performance emanating from such a source.   Fifteen years
before, the French Academy had crowned a similar performance ; now its equal was produced by a youth of twenty,
busily engaged in commercial pursuits, self-taught, and obliged
to snatch from sleep the hours devoted to study.   The paper
was immediately sent to Von Zach for publication, with a note

[1] *Briefwechsel mit Olbers*, p. xvi.

from Olbers explaining the circumstances of its author, and the name of Bessel became the common property of learned Europe.

He had, however, as yet no intention of adopting astronomy as his profession. For two years he continued to work in the counting-house by day, and to pore over the *Mécanique Céleste* and the Differential Calculus by night. But the post of assistant in Schröter's observatory at Lilienthal having become vacant by the removal of Harding to Göttingen in 1805, Olbers procured for him the offer of it. It was not without a struggle that he resolved to exchange the desk for the telescope. His reputation with his employers was of the highest; he had thoroughly mastered the details of the business, which his keen practical intelligence followed with lively interest; his years of apprenticeship were on the point of expiring, and an immediate, and not unwelcome prospect of comparative affluence lay before him. The love of science, however, prevailed; he chose poverty and the stars, and went to Lilienthal with a salary of a hundred thalers yearly. Looking back over his life's work, Olbers long afterwards declared that the greatest service which he had rendered to astronomy was that of having discerned, directed, and promoted the genius of Bessel.[1]

For four years he continued in Schröter's employment. At the end of that time the Prussian Government chose him to superintend the erection of a new observatory at Königsberg, which, after many vexatious delays caused by the prostrate condition of the country, was finished towards the end of 1813. Königsberg was the first really efficient German observatory. It became, moreover, a centre of improvement, not for Germany alone, but for the whole astronomical world. During two-and-thirty years it was the scene of Bessel's labours, and Bessel's labours had for their aim the reconstruction, on an amended and uniform plan, of the entire science of observation.

A knowledge of the places of the stars is the foundation of astronomy.[2] Their configuration lends to the skies their

[1] R. Wolf, *Gesch. der Astron.*, p. 518. [2] Bessel, *Pop. Vorl.*, p. 22.

distinctive features, and marks out the shifting tracks of more
mobile objects with relatively fixed, and generally unvarying
points of light.   A more detailed and accurate acquaintance
with the stellar multitude, regarded from a purely uranogra-
phical point of view, has accordingly formed at all times a
primary object of celestial science, and has, during the present
century, been cultivated with a zeal and success by which all
previous efforts are dwarfed into insignificance.   In Lalande's
Histoire Céleste, published in 1801, the places of no less than
47,390 stars were given, but in the rough, as it were, and consequently needing laborious processes of calculation to render
them available for exact purposes.   Piazzi set an example of
improved methods of observation, resulting in the publication,
in 1803 and 1814, of two catalogues of about 7000 stars, which
for their time were models of what such works should be.
Stephen Groombridge at Blackheath was similarly and most
beneficially active.   But something more was needed than the
diligence of individual observers.   A systematic reform was
called for ; and it was this which Bessel undertook and carried
through.

   Direct observation furnishes only what has been called the
"raw material" of the positions of the heavenly bodies.[1]   A
number of highly complex corrections have to be applied before
their *mean* can be disengaged from their *apparent* places on
the sphere.   Of these, the most considerable and familiar is
atmospheric refraction, by which objects seem to stand higher
in the sky than they in reality do, the effect being evanescent
at the zenith, and attaining, by gradations varying with conditions of pressure and temperature, a maximum at the horizon.
Moreover, the points from which measurements are taken are
themselves in motion, either continually in one direction, or
periodically to and fro.   The *precession* of the equinoxes is
slowly progressive, or rather retrogressive ; the *nutation* of the
pole oscillatory in a period of about eighteen years.   Added to
which, the successive transmission of light, combined with the

          [1] Bessel, *Pop. Vorl.*, p. 440.

movement of the earth in its orbit, causes a minute displacement known as *aberration*.

Now it is easy to see that any uncertainty in the application of these corrections saps the very foundations of exact astronomy. Extremely minute quantities, it is true, are concerned ; but the life and progress of modern celestial science depends upon the sure recognition of extremely minute quantities. In the early years of this century, however, no uniform system of "reduction" (so the complete correction of observational results is termed) had been established. Much was left to the individual caprice of observers, who selected for the several "elements" of reduction such values as seemed best to themselves. Hence arose much hurtful confusion, tending to hinder united action and mar the usefulness of laborious researches. For this state of things, Bessel, by the exercise of consummate diligence, sagacity, and patience, provided an entirely satisfactory remedy.

His first step was an elaborate investigation of the precious series of observations made by Bradley at Greenwich from 1750 until his death in 1762. The catalogue of 3222 stars which he extracted from them, gave the earliest example of the systematic reduction on a uniform plan of such a body of work. It is difficult, without entering into details out of place in a volume like the present, to convey an idea of the arduous nature of this task. It involved the formation of a theory of the errors of each of Bradley's instruments, and a difficult and delicate inquiry into the true value of each correction to be applied before the entries in the Greenwich journals could be developed into a finished and authentic catalogue. Although completed in 1813, it was not until five years later that the results appeared with the proud, but not inappropriate title of *Fundamenta Astronomiæ*. The eminent value of the work consisted in this, that by providing a mass of entirely reliable information as to the state of the heavens at the epoch 1755, it threw back the beginning of *exact* astronomy almost half a century. By comparison with Piazzi's catalogues the amount of precession was more accurately determined, the proper motions of a considerable number of stars became known with certainty, and definite prediction—the certificate of initiation into the secrets of Nature—at last became possible as regards the places of the stars. Bessel's final improvements in the methods of reduction were published in 1830 in his *Tabulæ Regiomontanæ*. They not only constituted an advance in accuracy, but afforded a vast increase of facility in application, and were at once and everywhere adopted. Thus astronomy became a truly universal science; uncertainties and disparities were banished, and observations made at all times and places rendered mutually comparable.[1]

More, however, yet remained to be done. In order to verify with greater strictness the results drawn from the Bradley and Piazzi catalogues, a third term of comparison was wanted, and this Bessel undertook to supply. By a course of 75,011 observations, executed during the years 1821–33, with the utmost nicety of care, the number of accurately known stars was brought up to above 50,000, and an ample store of trustworthy facts laid up for the use of future astronomers. In this department Argelander, whom he attracted from finance to astronomy, and trained in his own methods, was his assistant and successor. The great " Bonn Durchmusterung," [2] in which 324,198 stars visible in the northern hemisphere are enumerated, and the corresponding " Atlas," published in 1857–63, constituting a picture of our sidereal surroundings of heretofore unapproached completeness, may be justly said to owe their origin to Bessel's initiative, and to form a sequel to what he commenced.

But his activity was not solely occupied with the promotion of a comprehensive reform in astronomy; it embraced special problems as well. The long-baffled search for a parallax of the fixed stars was resumed with fresh zeal as each mechanical or optical improvement held out fresh hopes of a successful

[1] Durège, *Bessel's Leben und Wirken*, p. 28.
[2] *Bonner Beobachtungen*, Bd. iii.-v. 1859-62.

issue. Illusory results abounded. Piazzi in 1805 perceived, as he supposed, considerable annual displacements in Vega, Aldebaran, Sirius, and Procyon; the truth being that his instruments were worn out with constant use, and could no longer be depended upon.[1] His countryman, Calandrelli, was similarly deluded. The celebrated controversy between the Astronomer Royal and Dr. Brinkley, director of the Dublin College Observatory, turned on the same subject. Brinkley, who was in possession of a first-rate meridian-circle, believed himself to have discovered relatively large parallaxes for four of the brightest stars; Pond, relying on the testimony of the Greenwich instruments, asserted their nullity. The dispute was protracted for fourteen years, from 1810 to 1824, and was brought to no definite conclusion; but the strong presumption on the negative side was abundantly justified in the event.

There was good reason for incredulity in the matter of parallaxes. Announcements of their detection had become so frequent as to be discredited before they were disproved; and Struve, who made an investigation of the subject at Dorpat in 1818–21, had clearly shown that the quantities concerned were so small as to lie beyond the reliable measuring powers of any instrument then in use. Already, however, the means were being prepared of giving to those powers a large increase.

On the 21st July 1801, two old houses in an alley of Munich tumbled down, burying in their ruins the occupants, of whom one alone was extricated alive, though seriously injured. This was an orphan lad of fourteen, named Joseph Fraunhofer. The Elector Maximilian Joseph was witness of the scene, became interested in the survivor, and consoled his misfortune with a present of eighteen ducats. Seldom was money better bestowed or better employed. Part of it went to buy books and a glass-polishing machine, with the help of which young Fraunhofer studied mathematics and optics, and secretly exercised himself in the shaping and finishing of lenses; the

[1] Bessel, *Pop. Vorl.*, p. 238.

remainder purchased his release from the tyranny of one Weichselberger, a looking-glass maker by trade, to whom he had been bound apprentice on the death of his parents. A period of struggle and privation followed, during which, however, he rapidly extended his acquirements; and was thus eminently fitted for the task awaiting him, when, in 1806, he entered the optical department of the establishment founded two years previously by Von Reichenbach and Utzschneider. He now zealously devoted himself to the improvement of the achromatic telescope; and after prolonged study of the theory of lenses, and many toilsome experiments in the manufacture of flint-glass, he succeeded in perfecting, December 12, 1817, an object-glass of exquisite quality and finish, $9\frac{1}{2}$ inches in diameter, and of fourteen feet focal length.

This (as it was then considered) gigantic lens was secured by Struve for the Russian Government, and the "great Dorpat refractor"—the first of the large achromatics which have played such an important part in modern astronomy—was, late in 1824, set up in the place which it still occupies. By ingenious improvements in mounting and fitting, it was adapted to the finest micrometrical work, and thus offered unprecedented facilities both for the examination of double stars (in which Struve chiefly employed it), and for such subtle measurements as might serve to reveal or disprove the existence of a sensible stellar parallax. Fraunhofer, moreover, constructed for the observatory of Königsberg the first really available heliometer. The principle of this instrument (termed with more propriety a "divided object-glass micrometer") is the separation, by a strictly measurable amount, of two distinct images of the same object. If a double star, for instance, be under examination, the two half-lenses into which the object-glass is divided, are shifted until the upper star (say) in one image is brought into coincidence with the lower star in the other, when their distance apart becomes known by the amount of motion employed.[1]

[1] The heads of the screws applied to move the halves of the object-glass in the Königsberg heliometer are of so considerable a size that a

This virtually new engine of research was delivered and mounted in 1829, three years after the termination of the life of its deviser. The Dorpat lens had brought to Fraunhofer a title of nobility and the sole management of the Munich Optical Institute (completely separated since 1814 from the mechanical department). What he had achieved, however, was but a small part of what he meant to achieve. He saw before him the possibility of nearly quadrupling the light-gathering capacity of the great achromatic acquired by Struve ; he meditated improvements in reflectors as important as those he had already effected in refractors ; and was besides eagerly occupied with investigations into the nature of light, the momentous character of which we shall by-and-bye have an opportunity of estimating. But his health was impaired, it is said, from the weakening effects of his early accident combined with excessive and unwholesome toil, and, still hoping for its restoration from a projected journey to Italy, he died of consumption, June 7, 1826, aged thirty-nine years. His tomb in Munich bears the concise eulogy, *Approximavit sidera.*

Bessel had no sooner made himself acquainted with the exquisite defining powers of the Königsberg heliometer, than he resolved to employ them in an attack upon the now secular problem of star-distances. But it was not until 1837 that he found leisure to pursue the inquiry. In choosing his test-star he adopted a new principle. It had hitherto been assumed that our nearest neighbours in space must be found amongst the brightest ornaments of our skies. The knowledge of stellar proper motions afforded by the critical comparison of recent with earlier star-places, suggested a different criterion of distance. It is impossible to escape from the conclusion that the apparently swiftest-moving stars are, *on the whole*, also the nearest to us, however numerous the individual exceptions to the rule. Now, as early as 1792,[1] Piazzi had noted as an

---

thousandth part of a revolution, equivalent to $\frac{1}{10}$th of a second of arc, can be measured with the utmost accuracy. Main in *R. A. S. Mem.*, vol. xii. p. 53.

[1] *Specola Astronomica di Palermo*, lib. vi. p. 10, *note.*

indication of relative vicinity to the earth, the unusually large proper motion (5.2″ annually) of a double star of the fifth magnitude in the constellation of the Swan.   Still more emphatically in 1812 [1] Bessel drew the attention of astronomers to the fact, and 61 Cygni became known as the "flying star."   The *seeming* rate of its flight, indeed, is of so leisurely a kind, that in a thousand years it will have shifted its place by less than 3½ lunar diameters, and that a quarter of a million would be required to carry it round the entire circuit of the visible heavens.   Nevertheless, it has few rivals in rapidity of movement, the apparent displacement of the vast majority of stars being, by comparison, almost insensible.

This interesting, though inconspicuous object, then, was chosen by Bessel to be put to the question with his heliometer, while Struve made a similar, and somewhat earlier trial with the bright gem of the Lyre, whose Arabic title of the "Falling Eagle" survives as a time-worn remnant in "Vega."   Both astronomers agreed to use the "differential" method, for which their instruments and the vicinity to their selected stars of minute, physically detached companions offered special facilities.   In the last month of 1838 Bessel made known the result of one year's observations, showing for 61 Cygni a parallax of about a third of a second (0.3136″).[2]   He then had his heliometer taken down and repaired, after which he resumed the inquiry, and finally terminated a series of 402 measures in March 1840.[3]   The resulting parallax of 0.3483″ (corresponding to a distance about 600,000 times that of the earth from the sun), seemed to be ascertained beyond the possibility of cavil, and is memorable as the first *published* instance of the fathom-line, so industriously thrown into celestial space, having really and indubitably *touched bottom*.   It

[1] *Monatliche Correspondenz*, vol. xxvi. p. 162.

[2] *Astronomische Nachrichten*, Nos. 365–366.   It should be explained that what is called the "annual parallax" of a star is only half its apparent displacement.   In other words, it is the angle subtended at the distance of that particular star by the *radius* of the earth's orbit.

[3] *Astr. Nach.*, Nos. 401–402.

was confirmed in 1842–43 with curious exactness by Peters at Pulkowa ; but the latest researches show that it requires to be increased to just half a second.[1]

Struve's measurements inspired less confidence. They extended over three years (1835–38), but were comparatively few, and were frequently interrupted. Nevertheless the parallax of about a quarter of a second (0.2613″) which he derived from them for α Lyræ, and announced in 1840,[2] has proved real, though somewhat excessive.[3]

Meanwhile a result of the same kind, but of a more striking character than either Bessel's or Struve's, had been obtained, one might almost say casually, by a different method and in a distant region. Thomas Henderson, originally an attorney's clerk in his native town of Dundee, had become known for his astronomical attainments, and was appointed in 1831 to direct the recently completed observatory at the Cape of Good Hope. He began observing in April 1832, and the serious shortcomings of his instrument notwithstanding, executed during the thirteen months of his tenure of office a surprising amount of first-rate work. With a view to correcting the declination of the lustrous double star α Centauri (which ranks after Sirius and Canopus as the third brightest orb in the heavens), he effected a number of successive determinations of its position, and on being informed of its very considerable proper motion (3.6″ annually), he resolved to examine the observations already made for possible traces of parallactic displacement. This was done on his return to Scotland, where he filled the office of Astronomer Royal from 1834 until his premature death in 1844. The result justified his expectations. From the declination measurements made at the Cape and duly reduced, a parallax of about one second of arc clearly emerged (diminished by Gill's and Elkin's observations, 1882–

---

[1] Dr. Ball's measurements at Dunsink give a parallax of 0.47″ for 61 Cygni ; Professor A. Hall's at Washington, 0.48″.

[2] *Additamentum in Mensuras Micrometricas,* p. 28.

[3] Professor Hall in 1881 found the parallax of Vega = 0.18″.

1883, to o.75″), but, by perhaps an excess of caution, was withheld from publication until fuller certainty was afforded by the concurrent testimony of Lieutenant Meadows' determinations of the same star's right ascension.[1] When at last, January 9, 1839, Henderson communicated his discovery to the Astronomical Society, he could no longer claim the priority which was his due. Bessel had anticipated him with the parallax of 61 Cygni by just two months.

Thus from three different quarters, three successful, and almost simultaneous assaults were delivered upon a long-beleaguered citadel of celestial secrets. The same work has since been steadily pursued, with the general result of showing that, as regards their overwhelming majority, the stars are far too remote to show even the slightest trace of optical shifting from the revolution of the earth in its orbit. In about a score of cases, however, small parallaxes have been determined, some certainly (that is, within moderate limits of error), others more or less precariously. The list is an instructive one, in its omissions no less than in its contents. It includes stars of many degrees of brightness, from Sirius down to a nameless telescopic star in the Great Bear;[2] yet the vicinity of this minute object is so much greater than that of the brilliant Arcturus, that the latter transported to its place would increase in lustre fifteen times. Moreover, by far the greater number of the brightest stars are found to have no sensible parallax, while most of those ascertained to be nearest to the earth are of fifth, sixth, even ninth magnitudes. The obvious conclusions follow that the range of variety in the sidereal system is enormously greater than had been supposed, and that estimates of distance based upon apparent magnitude must be wholly futile. To which we may add the probable inference of a *real* preponderance of small stars over large—that is, of bodies

---

[1] *Mem. Roy. Astr. Soc.*, vol. xi. p. 61.

[2] That numbered 21,185 in Lalande's *Hist. Cél.*, found by Argelander to have a proper motion of 4.734″, and by Winnecke a parallax of 0.511″. *Month. Not.*, vol. xviii. p. 289.

inferior to our sun in size and lustre over such giants as Sirius, Arcturus, Aldebaran, and Capella. At the same time, both the so-called "optical" and "geometrical" methods of relatively estimating star-distances are seen to have a foundation of fact, although so disguised by complicated relations as to be of very doubtful individual application. On the whole, the chances are in favour of the superior vicinity of a bright star over a faint one; and, on the whole, the stars in swiftest *apparent* motion are amongst those whose *actual* remoteness is least. Indeed, there is no escape from either conclusion, unless on the supposition of special arrangements in themselves highly improbable, and, we may confidently say, non-existent.

The distances even of the few stars found to have measure-able parallaxes are on a scale entirely beyond the powers of the human mind to conceive. In the attempt both to realise them distinctly, and to express them conveniently, a new unit of length, itself of bewildering magnitude, has originated. This is what we may call the *light-journey* of one year. The subtle vibrations of the ether, propagated on all sides from the surface of luminous bodies, travel at the rate of 186,300 miles a second, or (in round numbers) six billions of miles a year. Four and a third such measures are needed to span the abyss that separates us from the nearest fixed star. In other words, light takes four years and four months to reach the earth from α Centauri; yet α Centauri lies some ten billions of miles nearer to us (so far as is yet known) than any other member of the sidereal system !

The determination of parallax leads, in the case of binary systems, to the determination of mass ; for the distance from the earth of the two bodies forming such a system being ascer-tained, the seconds of arc apparently separating them from each other can be at once translated into millions of miles ; and we only need to add a knowledge of their period to enable us, by an easy sum in proportion, to find their combined mass in terms of that of the sun. Thus, since—according to the elements published by Dr. Elkin in 1880—the two stars form-

D

ing α Centauri revolve round their common centre of gravity
at a mean distance 23⅓ times the radius of the earth's orbit, in
a period of 77½ years, the attractive force of the two together
must be fully twice the solar.   We may gather some idea of
their relations by placing in imagination a second luminary
like our sun in circulation between the orbits of Uranus and
Neptune.   But systems of still more majestic proportions lie
buried in distance impenetrable by our unaided sight.   A
minute double star in the constellation Eridanus, for which
Dr. Gill has detected a small parallax, appears to be more
than thrice as massive as the central orb of our world, while
61 Cygni affords an instance of a binary combination in
which only a fractional part of the solar gravitating power
resides.

Further, the actual rate of proper motions, so far as regards
that part of them which is projected upon the sphere, can be
ascertained for stars at known distances.   The annual journey,
for instance, of 61 Cygni *across the line of sight* amounts to
1000, and that of α Centauri to 436 millions of miles.   A
small star, numbered 1830 in Groombridge's Circumpolar Catalogue, "devours the way" at the rate of 200 miles a second—
a speed, in Newcomb's opinion, beyond the gravitating power
of the entire sidereal system to control; and ζ Tucanæ possesses, according to Dr. Gill, just half that amazing velocity,
besides whatever movement each may have towards or from
the earth, of which the spectroscope may eventually give an
account.

Herschel's conclusion as to the movement of the sun among
the stars was not admitted as valid by the most eminent of his
successors.   Bessel maintained that there was absolutely no
preponderating evidence in favour of its supposed direction
towards a point in the constellation Hercules.[1]   Biot, Burck-
hardt, even Herschel's own son, shared his incredulity.   But
the appearance of Argelander's prize-essay in 1837 [2] changed
the aspect of the question.   Herschel's first memorable solution

[1] *Fund. Astr.*, p. 309.   [2] *Mem. Prés. à l'Ac. de St. Pétersb.*, t. iii.

in 1783 was based upon the proper motions of thirteen stars, imperfectly known ; his second, in 1805, upon those of no more than six. Argelander now obtained an entirely concordant result from the large number of 390, determined with the scrupulous accuracy characteristic of Bessel's work and his own. The reality of the fact thus persistently disclosed could no longer be doubted ; it was confirmed five years later by the younger Struve, and still more strikingly in 1847 [1] by Galloway's investigation, founded exclusively on the apparent displacements of southern stars. In 1860, Mr. (now Sir George) Airy and Mr. Dunkin,[2] employing all the resources of modern science, and commanding the wealth of material furnished by 1167 proper motions carefully determined by Mr. Main, reached a conclusion closely similar to that indicated nearly eighty years previously by the first great sidereal astronomer ; which Mr. Plummer's reinvestigation of the subject in 1883 served but slightly to modify. The general direction of the solar movement may thus be regarded as known ; but as to its *rate*, the grounds of inference are much less satisfactory. Otto Struve's estimate of 154 million miles a year is based upon the assumption of an average annual parallax, for stars of the first magnitude, of about a quarter of a second ; and since only five out of eighteen stars of the first magnitude appear to have *any* measurable parallax, it is obvious that it merits a very restricted confidence.

As might have been expected, speculation has not been idle regarding the purpose and goal of the strange voyage of discovery through space upon which our system is embarked ; but altogether fruitlessly. The variety of the conjectures hazarded in the matter is in itself a measure of their futility. Long ago, before the construction of the heavens had as yet been made the subject of methodical inquiry, Kant was disposed to regard Sirius as the "central sun" of the Milky Way ; while Lambert surmised that the vast Orion nebula

---

[1] *Phil. Trans.*, vol. cxxxvii .p. 79.
[2] *Mem. Roy. Astr. Soc.*, vol. xxviii. 1860, and *Month. Not.*, vol. xxiii. p. 168.

might serve as the regulating power of a subordinate group including our sun. Herschel threw out the hint that the great cluster in Hercules (estimated to include 14,000 stars) might prove to be the supreme seat of attractive force ;[1] Argelander placed his central body in the constellation Perseus ;[2] Fomalhaut, the brilliant of the Southern Fish, was set in the post of honour by Boguslawski of Breslau. Mädler (who succeeded Struve at Dorpat in 1839) concluded from a more formal inquiry that the ruling power in the sidereal system resided, not in any single preponderating mass, but in the centre of gravity of the self-controlled revolving multitude.[3] In the former case (as we know from the example of the planetary scheme), the stellar motions would be most rapid near the centre ; in the latter, they would become accelerated with remoteness from it.[4] Mädler showed that no part of the heavens could be indicated as a region of exceptionally swift movements, such as would result from the vicinity of a gigantic (though possibly obscure) ruling body ; but that a community of extremely sluggish movements undoubtedly existed in, and in the neighbourhood of the group of the Pleiades, where, accordingly, he placed the centre of gravity of the Milky Way.[5] The bright star Alcyone thus became the " central sun," but in a purely passive sense, its headship being determined by its situation at the point of neutralisation of opposing tendencies, and of consequent rest. The solar period of revolution round this point was, by an avowedly conjectural method, fixed at 18,200,000 years, imply-ing, on the extremely hazardous supposition that the distance of Alcyone is thirty-four million times that of the earth from the sun, a velocity for our system of about thirty miles a second.

---

[1] *Phil. Trans.*, vol. xcvi. p. 230.

[2] *Mém. Prés. à l'Ac. de St. Pétersbourg*, t. iii. p. 603 (read Feb. 5, 1837).

[3] *Die Centralsonne, Astr. Nach.*, Nos. 566–567, 1846.

[4] Sir J. Herschel, note to *Treatise on Astronomy*, and *Phil. Trans.*, vol. cxxiii. part ii. p. 502.

[5] The position is (as Sir J. Herschel pointed out, *Outlines of Astronomy*, p. 631, 10th ed.) placed beyond the range of reasonable probability by its remoteness (fully 26°) from the galactic plane.

The scheme of sidereal government framed by the Dorpat astronomer was, it may be observed, of the most approved constitutional type ; deprivation, rather than increase of influence accompanying the office of chief dignitary. But while we are still ignorant, and shall perhaps ever remain so, of the fundamental plan upon which the Galaxy is organised, recent investigations tend more and more to exhibit it, not as monarchical (so to speak), but as federative. The community of proper motions detected by Mädler in the vicinity of the Pleiades may accordingly possess a significance altogether different from what he imagined.

Bessel's so-called "foundation of an Astronomy of the Invisible" now claims attention.[1] His prediction regarding the planet Neptune does not belong to the present division of our subject ; a strictly analogous discovery in the sidereal system was, however, also very clearly foreshadowed by him. His earliest suspicions of non-uniformity in the proper motion of Sirius dated from 1834 ; they extended to Procyon in 1840 ; and after a series of refined measurements with the new Repsold circle, he announced, in 1844, his conclusion that these irregularities were due to the presence of obscure bodies round which the two bright Dog-stars revolved as they pursued their way across the sphere.[2] He even assigned to each an approximate period of half a century. "I adhere to the conviction," he wrote later to Humboldt, "that Procyon and Sirius form real binary systems, consisting of a visible and an invisible star. There is no reason to suppose luminosity an essential quality of cosmical bodies. The visibility of countless stars is no argument against the invisibility of countless others."[3]

An inference so contradictory to received ideas obtained little credit, until Peters found, in 1851,[4] that the apparent anomalies in the movements of Sirius could be completely

[1] Mädler in *Westermann's Jahrbuch*, 1867, p. 615.
[2] Letter from Bessel to Sir J. Herschel, *Month. Not.*, vol. vi. p. 139.
[3] Wolf, *Gesch. d. Astr.*, p. 743, *note.*   [4] *Astr. Nach.*, Nos. 745–748.

explained by an orbital revolution in a period of fifty years.
Bessel's prevision was destined to be still more triumphantly
vindicated.   On the 31st of January 1862, while in the act of
trying a new 18-inch refractor, Alvan Clark, jun. (one of the
celebrated firm of American opticians), actually discovered
the hypothetical Sirian companion in the precise position required by theory.   It has now been watched through just half a
revolution (period 49.4 years), and—unless there should prove to
be other bodies concerned in disturbing the motion of Sirius—
must be very slightly luminous in proportion to its mass.   Its
attractive power, in fact, is about half that of its primary, while
it emits only $\frac{1}{10,000}$th of its light.   Sirius itself, on the other
hand, possesses a far higher radiative intensity than our sun.
It gravitates—admitting Dr. Gill's parallax of 0.38″ to be exact—
like three suns, but shines like seventy.   Possibly it is enormously distended by heat, and undoubtedly its atmosphere
intercepts a very much smaller proportion of its light than in
stars of the solar class.   As regards Procyon, visual verification is still wanting, but to the mental eye the presence of a
considerable disturbing mass is fully assured by the inquiry
instituted by Auwers in 1862.[1]   A period of forty years is
assigned by him to the system.

But Bessel was not destined to witness the recognition of
"the invisible" as a legitimate and profitable field for astronomical research.   He died March 17, 1846, just six months
before the discovery of Neptune, of an obscure disease,
eventually found to be occasioned by an extensive fungus-
growth in the stomach.   The place which he left vacant was
not one easy to fill.   Rarely indeed shall we find one who
reconciled with the same success the claims of theoretical and
practical astronomy, or surveyed the science which he had
made his own with a glance equally comprehensive, practical,
and profound.

The career of Friedrich Georg Wilhelm Struve illustrates
the maxim that science *differentiates* as it develops.   He

[1] *Astr. Nach.*, Nos. 1371-1373.

might be called a specialist in double stars. His earliest recorded use of the telescope was to verify Herschel's conclusion
as to the revolving movement of Castor, and he never varied
from the predilection which this first observation at once indicated and determined. He was born at Altona, of a respect-
able yeoman family, April 15, 1793, and in 1811 took a degree
in philology at the new Russian University of Dorpat. He
then turned to science, was appointed in 1813 to a professor-
ship of astronomy and mathematics, and began regular work
in the Dorpat Observatory just erected by Parrot for Alexander I.
It was not, however, until 1819 that the acquisition of a
5-foot refractor by Troughton enabled him to take the position-
angles of double stars with regularity and tolerable precision.
The resulting catalogue of 795 stellar systems gave the signal
for a general resumption of the Herschelian labours in this
branch. The extraordinary facilities for observation afforded
by the Fraunhofer achromatic encouraged him to undertake,
February 11, 1825, a review of the entire heavens down to 15°
south of the celestial equator, which occupied more than two
years, and yielded, from an examination of above 120,000 stars,
a harvest of about 2200 previously unnoticed composite objects.
The ensuing ten years were devoted to delicate and patient
measurements, the results of which were embodied in *Mensuræ
Micrometricæ*, published at St. Petersburg in 1837. This monumental work gives the places, positions, distances, colours,
and relative brightness of 3112 double and multiple stars, all
determined with the utmost skill and care. The record is one
which gains in value with the process of time, and will for
ages serve as a standard of reference by which to detect change
or confirm discovery.

It appears from Struve's researches that about one in forty
of all stars down to the ninth magnitude is composite, but
that the proportion is doubled in the brighter orders.[1] This
he attributed to the difficulty of detecting the faint companions
of very remote orbs. It was also noticed, both by him and

[1] *Ueber die Doppelsterne*, Bericht, 1827, p. 22.

Bessel, that double stars are in general remarkable for large proper motions. Struve's catalogue included no star of which the components were more than 32″ apart, because beyond that distance the chances of merely optical juxtaposition become appreciable; but the immense preponderance of extremely close over (as it were) loosely yoked bodies is such as to demonstrate their physical connection, even if no other proof were forthcoming. Many stars previously believed to be single divided under the scrutiny of the Dorpat refractor; while in some cases, one member of a (supposed) binary system revealed itself as double, thus placing the surprised observer in the unexpected presence of a triple group of suns. Five instances were noted of two pairs lying so close together as to induce a conviction of their mutual dependence;[1] besides which, 124 examples occurred of triple, quadruple, and multiple combinations, the reality of which was open to no reasonable doubt.[2]

It was first pointed out by Bessel that the fact of stars exhibiting a common proper motion might serve as an unfailing test of their real association into systems. This was, accordingly, one of the chief criteria employed by Struve to distinguish true binaries from merely optical couples. On this ground alone, 61 Cygni was admitted to be a genuine double star; and it was shown that, although its components appear to follow almost strictly rectilinear paths, yet the probability of their forming a connected pair is actually greater than that of the sun rising to-morrow morning.[3] Moreover, this tie of an identical movement was discovered to unite bodies[4] far beyond the range of distance ordinarily separating the members of binary systems, and to prevail so extensively

---

[1] *Ueber die Doppelsterne*, p. 25.   [2] *Mensuræ Micr.*, p. xcix.

[3] *Stellarum Fixarum imprimis Duplicium et Multiplicium Positiones Mediæ*, pp. cxc., cciii.

[4] For instance, the southern stars 36A Ophinchi (itself double) and 30 Scorpii, which are 12′ 10″ apart. *Ibid.*, p. cciii. Recent investigations have vastly enlarged the area over which this species of connection extends.

as to lead to the conclusion that single do not outnumber conjoined stars more than twice or thrice.[1]

In 1835 Struve was summoned by the Emperor Nicholas to superintend the erection of a new observatory at Pulkowa, near St. Petersburg, destined for the special cultivation of sidereal astronomy. Boundless resources were placed at his disposal, and the institution created by him was acknowledged to surpass all others of its kind in splendour, efficiency, and completeness. Its chief instrumental glory was a refractor of fifteen inches aperture by Merz and Mahler (Fraunhofer's successors), which left the famous Dorpat telescope far behind, and remained long without a rival. On the completion of this model establishment, August 19, 1839, Struve was installed as its director, and continued to fulfil the important duties of the post with his accustomed vigour until 1858, when illness compelled his virtual resignation in favour of his son Otto Struve, born at Dorpat in 1819. He died November 23, 1864.

An inquiry into the laws of stellar distribution, undertaken during the early years of his residence at Pulkowa, led Struve to confirm, in the main, the inferences arrived at by Herschel as to the construction of the heavens. According to his view, the appearance known as the Milky Way is produced by a collection of (for the most part) irregularly condensed star-clusters, within which the sun is somewhat eccentrically placed. The nebulous ring which thus integrates the light of countless worlds, he found to differ slightly from the form of a great circle, and he accounted for this deviation from symmetry by supposing the stars composing it to be scattered over a bent or "broken plane," or to lie in two planes slightly inclined to each other, our system occupying a position near their intersection.[2] He further attempted to show that the limits of this vast assemblage must remain for ever shrouded from human discernment, owing to the gradual extinction of light

[1] *Stellarum Fixarum, &c.,* p. ccliii.
[2] *Études d'Astronomie Stellaire,* 1847, p. 82.

in its passage through space,[1] and sought to confer upon this
celebrated hypothesis a definiteness and certainty far beyond
the aspirations of its earlier advocates, Chéseaux and Olbers ;
but arbitrary assumptions vitiated his reasonings on this, as
well as on some other points.[2]

In his special line as a celestial explorer of the most comprehensive type, Sir William Herschel had but one legitimate
successor, and that successor was his son. John Frederick
William Herschel was born at Slough, March 17, 1792,
graduated with the highest honours from St. John's College,
Cambridge, in 1813, and entered upon legal studies with a
view to being called to the Bar. But his share in an early
compact with Peacock and Babbage, "to do their best to
leave the world wiser than they found it," was not thus to be
fulfilled. The acquaintance of Dr. Wollaston decided his scientific vocation. Already, in 1816, we find him reviewing some
of his father's double stars ; and he completed in 1820 the
18-inch speculum which was to be the chief instrument of his
investigations. Soon after he undertook, in conjunction with
Mr. (afterwards Sir James) South, a series of observations,
issuing in the presentation to the Royal Society of a paper[3]
containing micrometrical measurements of 380 binary stars, by
which the elder Herschel's inferences of orbital motion were,
in many cases, strikingly confirmed. A star in the Northern
Crown, for instance (η Coronæ), was found to have completed
more than one entire circuit since its first discovery ; another,
τ Serpentarii, had *closed up* into apparent singleness ; while in
a third, ζ Orionis, the converse change had taken place, and deceptive singleness had been transformed into obvious duplicity.[4]

It was from the first confidently believed that the force
retaining double stars in curvilinear paths was identical with
that governing the planetary revolutions. But that identity

---

[1] *Études d'Astr. Stellaire*, p. 86.
[2] See Encke's criticism in *Astr. Nach.*, No. 622.
[3] *Phil. Trans.*, vol. cxiv. part iii. 1824.
[4] Grant, *Hist. Phys. Astr.*, p. 560.

was not ascertained until Savary of Paris showed, in 1827,[1] that the movements of a well-known binary in the hind-paw of the Great Bear (ξ Ursæ) could be represented with all attainable accuracy by an ellipse calculated on orthodox gravitational principles with a period of $58\frac{1}{4}$ years. Encke followed at Berlin with a still more elegant method; and Sir John Herschel, pointing out the uselessness of analytical refinements where the data were necessarily so imperfect, described in 1831 a graphical process by which " the aid of the eye and hand " was brought in " to guide the judgment in a case where judgment only, and not calculation, could be of any avail." [2] The subject has since been cultivated with diligence, and not without success; but our acquaintance with stellar orbits can hardly yet be said to have emerged from the tentative stage.

In 1825 Herschel undertook, and executed with great assiduity during the ensuing eight years, a general survey of the northern heavens, directed chiefly towards the verification of his father's nebular discoveries. The outcome was a catalogue of 2306 nebulæ and clusters, of which 525 were observed for the first time, besides 3347 double stars discovered almost incidentally.[3] " Strongly invited," as he tells us himself, " by the peculiar interest of the subject, and the wonderful nature of the objects which presented themselves," he resolved to attempt the completion of the survey in the southern hemisphere. With this noble object in view, he embarked his family and instruments on board the *Mount Stewart Elphinstone*, and, after a prosperous voyage, landed at Cape Town on the 16th of January 1834. Choosing as the scene of his observations a rural spot under the shelter of Table Mountain, he began regular " sweeping " on the 5th of March. The site of his great reflector is now marked with an obelisk, and the name of Feldhausen has become memorable in the history of science; for the four years' work done there may truly be said to open the chapter of our knowledge as regards the southern skies.

[1] *Conn. d. Temps*, 1830.   [2] *R. A. S. Mem.*, vol. v. 1833, p. 178.
[3] *Phil. Trans.*, vol. cxxiii., and *Results, &c.*, Introd.

The full results of Herschel's journey to the Cape were not made public until 1847, when a splendid volume [1] embodying them was brought out at the expense of the Duke of Northumberland. They form a sequel to his father's labours such as the investigations of one man have rarely received from those of another. What the elder observer did for the northern heavens, the younger did for the southern, and with generally concordant results. Reviving the paternal method of "star-gauging," he showed, from a count of 2299 fields, that the Milky Way surrounds the solar system as a complete annulus of minute stars; not, however, quite symmetrically, since it appears that the sun lies somewhat *above* its medial plane, as well as somewhat nearer to those portions visible in the southern hemisphere, which accordingly display a brighter lustre and a more complicated structure than the northern branches. The singular cosmical agglomerations known as the "Magellanic Clouds" were now, for the first time, submitted to a detailed, though admittedly incomplete, examination, the almost inconceivable richness and variety of their contents being such that a lifetime might with great profit be devoted to their study. In the Greater Nubecula, within a compass of forty-two square degrees, Herschel reckoned 278 distinct nebulæ and clusters, besides fifty or sixty outliers, and a large number of stars intermixed with diffused nebulosity,—in all, 919 catalogued objects, and, for the Lesser Cloud, 244. Yet this was only the most conspicuous part of what his twenty-foot revealed. Such an extraordinary concentration of bodies so various led him to the inevitable conclusion that "the Nubeculæ are to be regarded as systems *sui generis*, and which have no analogues in our hemisphere." [2] He noted also the blankness of surrounding space, especially in the case of Nubecula Minor, "the access to which on all sides," he remarked, "is through a desert;" as if the cosmical material in the neigh-

---

[1] *Results of Astronomical Observations made during the years* 1834-8 *at the Cape of Good Hope.*   [2] *Results, &c.,* p. 147.

bourhood had been swept up and garnered in these mighty groups.[1]

Of southern double stars, he discovered and gave careful measurements of 2102, and described 1708 nebulæ, of which at least 300 were new. The list was illustrated with a number of drawings, some of them extremely beautiful and elaborate.

Sir John Herschel's views as to the nature of nebulæ were considerably modified by Lord Rosse's success in "resolving" a crowd of these objects into stars with his great reflectors. His former somewhat hesitating belief in the existence of phosphorescent matter, "disseminated through extensive regions of space in the manner of a cloud or fog,"[2] was changed into a conviction that no valid distinction could be established between the faintest wisp of cosmical vapour just discernible in a powerful telescope, and the most brilliant and obvious cluster. He admitted, however, an immense range of possible variety in the size and mode of aggregation of the stellar constituents of various nebulæ. Some might appear nebulous from the closeness of their parts ; some from their smallness. Others, he suggested, might be formed of "discrete luminous bodies floating in a non-luminous medium ;"[3] while the annular kind probably consisted of "hollow shells of stars."[4] That a physical, and not merely an optical, connection unites nebulæ with the *embroidery* (so to speak) of small stars with which they are in many instances profusely decorated, was evident to him, as it must be to all who look as closely and see as clearly as he did. His description of No. 2093 in his northern catalogue as "a network or tracery of nebula following the lines of a similar network of stars,"[5] would alone suffice to dispel the idea of accidental scattering ; and many other examples of a like import might be quoted. The remarkably frequent occurrence of one or more minute stars in the close vicinity of "planetary" nebulæ led him to infer

---

[1] See Proctor's *Universe of Stars*, p. 92.
[2] *A Treatise of Astronomy*, 1833, p. 406.   [3] *Results, &c.*, p. 139.
[4] *Ibid.*, pp. 24, 142.   [5] *Phil. Trans.*, vol. cxxiii. p. 503.

their dependent condition ; and he advised the maintenance
of a strict watch for evidences of circulatory movements, not
only over these supposed stellar satellites, but also over the
numerous "double nebulæ," in which, as he pointed out, "all
the varieties of double stars as to distance, position, and
relative brightness, have their counterparts." He, moreover,
investigated the subject of nebular distribution by the simple
and effectual method of graphic delineation or "charting," and
succeeded in showing that while a much greater uniformity
of scattering prevails in the southern heavens than in the
northern, a condensation is nevertheless perceptible about the
constellations Pisces and Cetus, roughly corresponding to the
"nebular region" in Virgo by its vicinity (within 20° or 30°) to
the opposite pole of the Milky Way. He concluded "that the
nebulous system is distinct from the sidereal, though involving,
and perhaps to a certain extent intermixed with, the latter."[1]

Towards the close of his residence at Feldhausen, Herschel
was fortunate enough to witness one of those singular changes
in the aspect of the firmament which occasionally challenge
the attention even of the incurious, and excite the deepest
wonder of the philosophical observer. Immersed apparently
in the Argo nebula is a large star denominated η Argûs. When
Halley visited St. Helena in 1677, it seemed of the fourth
magnitude ; but Lacaille in the middle of the following century,
and others after him, classed it as of the second. In 1827 the
traveller Burchell, being then at St. Paul, near Rio Janeiro,
remarked that it had unexpectedly assumed the first rank,—a
circumstance the more surprising to him because he had frequently, when in Africa during the years 1811 to 1815, noted
it as of only fourth magnitude. This observation, however,
did not become generally known until later. Herschel, on his
arrival at Feldhausen, registered the star as a bright second,
and had no suspicion of its unusual character until December
16, 1837, when he suddenly perceived it with its light almost
tripled. It then far outshone Rigel in Orion, and on the 2d

[1] *Results, &c.,* p. 136.

of January following it very nearly matched α Centauri. From that date it declined; but a second and even brighter maximum occurred in April 1843, when Maclear, then director of the Cape Observatory, saw it blaze out with a splendour approaching that of Sirius. Its waxings and wanings were marked by curious "trepidations" of brightness extremely perplexing to theory. In 1863 it had sunk below the fifth magnitude, and in 1869 was barely visible to the naked eye; but it has now regained so much of its light as to shine with nearly the same lustre as when Halley observed it two centuries ago. There is some reason to believe that its fluctuations are included in a cycle of about seventy years,[1] confused probably by the superposition of more than one secondary period; but the extent and character of the vicissitudes to which it is subject, stamp it as a species of connecting link between regularly periodic and (so-called) "temporary stars."

Among the numerous topics which engaged Herschel's attention at the Cape was that of relative stellar brightness. Having contrived an "astrometer" in which an "artificial star," formed by the total reflection of moonlight from the base of a prism, served as a standard of comparison, he was able to estimate the lustre of the *natural* stars examined by the distances at which the artificial object appeared equal respectively to each. He thus constructed a table of 191 of the principal stars,[2] both in the northern and southern hemispheres, setting forth the numerical values of their apparent brightness relatively to that of α Centauri, which he selected as a unit of measurement. Further, the light of the full moon being found by him to exceed that of his standard star 27,408 times, and Dr. Wollaston having shown that the light of the full moon is to that of the sun as 1:801,072[3] (Zöllner finds the ratio 1:618,000), it became possible to compare stellar with solar radiance. Hence was derived, in the case of the few stars at ascertained distances, a knowledge of real lustre. Alpha Centauri, for example,

---

[1] Loomis in *Month. Not.*, vol. xxix. p. 298.
[2] *Outlines of Astr.*, App. I.   [3] *Phil. Trans.*, vol. cxix. p. 27.

is found to emit four times, Vega nearly forty times as much light as our sun; while Arcturus (if its measured parallax of 0.13″ can be depended upon) displays the splendour of fully 200 such luminaries.

Herschel returned to England in the spring of 1838, bringing with him a wealth of observation and discovery such as had perhaps never before been amassed in so short a time. Deserved honours awaited him. He was created a baronet on the occasion of the Queen's coronation (he had been knighted in 1831); universities and learned societies vied with each other in showering distinctions upon him; and the success of an enterprise in which scientific zeal was tinctured with an attractive flavour of adventurous romance, was justly regarded as a matter of national pride. His career as an observing astronomer was now virtually closed, and he devoted his leisure to the collection and arrangement of the abundant trophies of his father's and his own activity. The resulting great catalogue of 5079 nebulæ (including all then certainly known), published in the *Philosophical Transactions* for 1864, is, and will probably long remain, the leading source of information on the subject;[1] but he unfortunately did not live to finish the companion work on double stars, for which he had accumulated a vast store of materials.[2] He died at Collingwood in Kent, May 11, 1871, in the eightieth year of his age, and was buried in Westminster Abbey, close beside the grave of Sir Isaac Newton.

The consideration of Sir John Herschel's Cape observations brings us to the close of the period we are just now engaged in studying. They were given to the world, as already stated, three years before the middle of the century, and accurately

---

[1] Dr. Dreyer published in 1878 a supplement to the work, giving the places of 1880 new nebulæ.

[2] A list of 10,320 composite stars was drawn out by him in order of right ascension, and has been published in vol. xl. of *Mem. R. A. S.;* but the data requisite for their formation into a catalogue were not forthcoming. See Main's and Pritchard's *Preface* to above, and Dunkin's *Obituary Notices*, p. 73.

represent the condition of sidereal science at that date. Looking back over the fifty years traversed, we can see at a glance
how great was the stride made in the interval. Not alone was
acquaintance with individual members of the cosmos vastly
extended, but their mutual relations, the laws governing their
movements, their distances from the earth, masses, and intrinsic
lustre, had begun to be successfully investigated. *Begun to be ;*
for only regarding a scarcely perceptible minority had even
approximate conclusions been arrived at. Nevertheless the
whole progress of the future lay in that beginning ; it was
the thin end of the wedge of exact knowledge. The principle
of measurement had been substituted for that of probability ;
a basis had been found large and strong enough to enable
calculation to ascend from it to the sidereal heavens ; and
refinements had been introduced, fruitful in performance, but
still more in promise. Thus, rather the kind than the amount
of information collected was significant for the time to come—
rather the methods employed than the results actually secured
rendered the first half of the nineteenth century of epochal
importance in the history of our knowledge of the stars.

E

CHAPTER III.

PROGRESS OF KNOWLEDGE REGARDING THE SUN.

THE discovery of sun-spots in 1610 by Fabricius and Galileo first opened a way for inquiry into the solar constitution; but it was long before that way was followed with system or profit. The seeming irregularity of the phenomena discouraged continuous attention; casual observations were made the basis of arbitrary conjectures, and real knowledge received little or no increase. In 1620 we find Jean Tarde, canon of Sarlat, arguing that because the sun is "the eye of the world," and the eye of the world *cannot suffer from ophthalmia*, therefore the appearances in question must be due, not to actual specks or stains on the bright solar disc, but to the transits of a number of small planets across it! To this new group of heavenly bodies he gave the name of " Borbonia Sidera," and they were claimed in 1633 for the House of Hapsburg, under the title of " Austriaca Sidera," by Father Malapertius, a Belgian Jesuit.[1] A similar view was temporarily maintained against Galileo by the celebrated Father Scheiner of Ingolstadt, and later by William Gascoigne, the inventor of the micrometer; but most of those who were capable of thinking at all on such subjects (and they were but few) adhered either to the *cloud theory* or to the *slag theory* of sun-spots. The first was championed by Galileo, the second by Simon Marius, "astronomer and physician " to the brother Margraves of Branden-

[1] *Kosmos*, Bd. iii. p. 409 ; Lalande, *Bibliographie Astronomique*, pp. 179, 202.

burg.   The latter opinion received a further notable development from the fact that in 1618, a year remarkable for the
appearance of three bright comets, the sun was almost free
from spots; whence it was inferred that the cindery refuse
from the great solar conflagration, which usually appeared as
dark blotches on its surface, was occasionally thrown off in
the form of comets, leaving the sun, like a snuffed taper, to
blaze with renewed brilliancy.[1]

In the following century, Derham gathered from observations
carried on during the years 1703–11, " That the spots on the
sun are caused by the eruption of some new volcano therein,
which at first pouring out a prodigious quantity of smoke and
other opacous matter, causeth the spots ; and as that fuliginous
matter decayeth and spendeth itself, and the volcano at last
becomes more torrid and flaming, so the spots decay, and grow
to umbræ, and at last to faculæ." [2]

The view, confidently upheld by Lalande,[3] that spots were
rocky elevations uncovered by the casual ebbing of a luminous
ocean, the surrounding penumbræ representing shoals or sand-
banks, had even less to recommend it than Derham's volcanic
theory.   Both were, however, significant of a growing tendency
to bring solar phenomena within the compass of terrestrial
analogies.

For 164 years, then, after Galileo first levelled his telescope
at the setting sun, next to nothing was learned as to its nature ;
and the facts immediately ascertained of its rotation on an

[1] R. Wolf, *Die Sonne und ihre Flecken*, p. 9.   Marius himself, however,
seems to have held the Aristotelian terrestrial-exhalation theory of cometary origin.   See his curious little tract, *Astronomische und Astrologische
Beschreibung des Cometen*, Nürnberg, 1619.

[2] *Phil. Trans.*, vol. xxvii. p. 274.   *Umbræ* (now called *penumbræ*) are
spaces of half-shadow which usually encircle spots.   *Faculæ* ("little
torches," so named by Scheiner) are bright streaks or patches closely
associated with spots.

[3] *Mém. Ac. Sc.*, 1776 (pub. 1779), p. 507.   The merit, however (if merit
it be), of having first put forward (about 1671) the hypothesis alluded to
in the text, belongs to D. Cassini.   See Delambre, *Hist. de l'Astr. Mod.*,
t. ii. p. 694, and *Kosmos*, Bd. iii. p. 410.

axis nearly erect to the plane of the ecliptic, in a period of between twenty-five and twenty-six days, and of the virtual limitation of the spots to a so-called "royal" zone extending some thirty degrees north and south of the solar equator, gained little either in precision or development from five generations of astronomers.

But in November 1769 a spot of extraordinary size engaged the attention of Alexander Wilson, professor of astronomy in the University of Glasgow. He watched it day by day, and to good purpose. As the great globe slowly revolved, carrying the spot towards its western edge, he was struck with the gradual contraction and final disappearance of the penumbra *on the side next the centre of the disc;* and when, on the 6th of December, the same spot re-emerged on the eastern limb, he perceived, as he had anticipated, that the shady zone was now deficient *on the opposite side*, and resumed its original completeness as it returned to a central position. Similar perspective effects were visible in numerous other spots subsequently examined by him, and he was thus in 1774 [1] able to prove by strict geometrical reasoning that such appearances were, as a matter of fact, produced by vast excavations in the sun's substance. It was not, indeed, the first time that such a view had been suggested. Father Scheiner's later observations plainly foreshadowed it; [2] a conjecture to the same effect was emitted by Leonhard Rost of Nuremberg early in the eighteenth century; [3] both by Lahire in 1703 and by J. Cassini in 1719 spots had been seen to form actual notches on the solar limb; while Pastor Schülen of Essingen convinced himself in 1770, by the careful study of appearances similar to those noted by Wilson, of the fact detected by him. [4] Nevertheless, Wilson's demonstration came with all the surprise of novelty, as well as with all the force of truth.

[1] *Phil. Trans.*, vol. lxiv. part I, pp. 7-11.
[2] *Rosa Ursina*, lib. iv. p. 507.
[3] R. Wolf, *Die Sonne und ihre Flecken*, p. 12.
[4] Schellen, *Die Spectralanalyse*, Bd. ii. p. 56 (3d ed.)

The general theory by which it was accompanied rested on a very different footing. It was avowedly tentative, and was set forth in the modest shape of an interrogatory. "Is it not reasonable to think," he asks, "that the great and stupendous body of the sun is made up of two kinds of matter, very different in their qualities ; that by far the greater part is solid and dark, and that this immense and dark globe is encompassed with a thin covering of that resplendent substance from which the sun would seem to derive the whole of his vivifying heat and energy?"[1] He further suggests that the excavations or spots may be occasioned "by the working of some sort of elastic vapour which is generated within the dark globe," and that the luminous matter being in some degree fluid, and being acted upon by gravity, tends to flow down and cover the nucleus. From these hints, supplemented by his own diligent observations and sagacious reasonings, Herschel elaborated a scheme of solar constitution which held its ground until the physics of the sun were revolutionised by the spectroscope.

A cool, dark, solid globe, its surface diversified with mountains and valleys, clothed with luxuriant vegetation, and "richly stored with inhabitants," protected by a heavy cloud-canopy from the intolerable glare of the upper luminous region, where the dazzling coruscations of a solar aurora some thousands of miles in depth evolved the stores of light and heat which vivify our world—such was the central luminary which Herschel constructed with his wonted ingenuity, and described with his wonted eloquence.

"This way of considering the sun and its atmosphere," he says,[2] "removes the great dissimilarity we have hitherto been used to find between its condition and that of the rest of the great bodies of the solar system. The sun, viewed in this light, appears to be nothing else than a very eminent, large, and lucid planet, evidently the first, or, in strictness of speaking, the only primary one of our system; all others being truly

[1] *Phil. Trans.*, vol. lxiv. p. 20.   [2] *Ibid.*, vol. lxxxv. 1795, p. 63.

secondary to it.   Its similarity to the other globes of the solar
system with regard to its solidity, its atmosphere, and its
diversified surface, the rotation upon its axis, and the fall of
heavy bodies, leads us on to suppose that it is most probably
also inhabited, like the rest of the planets, by beings whose
organs are adapted to the peculiar circumstances of that vast
globe."

   We smile at conclusions which our present knowledge con-
demns as extravagant and impossible, but such incidental
flights of fancy in no way derogate from the high value of
Herschel's contributions to solar science.   The cloud-like character which he attributed to the radiant shell of the sun (first
named by Schröter the "photosphere") is borne out by all
recent investigations ; he observed its mottled or corrugated
aspect, resembling, as he described it, the roughness on the
rind of an orange ; showed that "faculæ" are elevations or
heaped-up ridges of the disturbed photospheric matter ; and
threw out the idea that spots may be caused by an excess of
the ordinary luminous emissions.   A certain "empyreal" gas
was, he supposed (very much as Wilson had done), generated
in the body of the sun, and rising everywhere by reason of its
lightness, made for itself, when in moderate quantities, small
openings or "pores,"[1] abundantly visible as dark points on
the solar disc.   But should an uncommon quantity be formed,
" it will," he maintained, "burst through the planetary[2] regions
of clouds, and thus will produce great openings ; then, spreading itself above them, it will occasion large shallows (penumbræ),
and, mixing afterwards gradually with other superior gases, it
will promote the increase, and assist in the maintenance of the
general luminous phenomena."[3]

   This partial anticipation of the modern view that the solar
radiations are maintained by some process of circulation
within the solar mass, was reached by Herschel through pro-

---

[1] *Phil. Trans.*, vol. xci. 1801, p. 303.
[2] The supposed opaque or protective stratum was named by him "planetary," from the analogy of terrestrial clouds.   [3] *Ibid.*, p. 305.

longed study of the phenomena in question. The novel and important idea contained in it, however, it was at that time premature to attempt to develop. But though many of the subtler suggestions of Herschel's genius passed unnoticed by his contemporaries, the main result of his solar researches was an unmistakable one. It was nothing less than the definitive introduction into astronomy of the paradoxical conception of the central fire and hearth of our system as a cold, dark, terrestrial mass, wrapt in a mantle of innocuous radiance—an earth, so to speak, within—a sun without.

Let us pause for a moment to consider the value of this remarkable innovation. It certainly was not a step in the direction of truth. On the contrary, the crude notions of Anaxagoras and Xeno approached more nearly to what we now know of the sun, than the complicated structure devised for the happiness of a nobler race of beings than our own, by the benevolence of eighteenth-century astronomers. And yet it undoubtedly constituted a very important advance in science. It was the first earnest attempt to bring solar phenomena within the compass of a rational system ; to put together into a consistent whole the facts ascertained ; to fabricate, in short, a solar machine that would in some fashion work. It is true that the materials were inadequate and the design faulty. The resulting construction has not proved strong enough to stand the wear and tear of time and discovery, but has had to be taken to pieces and remodelled on a totally different plan. But the work was not therefore done in vain. None of Bacon's aphorisms show a clearer insight into the relations between the human mind and the external world than that which declares " Truth to emerge sooner from error than from confusion." [1] A definite theory (even if a false one) gives holding-ground to thought. Facts acquire a meaning with reference to it. It affords a motive for accumulating them and a means of co-ordinating them ; it provides a framework for their arrangement and a receptacle for their preservation, until they

[1] *Novum Organum*, lib. ii. aph. 20.

become too strong and numerous to be any longer included
within arbitrary limits, and shatter the vessel originally framed
to contain them.

Such was the purpose subserved by Herschel's theory of the
sun. It helped to *clarify* ideas on the subject. The turbid
sense of groping and viewless ignorance gave place to the
lucidity of a plausible scheme. The persuasion of knowledge
is a keen incentive to its increase. Few men care to investigate
what they are obliged to admit themselves entirely ignorant
of; but once started on the road of knowledge (real or supposed), they are eager to pursue it. By the promulgation of
a confident and consistent view regarding the nature of the
sun, accordingly, research was encouraged, because it was
rendered hopeful, and inquirers were shown a path leading
indefinitely onwards where an impassable thicket had before
seemed to bar the way.

We have called the "terrestrial" theory of the sun's nature
an innovation, and so, as far as its general acceptance is concerned, it may justly be termed; but, like all successful in-
novations, it was a long time brewing. It is extremely curious
to find that Herschel had a predecessor in its advocacy who
never looked through a telescope (nor, indeed, imagined the
possibility of such an instrument), who knew nothing of sun-
spots, was still (mistaken assertions to the contrary notwithstanding) in the bondage of the geocentric system, and regarded Nature from the lofty standpoint of an idealist
philosophy. This was the learned and enlightened Cardinal
Cusa, a fisherman's son from the banks of the Moselle, whose
distinguished career in the Church and in literature extended
over a considerable part of the fifteenth century (1401–64).
In his singular treatise *De Doctâ Ignorantiâ*, one of the most
notable literary monuments of the early Renaissance, the
following passage occurs :—" To a spectator on the surface of
the sun, the splendour which appears to us would be invisible,
since it contains, as it were, an earth for its central mass, with
a circumferential envelope of light and heat, and between the

two an atmosphere of water and clouds and translucent air."
The luminary of Herschel's fancy could scarcely be more
clearly portrayed; some added words, however, betray the
origin of the Cardinal's idea. "The earth also," he says,
"would appear as a shining star to any one outside the fiery
element." It was, in fact, an extension to the sun of the
ancient elemental doctrine; but an extension remarkable at
that period, as premonitory of the tendency, so powerfully
developed by subsequent discoveries, to assimilate the orbs of
heaven to the model of our insignificant planet, and to extend
the brotherhood of our system and our species to the farthest
limit of the visible or imaginable universe.

In later times we find Flamsteed communicating to Newton,
March 7, 1681, his opinion "that the substance of the sun is
terrestrial matter, his light but the liquid menstruum encompassing him."[1] Bode in 1776 arrived independently at the
conclusion that "the sun is neither burning nor glowing, but
in its essence a dark planetary body, composed like our earth
of land and water, varied by mountains and valleys, and enveloped in a vaporous atmosphere;[2]" and the learned in general
applauded and acquiesced. The view, however, was in 1787
still so far from popular, that the holding of it was alleged as
a proof of insanity in Dr. Elliot when accused of a murder-
ous assault on Miss Boydell. His friend Dr. Simmons stated
on his behalf that he had received from him in the preceding
January a letter giving evidence of a deranged mind, wherein
he asserted "that the sun is not a body of fire, as hath been
hitherto supposed, but that its light proceeds from a dense
and universal aurora, which may afford ample light to the
inhabitants of the surface beneath, and yet be at such a distance
aloft as not to annoy them. No objection, he saith, ariseth to
that great luminary's being inhabited; vegetation may obtain
there as well as with us. There may be water and dry land,
hills and dales, rain and fair weather; and as the light, so the

---

[1] Brewster's *Life of Newton*, vol. ii. p. 103.
[2] *Beschäftigungen d. Berl. Ges. Naturforschender Freunde*, Bd. ii. p. 233.

season must be eternal, consequently it may easily be conceived to be by far the most blissful habitation of the whole
system!" The Recorder, however, we are told, objected that
if an extravagant hypothesis were to be adduced as proof of
insanity, the same might hold good with regard to some other
speculators, and desired Dr. Simmons to tell the court what
he thought of the theories of Burnet and Buffon.[1]

Eight years later, this same "extravagant hypothesis," backed
by the powerful recommendation of Sir William Herschel,
obtained admittance to the venerable halls of science, there to
abide undisturbed for nearly seven decades. It is true there
were individual objectors, but their arguments made little
impression on the general body of opinion. Ruder blows
were required to shatter an hypothesis flattering to human
pride of invention in its completeness, in the plausible detail
of observations by which it seemed to be supported, and in its
condescension to the natural pleasure in discovering resem·
blance under all but total dissimilarity.

Sir John Herschel included among the results of his multifarious labours at the Cape of Good Hope a careful study of
the sun-spots conspicuously visible towards the end of the
year 1836 and in the early part of 1837. They were remarkable, he tells us, for their forms and arrangement, as well as
for their number and size ; one group, measured on the 29th
of March in the latter year, covering (apart from what may be
called its outlying dependencies) the vast area of five square
minutes or 3780 million square miles.[2] We have at present
to consider, however, not so much these observations in themselves, as the chain of theoretical suggestions by which they
were connected. The distribution of spots, it was pointed out,
on two zones parallel to the equator, showed plainly their
intimate connection with the solar rotation, and indicated as
their cause fluid circulations analogous to those producing the
terrestrial trade- and anti-trade winds.

---

[1] *Gentleman's Magazine,* 1787, p. 636.   [2] *Results, &c.,* p. 432.

"The spots, in this view of the subject," he went on to say,[1] "would come to be assimilated to those regions on the earth's surface where, for the moment, hurricanes and tornadoes prevail; the upper stratum being temporarily carried downwards, displacing by its impetus the two strata of luminous matter beneath, the upper of course to a greater extent than the lower, and thus wholly or partially denuding the opaque surface of the sun below. Such processes cannot be unaccompanied by vorticose motions, which, left to themselves, die away by degrees and dissipate, with the peculiarity that their lower portions come to rest more speedily than their upper, by reason of the greater resistance below, as well as the remoteness from the point of action, which lies in a higher region, so that their centres (as seen in our waterspouts, which are nothing but small tornadoes) appear to retreat upwards. Now this agrees perfectly with what is observed during the obliteration of the solar spots, which appear as if filled in by the collapse of their sides, the penumbra closing in upon the spot and disappearing after it."

When, however, it comes to be asked whether a cause can be found by which a diversity of solar temperature might be produced corresponding with that which sets the currents of the terrestrial atmosphere in motion, we are forced to reply that we know of no such cause. For Sir John Herschel's hypothesis of an increased retention of heat at the sun's equator, due to the slightly spheroidal or bulging form of its outer atmospheric envelope, assuredly gives no sufficient account of such circulatory movements as he supposed to exist. Nevertheless, the view that the sun's rotation is intimately connected with the formation of spots is so obviously correct, that we can only wonder it was not thought of sooner, while we are even now unable to explain with any certainty *how* it is so connected.

Mere scrutiny of the solar surface, however, is not the only means of solar observation. We have a satellite, and that

---

[1] *Results, &c.,* p. 434.

satellite from time to time acts most opportunely as a screen, cutting off a part or the whole of those dazzling rays in which the master-orb of our system veils himself from over-curious regards. The importance of eclipses to the study of the solar surroundings is of comparatively recent recognition; nevertheless, much of what we know concerning them has been snatched, as it were, by surprise under favour of the moon. In former times, the sole astronomical use of such incidents was the correction of the received theories of the solar and lunar movements; the precise time of their occurrence was the main fact to be noted, and subsidiary phenomena received but casual attention. Now, their significance as a geometrical test of tabular accuracy is altogether overshadowed by the interest attaching to the physical observations for which they afford propitious occasions. This change may be said to date, in its pronounced form, from the great eclipse of 1842. Although a necessary consequence of the general direction taken by scientific progress, it remains associated in a special manner with the name of Francis Baily.

The "philosopher of Newbury" was by profession a London stockbroker, and a highly successful one. Nevertheless, his services to science were numerous and invaluable, though not of the brilliant kind which attracts popular notice. Born at Newbury in Berkshire, April 28, 1774, and placed in the City at the age of fourteen, he derived from the acquaintance of Dr. Priestley a love of science which never afterwards left him. It was, however, no passion such as flames up in the brain of the destined discoverer, but a regulated inclination, kept well within the bounds of an actively pursued commercial career. After travelling for a year or two in what were then the wilds of North America, he went on the Stock Exchange in 1799, and earned during twenty-four years of assiduous application to affairs a high reputation for integrity and ability, to which corresponded an ample fortune. In the meantime the Astronomical Society (largely through his co-operation) had been founded; he had for three years acted as its secretary, and he

now felt entitled to devote himself exclusively to a subject
which had long occupied his leisure hours. He accordingly
in 1825 retired from business, purchased a house in Tavistock
Place, and fitted up there a small observatory. He was, however, by preference a computator rather than an observer.
What Sir John Herschel calls the "archæology of practical
astronomy" found in him an especially zealous student. He
re-edited the star-catalogues of Ptolemy, Ulugh Beigh, Tycho
Brahe, Hevelius, Halley, Flamsteed, Lacaille, and Mayer;
calculated the eclipse of Thales and the eclipse of Agathocles,
and vindicated the memory of the first Astronomer Royal. But
he was no less active in meeting present needs than in revising
past performances. The subject of the reduction of observations, then, as we have already explained,[1] in a state of deplorable confusion, attracted his most earnest attention, and he was
close on the track of Bessel when made acquainted with the
method of simplification devised at Königsberg. Anticipated
as an inventor, he could still be of eminent use as a promoter
of these valuable improvements; and, carrying them out on a
large scale in the star-catalogue of the Astronomical Society
(published in 1827), "he put" (in the words of Herschel) "the
astronomical world in possession of a power which may be said,
without exaggeration, to have changed the face of sidereal
astronomy."[2]

His reputation was still further enhanced by his renewal,
with vastly improved apparatus, of the method, first used by
Henry Cavendish in 1797–98, for determining the density of
the earth. From a series of no less than 2153 delicate and
difficult experiments, conducted at Tavistock Place during
the years 1838–42, he concluded our planet to weigh 5.66 as
much as a globe of water of the same bulk; and this result
(slightly corrected) is still accepted as a very close approximation to the truth.

What we have thus glanced at is but a fragment of the truly

[1] See *ante*, p. 41.
[2] *Memoir of Francis Baily, Mem. R. A. S.,* vol. xv. p. 324.

surprising mass of work accomplished by Baily in the course
of a variously occupied life. A rare combination of qualities
fitted him for his task. Unvarying health, undisturbed equa-
nimity, methodical habits, the power of directed and sustained
thought, joined to form in him an intellectual toiler of the
surest, though not perhaps of the very highest quality. He
was in harness almost to the end. He was destined scarcely
to know the miseries of enforced idleness or of consciously
failing powers. In 1842 he completed the laborious reduction
of Lalande's great catalogue, undertaken at the request of the
British Association, and was still engaged in seeing it through
the press when he was attacked with what proved his last, as it
was probably his first, serious illness. He, however, recovered
sufficiently to attend the Oxford Commemoration of July 2,
1844, where an honorary degree of D.C.L. was conferred
upon him in company with Airy and Struve ; but sank rapidly
after the effort, and died on the 30th of August following, at
the age of seventy, lamented and esteemed by all who knew
him.

It is now time to consider his share in the promotion of solar
research. Eclipses of the sun, both ancient and modern, were
a speciality with him, and he was fortunate in those which came
under his observation. Such phenomena are of three kinds—
partial, annular, and total. In a partial eclipse, the moon,
instead of passing directly between us and the sun, slips by, as
it were, a little on one side, thus cutting off from our sight
only a portion of his surface. An annular eclipse, on the
other hand, takes place when the moon is indeed centrally
interposed, but falls short of the apparent size required for the
entire concealment of the solar disc, which consequently remains visible as a bright ring or annulus, even when the
obscuration is at its height. In a total eclipse, on the contrary, the sun completely disappears behind the dark body of
the moon. The difference of the two latter varieties is due to
the fact that the apparent diameters of the sun and moon are
so nearly equal as to gain alternate preponderance one over

the other, through the slight periodical changes in their respective distances from the earth.

Now on the 15th of May 1836, an annular eclipse was visible in the northern parts of Great Britain, and was observed by Baily at Inch Bonney near Jedburgh. It was here that he saw the phenomenon which obtained the name of " Baily's Beads " from the notoriety conferred upon it by his vivid description.

" When the cusps of the sun," he writes, " were about 40° asunder, a row of lucid points, like a string of bright beads, irregular in size and distance from each other, *suddenly* formed round that part of the circumference of the moon that was about to enter, or which might be considered as having just entered on the sun's disc. Its formation indeed was so rapid that it presented the appearance of having been caused by the ignition of a fine train of gunpowder." He expected every moment to see the thread of light completed round the moon, attributing the serrated aspect of its limb to the projection of lunar mountains. " My surprise however was great," he continues, " on finding that these luminous points as well as the dark intervening spaces increased in magnitude, some of the contiguous ones appearing to run into each other like drops of water. . . . Finally, as the moon pursued her course, these dark intervening spaces (which, at their origin, had the appearance of lunar mountains in high relief, and which still continued attached to the sun's border) were stretched out into long, black, thick, parallel lines, joining the limbs of the sun and moon ; when all at once they *suddenly* gave way, and left the circumference of the sun and moon in those points, as in the rest, comparatively smooth and circular, and the moon perceptibly advanced on the face of the sun." [1]

A lively interest was excited by the communication from which the above passages are taken. The curious appearances described in it were not, indeed, an absolute novelty, but they had previously received only transient or partial notice. Webber in 1791, and Von Zach in 1820, had seen the

[1] *Mem. R. A. S.*, vol. x. pp. 5–6.

"beads;" Van Swinden had described the "belts" or "threads."[1]   These last were, moreover (as Baily clearly perceived), completely analogous to the "black ligament" which formed so troublesome a feature in the transits of Venus in 1764 and 1769, and which, to the regret and confusion, though no longer to the surprise of observers, was renewed in that of 1874.   No completely satisfactory explanation of the entire phenomenon has yet been offered.   Fundamentally, no doubt, it is an effect of what is called *irradiation*, by which a bright object seems to encroach upon a dark one; but other circumstances, both instrumental and atmospheric, aid in its production;[2] while the inequalities of the moon's edge complicate the action of other causes.

The immediate result of Baily's observation at Jedburgh was powerfully to stimulate attention to solar eclipses in their *physical* aspect.   Never before had an occurrence of the kind been expected so eagerly or prepared for so actively as that which was total over Central and Southern Europe on the 8th of July 1842.   Astronomers hastened from all quarters to the favoured region.   The Astronomer Royal (Airy) repaired to Turin; Baily to Pavia; Otto Struve threw aside his work amidst the stars at Pulkowa, and went south as far as Lipeszk; Schumacher travelled from Altona to Vienna; Arago from Paris to Perpignan.   Nor did their trouble go unrewarded. The expectations of the most sanguine were outdone by the wonders disclosed.

Baily (whose narrative we again have recourse to) had set up his Dollond's achromatic ($3\frac{1}{2}$ feet focal length) in an upper room of the University of Pavia, and was eagerly engaged in noting a partial repetition of the singular appearances seen by him in 1836, when he was "astounded by a tremendous burst of applause from the streets below, and at the *same moment* was electrified at the sight of one of the most brilliant and splendid phenomena that can well be imagined.   For at that

[1] *Mem. R. A. S.,* vol. x. pp. 14–17.
[2] See Proctor, *Transits of Venus,* pp. 63–66.

instant the dark body of the moon was *suddenly* surrounded
with a corona, or kind of bright glory similar in shape and
relative magnitude to that which painters draw round the
heads of saints, and which by the French is designated an
auréole. Pavia contains many thousand inhabitants, the major
part of whom were, at this early hour, walking about the streets
and squares or looking out of windows, in order to witness
this long-talked-of phenomenon ; and when the total obscuration took place, which was *instantaneous*, there was an universal shout from every observer, which 'made the welkin
ring,' and, for the moment, withdrew my attention from the
object with which I was immediately occupied. I had indeed
anticipated the appearance of a luminous circle round the
moon during the time of total obscurity ; but I did not expect,
from any of the accounts of preceding eclipses that I had read,
to witness so magnificent an exhibition as that which took
place. . . . The breadth of the corona, measured from the
circumference of the moon, appeared to me to be nearly equal
to half the moon's diameter. It had the appearance of
brilliant rays. The light was most dense (indeed I may say
quite dense) close to the border of the moon, and became
gradually and uniformly more attenuate as its distance therefrom increased, assuming the form of diverging rays in a
rectilinear line, which at the extremity were more divided, and
of an unequal length ; so that in no part of the corona could I
discover the regular and well-defined shape of a ring at its
outer margin. It appeared to me to have the sun for its
centre, but I had no means of taking any accurate measures
for determining this point. Its colour was quite white, not
pearl-colour, nor yellow, nor red, and the rays had a vivid and
flickering appearance, somewhat like that which a gaslight
illumination might be supposed to assume if formed into a
similar shape. . . . Splendid and astonishing, however, as
this remarkable phenomenon really was, and although it could
not fail to call forth the admiration and applause of every
beholder, yet I must confess that there was at the same time

F

something in its singular and wonderful appearance that was appalling ; and I can readily imagine that uncivilised nations may occasionally have become alarmed and terrified at such an object, more especially at times when the true cause of the occurrence may have been but faintly understood, and the phenomenon itself wholly unexpected.

" But the most remarkable circumstance attending the phenomenon was the appearance of *three large protuberances*, apparently emanating from the circumference of the moon, but evidently forming a portion of the corona. They had the appearance of mountains of a prodigious elevation ; their colour was red tinged with lilac or purple ; perhaps the colour of the peach-blossom would more nearly represent it. They somewhat resembled the snowy tops of the Alpine mountains when coloured by the rising or setting sun. They resembled the Alpine mountains also in another respect, inasmuch as their light was perfectly steady, and had none of that flickering or sparkling motion so visible in other parts of the corona. All the three projections were of the same roseate cast of colour, and very different from the brilliant vivid white light that formed the corona ; but they differed from each other in magnitude. . . . The whole of these three protuberances were visible even to the last moment of total obscuration ; at least, I never lost sight of them when looking in that direction ; and when the first ray of light was admitted from the sun, they vanished, with the corona, altogether, and daylight was *instantaneously* restored." [1]

Notwithstanding unfavourable weather, the "red flames" were perceived with little less clearness and no less amazement from the Superga than at Pavia, and were even discerned by Mr. Airy with the naked eye. "Their form" (the Astronomer Royal wrote) "was nearly that of saw-teeth in the position proper for a circular saw turned round in the same direction in which the hands of a watch turn ; . . . their colour was

[1] *Mem. R. A. S.*, vol. xv. pp. 4–6.

a full lake red, and their brilliancy greater than that of any other part of the ring." [1]

The height of these extraordinary objects was estimated by Arago at two minutes of arc, representing, at the sun's distance, an actual elevation of 56,000 miles. When carefully watched, the rose-flush of their illumination was perceived to fade through violet to white as the light returned; the same changes in a reversed order having accompanied their first appearance. Their forms, however, during about three minutes of visibility, showed no change, although of so (apparently) unstable a character as to suggest to Arago " mountains on the point of crumbling into ruins " through topheaviness. [2]

The corona, both as to figure and extent, presented very different appearances at different stations. This was no doubt due to varieties in atmospheric conditions. At the Superga, for instance, all details of structure seem to have been effaced by the murky air, only a comparatively feeble ring of light being seen to encircle the moon. Elsewhere, a brilliant radiated formation was conspicuous, spreading, at four opposite points, into four vast luminous expansions, compared to feather-plumes or *aigrettes.* [3] Arago at Perpignan noticed considerable irregularities in the divergent rays ; some appeared curved and twisted ; a few lay *across* the others, in a direction almost tangential to the moon's limb; the general effect being described as that of a " hank of thread in disorder." [4] At Lipeszk, where the sun stood much higher above the horizon than in Italy or France, the corona showed with surprising splendour. Its apparent extent was judged by Struve to be no less than twenty-five minutes (more than six times Airy's estimate), while the great plumes spread their radiance to three or four degrees from the dark lunar edge. So dazzling was the light, that many well-instructed persons denied the totality of the eclipse. Nor was the error without precedent, although the appearances attending respectively a total and an

[1] *Mem. R. A. S.*, vol. xv. p. 16.   [2] *Annuaire*, 1846, p. 409.
[3] *Ibid.*, p. 317.   [4] *Ibid.*, p. 322.

annular eclipse are in reality wholly dissimilar.   In the latter
case, the surviving ring of sunlight becomes so much enlarged
by irradiation, that the interposed dark lunar body is reduced
to comparative insignificance, or even invisibility.   Maclaurin
tells us,[1] that during an eclipse of this character which he
observed at Edinburgh in 1737, "gentlemen by no means
shortsighted declared themselves unable to discern the moon
upon the sun without the aid of a smoked glass ; " and Baily
(who, however, *was* shortsighted) could distinguish, in 1836,
with the naked eye no trace of "the globe of purple velvet"
which the telescope revealed as projected upon the face of the
sun.[2]   Moreover, the diminution of light is described by him
as " little more than might be caused by a temporary cloud
passing over the sun ; " birds continued in full song ; and
"one cock in particular was crowing with all his might while
the annulus was forming."

Very different were the effects of the eclipse of 1842, as to
which some interesting particulars were collected by Arago.[3]
Beasts of burthen, he tells us, paused in their labour, and
could by no amount of punishment be induced to move until
the sun reappeared.   Birds and beasts abandoned their food ;
linnets were found dead in their cages ; even ants suspended
their toil.   Diligence-horses, on the other hand, seemed as
insensible to the phenomenon as locomotives.   The convolvulus
and some other plants closed their leaves ; but those of the
mimosa remained open.   The little light that remained was of
a livid hue.   One observer described the general coloration
as resembling the lees of wine, but human faces showed pale
olive or greenish.   We may, then, rest assured that none of
the remarkable obscurations recorded in history were due to
eclipses of the annular kind.

The existence of the corona is no modern discovery.
Indeed, it is too conspicuous an apparition to escape notice
from the least attentive, or least practised observer of a total

---

[1] *Phil. Trans.*, vol. xl. p. 192.   [2] *Mem. R. A. S.*, vol. x. p. 17.
[3] *Ann. du Bureau des Long.*, 1846, p. 309.

eclipse. Nevertheless, explicit references to it are rare in early times. Both Plutarch,[1] however, and Philostratus in his Life of Apollonius of Tyana,[2] are unmistakable in their allusions, the latter describing a " crown," or garland similar to the iris, by which the sun was encompassed and obscured during an eclipse. The first to take the phenomenon into scientific consideration was Kepler. He showed, from the positions in their orbits at the time of the sun and moon, that an eclipse observed by Clavius at Rome in 1567 could not have been annular,[3] as the dazzling coronal radiance visible during the obscuration had caused it to be believed. Although he himself never witnessed a total eclipse of the sun, he carefully collected and compared the remarks of those more fortunate, and concluded that the ring of " flame-like splendour" seen on such occasions was caused by the reflection of the solar rays from matter condensed in the neighbourhood either of the sun or moon.[4] To the solar explanation he gave his own decided preference, but, with one of those curious flashes of half-prophetic insight characteristic of his genius, declared that " it should be laid by ready for use, not brought into immediate requisition." [5] So literally was his advice acted upon, that the theory, which we now know to be (broadly speaking) the correct one, only emerged from the repository of anticipated truths after 236 years of almost complete retirement, and even then timorously and with hesitation.

The first eclipse of which the attendant phenomena were observed with tolerable exactness was that which was central in the South of France, May 12, 1706. Cassini then put forward the view that the " crown of pale light " seen round the lunar disc was caused by the illumination of the zodiacal light ; [6] but it failed to receive the attention which, as a step

---

[1] *Op. Mor. et Phil.*, vol. ix. p. 682, edit. Lips. 1778.

[2] Book viii. chap. xxiii. Both references are due to R. Grant, *Astr. Nach.*, No. 1838.   [3] *Astronomiæ Pars Optica, Op. omnia*, t. ii. p. 317.

[4] *De Stellâ Novâ, Op.*, t. ii. pp. 696–697.   [5] *Astr. Pars. Op.*, p. 320.

[6] *Mém. de l'Ac. des Sciences*, 1715, p. 119.

in the right direction, it undoubtedly merited. Nine years later we meet with Halley's comments on a similar event, the first which had occurred in London since March 20, 1140. By nine in the morning of April 22 (O.S.), 1715, the obscuration, he tells us, "was about ten digits,[1] when the face and colour of the sky began to change from perfect serene azure blue to a more dusky livid colour, having an eye of purple intermixt. . . . A few seconds before the sun was all hid, there discovered itself round the moon a luminous ring, about a digit or perhaps a tenth part of the moon's diameter in breadth. It was of a pale whiteness or rather pearl colour, seeming to me a little tinged with the colours of the iris, and to be concentric with the moon, whence I concluded it the moon's atmosphere. But the great height thereof, far exceeding our earth's atmosphere, and the observation of some, who found the breadth of the ring to increase on the west side of the moon as emersion approached, together with the contrary sentiments of those whose judgment I shall always revere" (Newton is most probably referred to), "makes me less confident, especially in a matter whereto I confess I gave not all the attention requisite." He concludes by declining to decide whether the " enlightened atmosphere," which the appearance "in all respects resembled," " belonged to sun or moon." [2]

A French Academician, who happened to be in London at the time, was less guarded in expressing an opinion. The Chevalier de Louville declared emphatically for the lunar atmospheric theory of the corona,[3] and his authority carried great weight. It was, however, much discredited by an observation made by Maraldi in 1724, to the effect that the luminous ring, instead of travelling *with* the moon, was traversed *by* it.[4] This was in reality decisive, though, as usual, belief lagged far behind

---

[1] A digit = $\frac{1}{12}$th of the solar diameter.

[2] *Phil. Trans.*, vol. xxix. pp. 247–249.

[3] *Mém de l'Ac. des Sciences*, 1715; *Histoire*, p. 49; *Mémoires*, pp. 93–98.

[4] *Ibid.*, 1724, p. 178.

demonstration. Moreover, the advantage accruing from this fresh testimony was adjudged to the wrong claimant. In 1715 a novel explanation had been offered by Delisle and Lahire,[1] supported by experiments regarded at the time as perfectly satisfactory. The aureola round the eclipsed sun, they argued, is simply a result of the *diffraction* or apparent bending of the sunbeams that graze the surface of the lunar globe—an effect of the same kind as the coloured fringes of shadows. And this view prevailed amongst men of science until (and even after) Brewster showed, with clear and simple decisiveness, that such an effect could by no possibility be appreciable at our distance from the moon.[2] Don Jose Joaquim de Ferrer, who observed a total eclipse of the sun at Kinderhook, in the State of New York, on June 16, 1806, seems to have been ignorant that such a refined optical *rationale* of the phenomenon was current in the learned world. Two alternative explanations alone presented themselves to his mind as possible. The bright ring round the moon must be due to the illumination either of a lunar or of a solar atmosphere. If the former, he calculated that it should have a height fifty times that of the earth's gaseous envelope. "Such an atmosphere," he rightly concluded, "cannot belong to the moon, but must without any doubt belong to the sun."[3] He, however, stood alone in this unhesitating assertion.

The importance of the problem was first brought fully home to astronomers by the eclipse of 1842. The brilliant and complex appearance which, on that occasion, challenged the attention of so many observers, demanded and received, no longer the casual attention hitherto bestowed upon it, but the most earnest study of those interested in the progress of science. Nevertheless, it was only by degrees and through a process of "exclusions" (to use a Baconian phrase) that the corona was put in its right place as a solar appendage. As

[1] *Mém de l'Ac. des Sciences*, 1715, p. 161, and pp. 166-169.
[2] *Ed. Ency.*, art. *Astronomy*, p. 635.
[3] *Trans. Am. Phil. Soc.*, vol. vi. p. 274.

every other available explanation proved inadmissible and
dropped out of sight, the broad presentation of Nature's fact
remained, which, though of sufficiently obvious interpretation,
was long and persistently misconstrued. Nor was it until 1869
that absolutely decisive evidence on the subject was forthcoming, as we shall see farther on.

Sir John Herschel, writing to his venerable aunt, relates that
when the brilliant red flames burst into view behind the dark
moon on the morning of the 8th July 1842, the populace of
Milan, with the usual inconsequence of a crowd, raised the
shout, " *Es leben die Astronomen !* " [1] In reality, none were less
prepared for their apparition than the class to whom the
applause due to the magnificent spectacle was thus adjudged.
And in some measure through their own fault ; for many partial
hints and some distinct statements from earlier observers had
given unheeded notice that some such phenomenon might be
expected to attend a solar eclipse.

What we now call the "chromosphere" is an envelope of
glowing gases, principally hydrogen, by which the sun is
completely covered, and from which the "prominences" are
emanations, eruptive or otherwise. Now, continual indications
of the presence of this fire-ocean had been detected during
eclipses in the eighteenth and nineteenth centuries. Captain
Stannyan, describing in a letter to Flamsteed an occurrence of
the kind witnessed by him at Berne on May 1 (O.S.), 1706,
says that the sun's "getting out of the eclipse was preceded
by a blood-red streak of light from its left limb." [2] A precisely similar appearance was noted by both Halley and De
Louville in 1715 ; during annular eclipses by Lord Aberdour
in 1737,[3] and by Short in 1748,[4] the tint of the ruby border
being, however, subdued to "brown" or "dusky red" by the
surviving sunlight ; while observations identical in character

---

[1] *Memoir of Caroline Herschel*, p. 327.
[2] *Phil. Trans.*, vol. xxv. p. 2240.   [3] *Ibid.*, vol. xl. p. 182.
[4] *Ibid.*, vol. xlv. p. 586.

were made at Amsterdam in 1820,[1] at Edinburgh (by Henderson) in 1836, and at New York in 1838.[2]

"Flames" or "prominences," if more conspicuous, are less
constant in their presence than the glowing stratum from
which they spring. The first to describe them was a Swedish
professor named Vassenius, who observed a total eclipse at
Gottenburg, May 2 (O.S.), 1733.[3] His astonishment equalled
his admiration when he perceived, just outside the edge of the
lunar disc, and suspended, as it seemed, in the coronal atmosphere, three or four reddish spots or clouds, one of which was
so large as to be detected with the naked eye. As to their
nature, he did not even offer a speculation, further than by
tacitly referring them to the moon, in which position they
appear to have remained so long as the observation was held
in mind. It was repeated in 1778 by a Spanish admiral, but
with no better success in directing efficacious attention to the
phenomenon. Don Antonio Ulloa was on board his ship the
Espagne in passage from the Azores to Cape St. Vincent
on the 24th of June in that year, when a total eclipse of the
sun occurred, of which he has left a valuable description.
His notices of the corona are full of interest ; but what just
now concerns us is the appearance of "a red luminous point"
"near the edge of the moon," which gradually increased in
size as the moon moved away from it, and was visible during
about a minute and a quarter.[4] He was satisfied that it
belonged to the sun because of its fiery colour and growth
in magnitude, and supposed that it was occasioned by some
crevice or inequality in the moon's limb, through which the
solar light penetrated.

---

[1] *Mem. R. A. S.*, vol. i. pp. 145, 148.

[2] *American Journal of Science*, vol. xlii. p. 396.

[3] *Phil. Trans.*, vol. xxxviii. p. 134. Father Secchi has, however,
pointed out a tolerably distinct mention of a prominence so far back as
1239 A.D. In a description of a total eclipse of that date it is added, "Et
quoddam foramen erat ignitum in circulo solis ex parte inferiori " (Mura-
tori, *Rer. It. Scriptores*, t. xiv. col. 1097). The " circulus solis " of course
signifies the corona. [4] *Phil. Trans.*, vol. lxix. p. 114

Allusions less precise, both prior and subsequent, which it is now easy to refer to similar objects (such as the "slender columns of smoke" seen by Ferrer),[1] might be detailed; but the evidence already adduced suffices to show that the prominences viewed with such amazement in 1842 were no unprecedented, or even unusual phenomenon.

It was more important, however, to decide what was their nature than whether their appearance might have been anticipated. They were generally, and not very incorrectly, set down as solar clouds. Arago believed them to shine by reflected light,[2] but the Abbé Peytal rightly considered them to be self-luminous. Writing in a Montpélier paper of July 16, 1842, he declared that we had now become assured of the existence of a third or outer solar envelope, composed of a glowing substance of a bright rose tint, forming mountains of prodigious elevation, analogous in character to the clouds piled above our horizons.[3] This first extant description of a very important feature of our great luminary was probably founded on an observation made by Bérard at Toulon during the then recent eclipse, "of a very fine red band, irregularly dentelated, or, as it were, crevassed here and there,"[4] encircling a large arc of the moon's circumference. It can hardly, however, be said to have obtained distinct recognition until the 28th of July 1851. On that day a total eclipse took place, which was observed with considerable success in various parts of Sweden and Norway by a number of English astronomers. Mr. Hind saw, on the south limb of the moon, "a long range of rose-coloured flames,"[5] described by Mr. Dawes as "a low ridge of red prominences, resembling in outline the tops of a very irregular range of hills."[6] Mr. Airy termed the portion of this "rugged line of projections." visible to him the *sierra*, and was struck with its brilliant

---

[1] *Trans. Am. Phil. Soc.*, vol. vi. 1809, p. 267.
[2] *Annuaire*, 1846, p. 460.   [3] *Ibid.*, p. 439, *note*.
[4] *Ibid.*, 1846, p. 416.   [5] *Mem. R. A. S.*, vol. xxi. p. 82.
[6] *Ibid.*, p. 90.

light and "nearly scarlet" colour.[1] Its true character of a
continuous solar envelope was inferred from these data by
(amongst others) Grant, Swan, and Littrow; and was by
Father Secchi formally accepted as established after the
great eclipse of 1860.[2]

Several prominences of remarkable forms, especially one
variously compared to a Turkish scimitar, a sickle, and a
boomerang, were seen in 1851. In connection with them
two highly significant circumstances were pointed out. First,
that of the approximate coincidence between their positions
and those of sun-spots previously observed.[3] Next, that "the
moon passed over them, leaving them behind, and revealing
successive portions as she advanced."[4] This latter fact (as to
which there could be no doubt, since it was separately noted
by at least four first-rate observers) was justly considered by
the Astronomer Royal and others as affording absolute certainty of the solar dependence of these singular objects.
Nevertheless sceptics were still found. M. Faye of the
French Academy inclined to a lunar origin for them;[5]
Professor von Feilitsch of Greifswald published in 1852 a
treatise for the express purpose of proving all the luminous
phenomena attendant on solar eclipses—corona, prominences,
and "sierra" (or chromosphere)—to be purely optical appearances.[6] Happily, however, the unanswerable arguments of the
photographic camera were soon to be made available against
such hardy incredulity.

Thus, the virtual discovery of the solar appendages, both
coronal and chromospheric, may be said to have been begun
in 1842, and completed in 1851. The current Herschelian

---

[1] *Mem. R. A. S.*, vol. xxi. pp. 7–8.   [2] *Le Soleil*, t. i. p. 386.

[3] By Williams and Stanistreet, *Mem. R. A. S.*, vol. xxi. pp. 54, 56.
Santini had made a similar observation at Padua in 1842. Grant, *Hist.
Astr.*, p. 401.

[4] Lassell in *Month. Not.*, vol. xii. p. 53.

[5] *Comptes Rendus*, t. xxxiv. p. 155.

[6] *Optische Untersuchungen*, and *Zeitschrift für populäre Mittheilungen*,
Bod. i. 186, p. 201.

theory of the solar constitution remained, however, for the time, intact. Difficulties, indeed, were thickening around it ; but their discussion was perhaps felt to be premature, and they were permitted to accumulate without debate, until fortified by fresh testimony into unexpected and overwhelming preponderance.

CHAPTER IV.

PLANETARY DISCOVERIES.

In the course of his early gropings towards a law of the planetary distances, Kepler tried the experiment of setting a planet, invisible by reason of its smallness, to revolve in the vast region of (seemingly) desert space separating Mars from Jupiter.[1] The disproportionate magnitude of the same interval was explained by Kant as due to the overweening size of Jupiter. The zone in which each planet moved was, according to the philosopher of Königsberg, to be regarded as the empty storehouse from which its materials had been derived. A definite relation should thus exist between the planetary masses and the planetary intervals.[2] Lambert, on the other hand, sportively suggested that the body or bodies (for it is noticeable that he speaks of them in the plural) which once bridged this portentous gap in the solar system, might, in some remote age, have been swept away by a great comet, and forced to attend its wanderings through space.[3]

These speculations were destined before long to assume a more definite form. Johann Daniel Titius, a professor at Wittenberg (where he died in 1796), pointed out in 1772, in a note to a translation of Bonnet's *Contemplation de la Nature,*[4]

---

[1] *Op.*, t. i. p. 107. He interposed, but tentatively only, another similar body between Mercury and Venus.

[2] *Allgemeine Naturgeschichte* (ed. 1798), pp. 118–119.

[3] *Cosmologische Briefe,* No. 1 (quoted by Von Zach, *Monat. Corr.,* vol. iii. p. 592).

[4] Second ed., p. 7. See Bode, *Von dem neuen Hauptplaneten,* p. 43, *note.*

the existence of a remarkable symmetry in the disposition of
the bodies constituting the solar system.   By a certain series of
numbers, increasing in regular progression,[1] he showed that
the distances of the six known planets from the sun might
be represented with a close approach to accuracy.   But with
one striking interruption.   The term of the series succeeding
that which corresponded to the orbit of Mars was without a
celestial representative.   The orderly flow of the sequence was
thus singularly broken.   The space where a planet should—in
fulfilment of the "Law"—have revolved, was, it appeared,
untenanted.   Johann Elert Bode, then just about to begin his
long career as leader of astronomical thought and work at
Berlin, marked at once the anomaly, and filled the vacant
interval with an hypothetical planet.   The discovery of Uranus
at a distance falling but slightly short of perfect conformity
to the law of Titius, lent weight to a seemingly hazardous
prediction, and Von Zach was actually at the pains, in 1785,
to calculate what he termed "analogical" elements[2] for this
unseen and (by any effect or influence) *unfelt* body.   The
search for it, though confessedly scarcely less chimerical than
that of alchemists for the philosopher's stone, he kept steadily
in view for fifteen years, and at length, September 21, 1800,
succeeded in organising, in combination with five other German astronomers assembled at Lilienthal, a force of what he
jocularly termed celestial police, for the express purpose of
tracking and intercepting the fugitive subject of the sun.   The
zodiac was accordingly divided for purposes of scrutiny into
twenty-four zones; their apportionment to separate observers
was in part effected, and the association was rapidly getting into
working order, when news arrived that the missing planet had

---

[1] The representative numbers are obtained by adding 4 to the following
series (irregular, it will be observed, in its first member, which should be
$1\frac{1}{2}$ instead of 0): 0, 3, 6, 12, 24, 48, &c.   The formula is a purely empirical one, and is, moreover, completely at fault as regards the distance of
Neptune.

[2] *Monat. Corr.*, vol. iii. p. 596.

been found, through no systematic plan of search, but by the diligent, though otherwise directed labours of a distant watcher of the skies.

Giuseppe Piazzi was born at Ponte in the Valtelline, July 16, 1746. He studied at various places and times under Tiraboschi, Beccaria, Jacquier, and Le Sueur; and having entered the Theatine order of monks at the age of eighteen, he taught philosophy, science, and theology in several of the Italian cities as well as in Malta until 1780, when the chair of mathematics in the University of Palermo was offered to and accepted by him. Prince Caramanico, then viceroy of Sicily, had scientific leanings, and was easily won over to the project of building an observatory, a commodious foundation for which was afforded by one of the towers of the viceregal palace. This architecturally incongruous addition to an ancient Saracenic edifice—once the abode of Kelbite and Zirite Emirs—was completed in February 1791. Piazzi, meanwhile, had devoted nearly three years to the assiduous study of his new profession, acquiring a practical knowledge of Lalande's methods at the École Militaire, and of Maskelyne's at the Royal Observatory; and returned to Palermo in 1789, bringing with him, in the great five-foot circle which he had prevailed upon Ramsden to construct, the most perfect measuring instrument hitherto employed by an astronomer.

He had been above nine years at work on his star-catalogue, and was still profoundly unconscious that a place amongst the Lilienthal band of astronomical detectives was being held in reserve for him, when, on the first evening of the nineteenth century, January 1, 1801, he noted the position of an eighth-magnitude star in a part of the constellation Taurus, towards which an error of Wollaston's had directed his special attention. On the following night, it seemed to him that the star had slightly shifted its position to the west; on the 3d, he assured himself of the fact, and believed that he had chanced upon a new kind of comet without tail or coma. The wandering body (whatever its nature) exchanged retrograde for direct motion on January

13,[1] and was carefully watched by Piazzi until February 11, when a dangerous illness interrupted his observations. He had, however, not omitted to give notice of his discovery, but so precarious were communications in those unpeaceful times, that his letter to Oriani of January 23 did not reach Milan until April 5, while a missive of one day later addressed to Bode came to hand at Berlin, March 20. The delay just afforded time for the publication, by a young philosopher of Jena named Hegel, of a "Dissertation" showing, by the clearest light of reason, that the number of the planets could not exceed seven, and exposing the folly of certain devotees of induction who sought a new celestial body merely to fill a gap in a numerical series.[2]

Unabashed by speculative scorn, Bode had scarcely read Piazzi's letter when he concluded that it referred to the precise body in question. The news spread rapidly, and created a profound sensation not unmixed with alarm lest this latest addition to the solar family should have been found only to be again lost. For by that time Piazzi's moving star was too near the sun to be any longer visible, and in order to redis-cover it after conjunction a tolerably accurate knowledge of its path was indispensable. But a planetary orbit had never before been calculated from such scanty data as Piazzi's ob-servations afforded;[3] and the attempts made by nearly every astronomer of note in Germany to compass the problem were manifestly inadequate, failing even to account for the positions in which the body had been actually seen, and *à fortiori* serv-ing only to mislead as to the places where, from September 1801, it ought once more to have become discernible. It was in this extremity that the celebrated mathematician Gauss came to the rescue. He was then in his twenty-fifth year, and

[1] Such reversals of direction in the apparent movements of the planets are a consequence of the earth's revolution in its orbit.

[2] *Dissertatio Philosophica de Orbitis Planetarum*, 1801. See Wolf, *Gesch. d. Astr.*, p. 685.

[3] Observations on Uranus, as a supposed fixed star, reached back to 1690.

was earning his bread by tuitions at Brunswick, with many
possibilities, but no settled career before him. The news
from Palermo may be said to have converted him from an
arithmetician into an astronomer. He was already in possession of a new and more general method of computing elliptical
orbits, and the system of "least squares," which he had devised though not published, enabled him to extract the utmost
amount of probable truth from a given set of observations.
Armed with these novel powers, he set to work, and the communication in November of his elements and ephemeris for
the lost object revived the drooping hopes of the little band
of eager searchers. Their patience, however, was to be still
further tried. Clouds, mist, and sleet seemed to have conspired to cover the retreat of the fugitive; but on the last
night of the year the sky cleared unexpectedly with the setting
in of a hard frost, and there, in the north-western part of Virgo,
nearly in the position assigned by Gauss to the runaway
planet, a strange star was discerned by Von Zach [1] at Gotha,
and on the subsequent evening—the anniversary of the original
discovery—by Olbers at Bremen. The name of Ceres (as the
tutelary goddess of Sicily) was, by Piazzi's request, bestowed
upon this first known of the numerous, and probably all but
innumerable, family of the minor planets.

The recognition of the second followed as the immediate
consequence of the detection of the first. Olbers had made
himself so familiar with the positions of the small stars along
the track of the long-missing body, that he was at once struck,
March 28, 1802, with the presence of an intruder near the
spot where he had recently identified Ceres. He at first
believed the newcomer to be a variable star usually inconspicuous, but just then at its maximum of brightness; but
within two hours he had convinced himself that it was no *fixed*
star, but a rapidly moving object. The aid of Gauss was

---

[1] He had caught a glimpse of it on December 7, but was prevented by
bad weather from verifying his suspicion. *Monat. Corr.*, vol. v. p.
171.

again invoked, and his prompt calculations showed that this
fresh celestial acquaintance (named "Pallas" by Olbers) revolved round the sun at nearly the same mean distance
as Ceres, and was beyond question of a strictly analogous
character.

This result was perplexing in the extreme. The symmetry
and simplicity of the planetary scheme appeared fatally com-
promised by the admission of many, where room could,
according to old-fashioned rules, only be found for one. A
daring hypothesis of Olbers' invention provided an exit from
the difficulty. He supposed that both Ceres and Pallas were
fragments of a primitive trans-Martian planet, blown to pieces
in the remote past, either by the action of internal forces or by
the impact of a comet; and predicted that many more such
fragments would be found to circulate in the same region. He,
moreover, pointed out that these numerous orbits, however
much they might differ in other respects, must all have a
common line of intersection,[1] and that the bodies moving in
them must consequently pass, at each revolution, through two
opposite points of the heavens, one situated in the Whale, the
other in the constellation of the Virgin, where already Pallas
had been found, and Ceres recaptured. The intimation that
fresh discoveries might be expected in those particular regions
was singularly justified by the detection of the two bodies now
known respectively as Juno and Vesta. The first was found
near the predicted spot in Cetus by Harding, Schröter's
assistant at Lilienthal, September 2, 1804; the second by
Olbers himself in Virgo, after three years of patient scrutiny,
March 29, 1807.

The theory of an exploded planet now seemed to have
everything in its favour. It required that the mean or average
distances of the newly discovered bodies should be nearly the

[1] Planetary fragments, hurled *in any direction*, and *with any velocity*
short of that which would for ever release them from the solar sway, would
continue to describe elliptic orbits round the sun, all passing through the
scene of the explosion, and thus possessing a common line of intersection.

same, but admitted a wide range of variety in the shapes and positions of their orbits, provided always that they preserved common points of intersection. These conditions were fulfilled with a striking approach to exactness. Three of the four "asteroids" (a designation introduced by Sir W. Herschel[1]) conformed with very approximate precision to "Bode's law" of distances; they all traversed, in their circuits round the sun, nearly the same parts of Cetus and Virgo; while the eccentricities and inclinations of their paths departed widely from the planetary type—that of Vesta, for example, making with the ecliptic an angle of no less than 35°. The minuteness of these bodies appeared further to strengthen the imputation of a fragmentary character. Herschel estimated the diameter of Ceres at 162, that of Pallas at 147 miles.[2] Juno is smaller than either; and even Vesta, which surpasses all the minor planets in size, and may, under favourable circumstances, be seen with the naked eye, has a diameter scarcely, if at all, exceeding 500 miles. A suspected variability of brightness in some of the asteroids, somewhat hazardously explained as due to the irregularities of figure to be expected in cosmical *potsherds* (so to speak), was added to the confirmatory evidence.[3] The strong point of the theory, however, lay not in what it explained, but in what it had predicted. It had been twice confirmed by actual exploration of the skies, and had produced, in the recognition of Vesta, the first recorded instance of the *premeditated* discovery of a heavenly body.

The view not only commended itself to the facile imagination of the unlearned, but received the sanction of the highest scientific authority. The great Lagrange bestowed upon it his analytical *imprimatur,* showing that the explosive forces required to produce the supposed catastrophe came well within

---

[1] *Phil. Trans.*, vol. xcii. part ii. p. 228.

[2] *Ibid.*, p. 218. In a letter to Von Zach of June 24, 1802, he speaks of Pallas as "almost incredibly small," and makes it only seventy English miles in diameter. *Monat. Corr.*, vol. vi. pp. 89–90.

[3] Olbers, *Monat. Corr.*, vol. vi. p. 88.

the bounds of possibility—a velocity of less than twenty times that of a cannon-ball leaving the gun's mouth sufficing, according to his calculation, to have launched the asteroidal fragments on their respective paths.   Indeed, he was disposed to regard the hypothesis of disruption as more generally available than its author had designed it to be, and proposed to supplement with it, as explanatory of the eccentric orbits of comets, the nebular theory of Laplace, thereby obtaining, as he said, "a complete view of the origin of the planetary system more conformable to Nature and mechanical laws than any yet proposed." [1]

Nevertheless the hypothesis of Olbers has not held its ground.   It seemed as if all the evidence available for its support had been produced at once and spontaneously, while the unfavourable items were elicited slowly, and, as it were, by cross-examination.   A more extended acquaintance with the group of bodies whose peculiarities it was framed to explain, has shown them, after all, as recalcitrant to any such explanation. Coincidences at the first view significant and striking have been swamped by contrary examples ; and a hasty general conclusion has, by a not uncommon destiny, at last perished under the accumulation of particulars.   Moreover, as has been remarked by Professor Newcomb,[2] mutual perturbations would rapidly efface all traces of a common disruptive origin, and the catastrophe, to be perceptible in its effects, should have been comparatively recent.

A new generation of astronomers had arisen before any additions were made to the little family of the minor planets.   Piazzi died in 1826, Harding in 1834, Olbers in 1840 ; all those who had prepared, or participated in the first discoveries passed away without witnessing their resumption.   In 1830, however, a certain Hencke, ex-postmaster in the Prussian town of Driessen, set himself to watch for new planets, and after fifteen long years his patience was rewarded.   The asteroid found by him December 8, 1845, received the name of Astræa,

[1] *Conn. d. Tems.* for 1814, p. 218.   [2] *Popular Astronomy*, p. 327.

and his further prosecution of the search resulted, July 1, 1847, in the discovery of Hebe. A few weeks later, August 13, Mr. Hind, after many months' exploration from Mr. Bishop's observatory in the Regent's Park, picked up Iris, and, October 18, Flora.[1] The next on the list was Metis, found by Mr. Graham, April 25, 1848, at Markree in Ireland.[2] At the close of the period to which our attention is at present limited, the number of these small bodies known to astronomy was thirteen; and the course of discovery has since proceeded still more rapidly and with less interruption.

Both in itself and in its consequences the recognition of the minor planets was of the highest importance to science. The traditional ideas regarding the constitution of the solar system were enlarged by the admission of a new class of bodies, strongly contrasted, yet strictly co-ordinate with the old-established planetary order; the profusion of resource, so conspicuous in the living kingdoms of Nature, was seen to prevail no less in the celestial spaces; and some faint preliminary notion was afforded of the indefinite complexity of relations underlying the apparent simplicity of the majestic scheme to which our world belongs. Theoretical and practical astronomy both derived profit from the admission of these apparently insignificant strangers to the rights of citizenship of the solar system. The disturbance of their motions by their giant neighbour afforded a more accurate knowledge of the Jovian mass, which Laplace had taken about $\frac{1}{50}$th too small; the anomalous character of their orbits presented geometers with highly stimulating problems in the theory of perturbations; while the exigencies of the first discovery had produced the *Theoria Motus*, and won Gauss over to the ranks of calculating astronomy. Moreover, the sure prospect of further detections powerfully incited to the exploration of the skies; observers became more numerous and more zealous in view of the prizes held out to them; star-maps were diligently constructed, and the sidereal multitude strewn along the great zodiacal belt acquired a fresh

[1] *Month. Not.*, vol. vii. p. 299; vol. viii. p. 1.   [2] *Ibid.*, vol. viii. p. 146.

interest when it was perceived that its least conspicuous
member might be a planetary shred or projectile in the majestic
disguise of a distant sun. Harding's " Celestial Atlas," designed
for the special purpose of facilitating asteroidal research, was
the first systematic attempt to represent to the eye the *telescopic*
aspect of the heavens. It was while engaged on its construction that the Lilienthal observer successfully intercepted Juno
on her passage through the Whale in 1804 ; whereupon promoted to Göttingen, he there completed, in 1822, the arduous
task so opportunely entered upon a score of years previously.
Still more important were the great star-maps of the Berlin
Academy, undertaken at Bessel's suggestion, with the same
object of distinguishing errant from fixed stars, and executed, under Encke's supervision, during the years 1830–59.
They have played a noteworthy part in the history of planetary
discovery, nor of the minor kind alone.

We have now to recount an event unique in scientific history.
The discovery of Neptune has been characterised as the result
of a " movement of the age," [1] and with some justice. It had
become necessary to the integrity of planetary theory. Until
it was accomplished, the phantom of an unexplained anomaly
in the orderly movements of the solar system must for ever
have haunted the brains of astronomers. Moreover, it was
prepared by many, suggested as possible by not a few, and
actually achieved, simultaneously, independently, and completely, by two investigators.

The position of the planet Uranus was recorded as that of
a fixed star no less than twenty times between 1690 and
the epoch of its final detection by Herschel. But these early
observations, far from affording the expected facilities for the
calculation of its orbit, proved a source of grievous perplexity.
The utmost ingenuity of geometers failed to combine them
satisfactorily with the later Uranian places, and it became
evident, either that they were widely erroneous, or that the
revolving body was wandering from its ancient track. The

    [1] Airy, *Mem. R. A. S.,* vol. xvi. p. 386.

simplest course was to reject them altogether, and this was done in the new Tables published in 1821 by Alexis Bouvard, the indefatigable computating partner of Laplace. But the trouble was not thus to be got rid of. After a few years fresh irregularities began to appear, and continued to increase until absolutely "intolerable." It may be stated, as illustrative of the perfection to which astronomy had been brought, that divergences regarded as menacing the very foundation of its theories never entered the range of unaided vision. In other words, if the theoretical and the real Uranus had been placed side by side in the sky, they would have seemed, to the sharpest eyes, to form a single body.[1]

The idea that these enigmatical disturbances were due to the attraction of an unknown exterior body was a tolerably obvious one ; and we accordingly find it suggested in many different quarters. Bouvard himself was perhaps the first to conceive it. He kept the possibility continually in view, and bequeathed to his nephew's diligence the inquiry into its reality when he felt that his own span was drawing to a close; but before any progress had been made with it, he had already (June 7, 1843) "ceased to breathe and to calculate." The Rev. T. J. Hussey actually entertained in 1834 the notion, but found his powers inadequate to the task, of assigning an approximate place to the disturbing body ; and Bessel, in 1840, laid his plans for an assault in form upon the Uranian difficulty, the triumphant exit from which fatal illness frustrated his hopes of effecting or even witnessing.

The problem was practically untouched when, in 1841, an undergraduate of St. John's College, Cambridge, formed the resolution of grappling with it. The projected task was an

[1] See Newcomb's *Pop. Astr.*, p. 359. The error of Uranus amounted, in 1844, to 2′ ; but even the tailor of Breslau, whose extraordinary powers of vision Humboldt commemorates (*Kosmos*, Bd. iii. p. 112), could only see Jupiter's first satellite at its greatest elongation, 2′ 15″. He might, however, possibly have distinguished two objects of *equal* lustre at a lesser interval. The components of the double star ε Lyræ, which Bessel, when a boy, could see separately with the naked eye, are 3½′ apart.

arduous one.   There were no guiding precedents for its
conduct.   Analytical obstacles had to be encountered so
formidable as to appear invincible even to such a mathematician as Airy.   John Couch Adams, however, had no sooner
taken his degree, which he did as senior wrangler in January
1843, than he set resolutely to work, and by October 1845
was able to communicate to the Astronomer Royal numerical estimates of the elements and mass of the unknown
planet, together with an indication of its actual place in the
heavens.

Sir George Biddell Airy had begun in 1835 his long and
energetic administration of Greenwich Observatory, and was
already in possession of data vitally important to the momentous inquiry then on foot.   At his suggestion, and under his
superintendence, the reduction of all the planetary observations
made at Greenwich from 1750 downwards had been undertaken in 1833.   The results, published in 1846, constituted a
permanent and universal stock of materials for the correction
of planetary theory.   But in the meantime, investigators, both
native and foreign, were freely supplied with the "places and
errors," which, clearly exhibiting the discrepancies between
observation and calculation — between what *was* and what
was *expected*—formed the very groundwork of future improvements.

Mr. Adams had no reason to complain of official discourtesy.
His labours were aided and encouraged; but they were *not*
fully believed in.   "I have always," Sir George Airy wrote,[1]
"considered the correctness of a distant mathematical result to
be a subject rather of moral than of mathematical evidence."
And, in the case actually before him, there was absolutely no
warrant for putting faith in the solution, by a young and
untried man, of a problem before the complexities of which
Laplace himself might have quailed.   Moreover, Mr. Adams
unaccountably neglected to answer (until too late) a question
regarded by Sir George Airy in the light of an *experimentum*

[1] *Mem. R. A. S.*, vol. xvi. p. 399.

crucis as to the soundness of the new theory. Nor did he himself take any steps to obtain a publicity which he was more anxious to merit than to secure. The investigation consequently remained buried in obscurity. It is now known that had a search been instituted in the autumn of 1845 for the remote body whose existence had been so marvellously foretold, it would have been found within *three and a half lunar diameters* (1° 49′) of the spot assigned to it by Mr. Adams.

A competitor, however, equally daring and more fortunate— *audax fortunâ adjutus*, as Gauss said of him—was even then entering the field. Urbain Jean Joseph Leverrier, the son of a small Government *employé* in Normandy, was born at Saint-Lô, March 11, 1811. He studied with brilliant success at the École Polytechnique, accepted the post of astronomical teacher there in 1837, and, "docile to circumstance," immediately concentrated the whole of his vast, though as yet undeveloped powers upon the formidable problems of celestial mechanics. He lost no time in proving to the mathematical world that the race of giants was not extinct. Two papers on the stability of the solar system, presented to the Academy of Sciences, September 16 and October 14, 1839, showed him to be the worthy successor of Lagrange and Laplace, and encouraged hopes destined to be abundantly realised. His attention was directed by Arago to the Uranian difficulty in 1845, when he cheerfully put aside certain intricate cometary researches upon which he happened to be engaged, in order to obey with dutiful promptitude the summons of the astronomical chief of France. In his first memoir on the subject (communicated to the Academy, November 10, 1845), he proved the inadequacy of all known causes of disturbance to account for the vagaries of Uranus; in a second (June 1, 1846), he demonstrated that only an exterior body occupying at a certain date a determinate position in the zodiac could produce the observed effects; in a third (August 31, 1846), he assigned the orbit of the disturbing body, and announced its visibility as an object with a sensible disc about as bright as a star of the eighth magnitude.

The question was now visibly approaching an issue. On September 10 Sir John Herschel declared to the British Association respecting the hypothetical new planet : " We see it as Columbus saw America from the coast of Spain. Its movements have been felt, trembling along the far-reaching line of our analysis with a certainty hardly inferior to that of ocular demonstration." Less than a fortnight later, September 23, Professor Galle, of the Berlin Observatory, received a letter from Leverrier requesting his aid in the telescopic part of the inquiry already analytically completed. He directed his refractor to the heavens that same night, and perceived within less than a degree of the spot indicated, an object with a measurable disc nearly three seconds in diameter. Its absence from Bremiker's recently completed map of that region of the sky showed it to be no star, and its movement in the predicted direction confirmed without delay the strong persuasion of its planetary nature.

In this remarkable manner the existence of the remote member of our system known as " Neptune " was ascertained. But the discovery, which faithfully reflected the duplicate character of the investigation which led to it, had been already secured at Cambridge before it was announced from Berlin. Sir George Airy's incredulity vanished in the face of the striking coincidence between the position assigned by Leverrier to the unknown planet in June, and that laid down by Mr. Adams in the previous October ; and on the 9th of July he wrote to Professor Challis, director of the Cambridge Observatory, recommending a search with the great Northumberland equatoreal. Had a good star-map been at hand, the process would have been a simple one ; but of Bremiker's " Hora XXI." no news had yet reached England, and there was no other sufficiently comprehensive to be available for an inquiry which, in the absence of such aid, promised to be both long and laborious. As the event proved, it might have been neither.

"After four days of observing," Professor Challis wrote, October 12, 1846, to Sir George Airy, "the planet was in my grasp if only I had examined or mapped the observations."[1] Had he done so, the first honours in the discovery, both theoretical and optical, would have fallen to the University of Cambridge. But Professor Challis had other astronomical avocations to attend to, and, moreover, his faith in the precision of the indications furnished to him was, by his own confession, a very feeble one. For both reasons he postponed to a later stage of the proceedings the discussion and comparison of the data nightly furnished to him by his telescope, and thus allowed to lie, as it were, latent in his observations the momentous result which his diligence had ensured, but which his delay suffered to be anticipated.[2]

Nevertheless it should not be forgotten that the Berlin astronomer had two circumstances in his favour apart from which his swift success could hardly have been achieved. The first was the possession of a good star-map; the second was the clear and confident nature of Leverrier's instructions. "Look where I tell you," he seemed authoritatively to say, "and you will see an object such as I describe."[3] And in fact, not only Galle on the 23d of September, but also Challis on the 29th, immediately after reading the French geometer's lucid and impressive treatise, picked out from among the stellar points strewing the zodiac, a small planetary disc, which eventually proved to be that of the precise body he had been in search of during two months.

Personal questions, however, vanish in the magnitude of the event they relate to. By it the last lingering doubts as to the universal validity of the Newtonian Law were dissipated. Recondite analytical methods received a confirmation brilliant

---

[1] *Mem. R. A. S.*, vol. xvi. p. 412.

[2] He had recorded the places of 3150 stars (three of which were different positions of the planet), and was preparing to map them, when, October 1, news of the discovery arrived from Berlin. Prof. Challis's *Report*, quoted in Obituary Notice, *Month. Not.*, Feb. 1803, p. 170.

[3] See Airy in *Mem. R. A. S.*, vol. xvi. p. 411.

and intelligible even to the minds of the vulgar, and emerged
from the patient solitude of the study to enjoy an hour of
clamorous triumph.   For ever invisible to the unaided eye of
man, a sister-globe to our earth was shown to circulate, in
perpetual frozen exile, at thirty times its distance from the sun.
Nay, the possibility was made apparent that the limits of our
system were not even thus reached, but that yet profounder
abysses of space might shelter obedient, though little favoured
members of the solar family, by future astronomers to be recognised through the sympathetic thrillings of Neptune, even
as Neptune himself was recognised through the tell-tale deviations of Uranus.

It is curious to find that the fruit of Adams' and Leverrier's
laborious investigations had been accidentally all but snatched
half a century before it was ripe to be gathered.   On the 8th,
and again on the 10th of May 1795, Lalande noted the position of Neptune as that of a fixed star, but perceiving that the
two observations did not agree, he suppressed the first as
erroneous, and pursued the inquiry no further.   An immor-
tality which he would have been the last to despise hung in
the balance; the feather-weight of his carelessness, however,
kicked the beam, and the discovery was reserved to be more
hardly won by later comers.

Bode's Law did good service in the quest for a trans-Uranian
planet by affording ground for a probable assumption as to
its distance.   A starting-point for approximation was provided by it; but it was soon found to be considerably at
fault.   Even Uranus is about 36 millions of miles nearer to
the sun than the order of progression requires; and Neptune's
vast distance of 2800 million should be increased by no less
than 800 million miles, and its period of 165 lengthened out to
225 years,[1] in order to bring it into conformity with the curious
and unexplained rule which planetary discoveries have alternately tended to confirm and to invalidate.

Within seventeen days of its identification with the Berlin

[1] Ledger, *The Sun, its Planets, and their Satellites*, p. 414.

achromatic, Neptune was found to be attended by a satellite. This discovery was the first notable performance of the celebrated two-foot reflector[1] erected by Mr. Lassell at his suggestively named residence of Starfield, near Liverpool. William Lassell was a brewer by profession, but by inclination an astronomer. Born at Bolton in Lancashire, June 18, 1799, he closed a life of eminent usefulness to science, October 5, 1880, thus spanning with his well-spent years almost the entire of the momentous period which we have undertaken to traverse. At the age of twenty-one, being without the means to purchase, he undertook to construct telescopes, and naturally turned his attention to the reflecting sort, as favouring amateur efforts by the comparative simplicity of its structure. His native ingenuity was remarkable, and was developed by the hourly exigencies of his successive enterprises. Their uniform success encouraged him to enlarge his aims, and in 1844 he visited Birr Castle for the purpose of inspecting the machine used in polishing the giant speculum of Parsonstown. In the construction of his new instrument, however, he eventually discarded the model there obtained, and worked on a method of his own, assisted by the supreme mechanical skill of James Nasmyth. The result was a Newtonian of exquisite definition, with an aperture of two, and a focal length of twenty feet, provided by a novel artifice with the equatoreal mounting, previously regarded as available only for refractors.

This beautiful instrument afforded to its maker, October 10, 1846, a cursory view of a Neptunian attendant. But the planet was then approaching the sun, and it was not until the following July that the observation could be verified, which it was completely, first by Lassell himself, and somewhat later by Otto Struve and Bond of Cambridge (U.S.) When it is considered that this remote object shines by reflecting sunlight reduced by distance to $\frac{1}{900}$th of the intensity with which it illuminates our moon, the fact of its visibility, even in the most perfect telescopes, is a somewhat surprising one.

[1] Lately presented by the Misses Lassell to the Greenwich Observatory.

It can only, indeed, be accounted for by attributing to it dimensions very considerable for a body of the secondary order. It shares with the moons of Uranus the peculiarity of retrograde motion ; that is to say, its revolutions, running counter to the grand current of movement in the solar system, are performed from east to west, in a plane inclined at an angle of 35° to that of the ecliptic. Their swiftness serves to measure the mass of the globe round which they are performed. For while our moon takes twenty-seven days and nearly eight hours to complete its circuit of the earth, the satellite of Neptune, at a distance not greatly inferior, sweeps round its primary in five days and twenty-one hours, showing (according to a very simple principle of computation) that it is urged by a force seventeen times greater than the terrestrial pull upon the lunar orb. Combining this result with that of measurements of the small telescopic disc of this farthest known planet, it is found that while in *mass* Neptune equals seventeen earths, in *bulk* it is equivalent to eighty-four. This is as much as to say that it is composed of relatively very light materials, or more probably of materials distended by internal heat, as yet unwasted by radiation into space, to about five times the volume they would occupy in the interior of our globe. The fact, at any rate, is fairly well ascertained that the average density of Neptune differs little from that of water.

We must now turn from this late-recognised member of our system to bestow some brief attention upon the still fruitful field of discovery offered by one of the immemorial five. The family of Saturn, unlike that of its brilliant neighbour, has been gradually introduced to the notice of astronomers. Titan, the sixth Saturnian moon in order of distance, led the way, being detected by Huygens, March 25, 1655 ; Cassini made the acquaintance of four more between 1671 and 1684 ; while Mimas and Enceladus, the two innermost, were caught by Herschel in 1789, as they threaded their lucid way along the edge of the almost vanished ring. In the distances of these seven revolving bodies from their primary, an order of pro-

gression analogous to that pointed out by Titius in the
planetary intervals was found to prevail ; but with one conspicuous interruption, similar to that which had first suggested
the search for new members of the solar system. Between
Titan and Japetus—the sixth and seventh reckoning outwards
—there was obviously room for another satellite. It was
discovered, on both sides of the Atlantic simultaneously, on
the 19th of September 1848. Mr. W. C. Bond, employing
the splendid 15-inch refractor of the Harvard Observatory,
noticed, September 16, a minute star situated in the plane
of Saturn's rings. The same object was discerned by Mr.
Lassell on the 18th. On the following evening, both observers
perceived that the problematical speck of light kept up with,
instead of being left behind by, the planet as it moved, and
hence inferred its true character.[1] Hyperion, the seventh by
distance and eighth by recognition of Saturn's attendant train,
is of so insignificant a size when compared with some of its
fellow-moons (Titan is but little inferior to the planet Mars),
as to have suggested to Sir John Herschel[2] the idea that it
might be only one of several bodies revolving very close
together—in fact, an *asteroidal satellite ;* but the conjecture has,
so far, not been verified.

The coincidence of its duplicate discovery was singularly
paralleled two years later. Galileo's amazement when his
" optic glass " revealed to him the " triple" form of Saturn—
planeta tergeminus—has proved to be, like the laughter of the
gods, "inextinguishable." It must revive in every one who
contemplates anew the unique arrangements of that world apart
known to us as the Saturnian system. The resolution of the
so-called *ansæ,* or " handles," into one encircling ring by
Huygens in 1655 ; the discovery by Cassini in 1675 of the
division of that ring into two concentric ones ; the closely concordant determination, theoretically by Laplace and optically
by Herschel, of their period of rotation,[3] constituted, with

[1] Grant, *Hist. of Astr.,* p. 271. [2] *Month. Not.,* vol. ix. p. 91.
[3] The computed period was 10h. 33m. 36s.; the observed period, 10h.
32m. 15s.

some minor observations, the sum of the knowledge obtained, up to the middle of the present century, on the subject of this remarkable formation. The first place in the discovery now about to be related belongs to an American astronomer.

William Cranch Bond, born in 1789 at Falmouth (now Portland), in the State of Maine, was a watchmaker whom the solar eclipse of 1806 attracted to study the wonders of the heavens. When, in 1815, the erection of an observatory in connection with Harvard College, Cambridge, was first contemplated, he undertook a mission to England for the purpose of studying the working of similar institutions there ; and, on his return, erected a private observatory at Dorchester, where he worked diligently for many years. Meanwhile, the time was approaching for the resumption of the long-postponed design of the Harvard authorities ; and on the completion of the new establishment in 1844, Bond, who had for some time been officially connected with the College, and had carried on his scientific labours within its precincts, was offered and accepted the post of its director. Placed thus in possession of one of the finest instruments in the world—a masterpiece of Merz and Mahler—he headed the now long list of distinguished Transatlantic observers. Like the elder Struve, he left an heir to his office and to his eminence ; but George Bond unfortunately died in 1865, at the early age of thirty-nine, having survived his father but six years.

On the night of November 15, 1850—the air, remarkably enough, being so hazy that only the brightest stars could be perceived with the naked eye—William Bond discovered a third dusky ring, extending about half-way between the inner bright one and the globe of Saturn. A fortnight later, but before the observation had been announced in England, the same appearance was seen by the Rev. W. R. Dawes with the comparatively small refractor of his observatory at Wateringbury, and on December 3 was described by Mr. Lassell (then on a visit with him) as "something like a crape veil covering

a part of the sky within the inner ring."[1] Next morning the
Times containing the report of Bond's discovery reached
Wateringbury. The most surprising circumstance in the
matter was that the novel appendage had remained so long
unrecognised. As the rings opened out to their full extent,
it became obvious with very moderate optical assistance; yet
some of the most acute observers who have ever lived, using
instruments of vast power, had heretofore failed to detect its
presence. It soon appeared, however, that Galle of Berlin[2]
had noticed, June 10, 1838, a veil-like extension of the lucid
ring across half the dark space separating it from the planet;
but the observation, although communicated at the time
to the Berlin Academy of Sciences, had remained barren.
Traces of the dark ring, moreover, were found in a drawing
executed by Campani in 1664;[3] and Picard (June 15, 1673),[4]
Hadley (spring of 1720),[5] and Herschel,[6] had all undoubtedly
seen it under the aspect of a dark bar or belt crossing the
Saturnian globe. It was, then, of no recent origin; but there
seemed reason to think that it had lately gained considerably
in brightness. The full meaning of this remarkable fact it was
reserved for later investigations to develop.

What we may, in a certain sense, call the closing result of
the race for discovery, in which several observers seemed at
that time to be engaged, was the establishment, on a satisfactory footing, of our acquaintance with the dependent system
of Uranus. Sir William Herschel, whose researches formed,
in so many distinct lines of astronomical inquiry, the starting-
points of future knowledge, detected, January 11, 1787,[7] two
Uranian moons, since called Oberon and Titania, and ascertained the curious circumstance of their motion in a plane

[1] *Month. Not.*, vol. xi. p. 21.
[2] *Astr. Nach.*, No. 756 (May 2, 1851).
[3] F. Secchi, *Month. Not.*, vol. xiii. p. 248.
[4] Hind, in *ibid.*, vol. xv. p. 32.
[5] Lynn, *Observatory*, Oct. 1, 1883; Hadley, *Phil. Trans.*, vol. xxxii. p. 385.
[6] Proctor, *Saturn and its System*, p. 64.
[7] *Phil. Trans.*, vol. lxxvii. p. 125.

almost at right angles to the ecliptic, in a direction contrary to that of all previously known denizens (other than cometary) of the solar kingdom.   He believed that he caught occasional glimpses of four more, but never succeeded in assuring himself of their substantial existence.   Even the two first remained unseen save by himself until 1828, when his son re-observed them with a 20-foot reflector, similar to that with which they had been originally discovered.   Thenceforward they were kept fairly within view, but their four questionable companions, in spite of some false alarms of detection, remained in the dubious condition in which Herschel had left them.   At last, on October 24, 1851,[1] after some years of fruitless watching, Mr. Lassell espied " Ariel" and " Umbriel," two Uranian attendants, interior to Oberon and Titania, and of about half their brightness ; so that their disclosure is still reckoned amongst the very highest proofs of instrumental power and perfection. In all probability they were then for the first time seen ; for although Professor Holden,[2] director of the Washburn Observatory (U.S.), has attempted to identify them with two of Herschel's doubtful quartette, Mr. Lassell's argument [3] that the glare of the planet in Herschel's great specula must have rendered almost impossible the perception of objects so minute and so close to its disc, appears tolerably decisive to the contrary. Uranus is thus attended by four moons, and—so far as present knowledge extends—by no more.   Amongst the most important of the "negative results"[4] secured by Mr. Lassell's observations at Malta during the years 1852–53 and 1861–65, were the convincing evidence afforded by them that, without great increase of optical power, no further Neptunian or Uranian satellites can be perceived, and the consequent relegation of Herschel's baffling four—notwithstanding the unquestioned place long assigned to them in astronomical textbooks—to the shadowy condition of telescopic "ghosts."

[1] *Month. Not.*, vol. xi. p. 248.   [2] *Ibid.*, vol. xxxv. pp. 16–22.
[3] *Ibid.*, p. 26.   [4] *Ibid.*, vol. xli. p. 190.

(115)

CHAPTER V.

COMETS.

NEWTON showed that the bodies known as "comets," or *hirsute* stars, obey the law of gravitation; but it was by no means certain that the individual of the species observed by him in 1680 formed a permanent member of the solar system. The velocity, in fact, of its rush round the sun was quite possibly sufficient to carry it off for ever into the depths of space, there to wander, a celestial casual, from star to star. With another comet, however, which appeared two years later, the case was different. . Edmund Halley, who afterwards succeeded Flamsteed as Astronomer Royal, calculated its orbit on Newton's principles, and found it such as to give a period of revolution of about seventy-six years. He accordingly announced its probable identity with the comets observed by Peter Apian in 1531 and by Kepler in 1607, and fixed its return for 1758–59. The prediction was one of the test-questions put by Science to Nature, on the replies to which largely depend both the development of knowledge and the conviction of its reality. In the present instance, the answer afforded may be said to have laid the foundation of this branch of astronomy. Halley's comet punctually reappeared on Christmas Day, 1758, and effected its perihelion passage on the 12th of March following, thus proving beyond dispute that some at least of these erratic bodies are domesticated within our system, and strictly conform, if not to its unwritten customs (so to speak), at any rate to its fundamental laws. Their movements, in short, were demonstrated by the most unanswerable of all arguments—that of verified calculation—to be *calculable*, and their investigation was erected into a legitimate department of astronomical science.

This notable advance was the chief *result* obtained in the field of inquiry just now under consideration during the eighteenth century. But before it closed, its cultivation had received a powerful stimulus through the invention of an improved *method*. The name of Olbers has already been brought prominently before our readers in connection with asteroidal discoveries; these, however, were but chance excursions from the path of cometary research which he steadily pursued through life. An early predilection for the stars was fixed in this particular direction by one of the happy inspirations of genius. As he was watching, one night in the year 1779, by the sick-bed of a fellow-student in medicine at Göttingen, an important simplification in the mode of computing the paths of comets occurred to him. Although not made public until 1797, "Olbers' method" was then universally adopted, and is still regarded as the most expeditious and convenient in cases where absolute rigour is not required. By its introduction, not only many a toilsome and thankless hour was spared, but workers were multiplied, and encouraged in the prosecution of labours more useful than attractive.

The career of Heinrich Olbers is a brilliant example of what may be done by an amateur in astronomy. He at no time did regular work in an observatory; he was never the possessor of a transit or any other fixed instrument; moreover, all the best years of his life were absorbed in the assiduous exercise of a toilsome profession. In 1781 he settled as a physician in his native town of Bremen (he was born in 1758 at Arbergen, a neighbouring village, of which his father was pastor), and continued in active practice for over forty years. It was thus only the hours which his robust constitution enabled him to spare from sleep that were available for his intellectual pleasures. Yet his recreation was, as Von Zach remarked,[1] no less

[1] *Allgemeine Geographische Ephemeriden,* vol. iv. p. 287.

prolific of useful results than the severest work of other men.
The upper part of his house in the Sandgasse was fitted up with
such instruments and appliances as restrictions of space permitted, and there, night after night during half a century and
upwards, he discovered, calculated, or observed the cometary
visitants of northern skies. Almost as effective in promoting
the interests of science as the valuable work actually done by
him, was the influence of his genial personality. He engaged
confidence by his ready and discerning sympathy; he inspired
affection by his benevolent disinterestedness; he quickened
thought and awakened zeal by the suggestions of a lively and
inventive spirit, animated with the warmest enthusiasm for the
advancement of knowledge. Nearly every astronomer in Germany enjoyed the benefits of a (frequently active) correspondence with him, and his communications to the scientific
periodicals of the time were numerous and striking. The
motive power of his mind was thus widely felt and continually
in action. Nor did it wholly cease to be exerted even when
the advance of age and the progress of infirmity rendered him
incapable of active occupation. He was, in fact, *alive* even to
the last day of his long life of eighty-one years; and his death,
which occurred March 2, 1840, left vacant a position which a
rare combination of moral and intellectual qualities had conspired to render unique.

Amongst the many younger men who were attracted and
stimulated by intercourse with him was Johann Franz Encke.
But while Olbers became a mathematician because he was an
astronomer, Encke became an astronomer because he was a
mathematician. A born geometer, he was naturally sent to
Göttingen and placed under the tuition of Gauss. But
geometers are men; and the contagion of patriotic fervour
which swept over Germany after the battle of Leipsic did not
spare Gauss's promising pupil. He took up arms in the Hanseatic Legion, and marched and fought until the oppressor of
his country was safely ensconced behind the ocean-walls of St.
Helena. In the course of his campaigning he met Lindenau,

the militant director of the Seeberg Observatory, and by his influence was appointed his assistant, and eventually, in 1822, became his successor.   Thence he was promoted in 1825 to Berlin, where he superintended the building of the new observatory, so actively promoted by Humboldt, and remained at its head until within some eighteen months of his death in August 1865.

On the 26th of November 1818, Pons of Marseilles discovered a comet, whose inconspicuous appearance gave little promise of its becoming one of the most interesting objects in our system. Encke at once took the calculation of its elements in hand, and brought out the unexpected result that it revolved round the sun in a period of about $3\frac{1}{4}$ years.[1]   He moreover detected its identity with comets seen by Méchain in 1786, by Caroline Herschel in 1795, by Pons, Huth, and Bouvard in 1805, and after six laborious weeks of research into the disturbances experienced by it from the planets during the entire interval since its first ascertained appearance, he fixed May 24, 1822, as the date of its next return to perihelion. Although on that occasion, owing to the position of the earth, invisible in the northern hemisphere, Sir Thomas Brisbane's observatory at Paramatta was fortunately ready equipped for its recapture, which Rümker effected quite close to the spot indicated by Encke's ephemeris.

The importance of this event can be better understood when it is remembered that it was only the second instance of the recognised return of a comet (that of Halley's, sixty-three years previously, having, as already stated, been the first) ; and that it moreover established the existence of a new class of celestial objects, somewhat loosely distinguished as " comets of short period."   These bodies (of which a dozen are known to circulate within the orbit of Saturn) are remarkable as showing certain planetary affinities in the manner of their motions not at all perceptible in the wider travelling members of their order.

[1] *Astr. Jahrbuch*, 1823, p. 217. The period (1208 days) of this body is considerably shorter than that of any other known comet.

They revolve, without exception, in the same direction as the planets—from west to east; they exhibit a marked tendency to conform to the zodiacal track which limits planetary excursions north and south; and their paths round the sun, although much more eccentric than the approximately circular planetary orbits, are far less so than the extravagantly long ellipses in which comets comparatively untrained (as it were) in the habits of the solar system ordinarily perform their revolutions.

No *great* comet is of the "planetary" kind. These are, indeed, only by exception visible to the naked eye; they display extremely feeble tail-producing powers, and give small signs of central condensation. Thin wisps of cosmical cloud, they flit across the telescopic field of view without sensibly obscuring the smallest star. Their appearance, in short, suggests—what some notable facts in their history will presently be shown to confirm—that they are bodies already effete, and verging towards dissolution. . If it be asked what possible connection can be shown to exist between the shortness of period by which they are essentially characterised, and what we may call their *superannuated* condition, we are not altogether at a loss for an answer. Kepler's remark,[1] that comets are consumed by their own emissions, has undoubtedly a measure of truth in it. The substance ejected into the tail must, in overwhelmingly large proportion, be for ever lost to the central mass from which it issues. True, it is of a nature inconceivably tenuous; but unrepaired waste, however small in amount, cannot be persisted in with impunity. The incitement to such self-spoliation proceeds from the sun; it accordingly progresses more rapidly the more numerous are the returns to the solar vicinity. Comets of short period may thus reasonably be expected to *wear out* quickly.

They are, moreover, bodies subject to many adventures and vicissitudes. Their aphelia—or the farthest points of their

[1] "Sicut bombyces filo fundendo, sic cometas cauda exspiranda consumi et denique mori."—*De Cometis, Op.*, vol. vii. p. 110.

orbits from the sun—are all situated so near to the path either
of Jupiter or of Saturn, as to permit these giant planets to act
as secondary rulers of their destinies. By their influence they
were, in all probability, originally fixed in their present tracks ;
and by their influence, exerted in an opposite sense, they may,
in some cases, be eventually ejected from them. A curious
instance of such capricious dealing on the part of Jupiter, was
afforded by the comet of 1770, found by Lexell of St. Petersburg to perform its circuit of the sun in $5\frac{1}{2}$ years, but which
had never previously, and has never since been seen. The
explanation of this anomaly, suggested by Lexell, and fully confirmed by the analytical inquiries both of Laplace and Leverrier, was that a very close approach to Jupiter in 1767 had
completely changed the character of its orbit, and brought it
within the range of terrestrial observation ; while in 1779,
after having only twice traversed its new path (at its second
return it was so circumstanced as to be invisible from the
earth), it was, by a fresh encounter, diverted into one entirely
different.[1]

It can easily be imagined that careers so varied are likely to
prove instructive, and astronomers have not been backward
in extracting from them the lessons they are fitted to convey.
Encke's comet, above all, has served as an index to much
curious information, and it may be hoped that its function in
that respect is by no means at an end. The great extent of
the solar system traversed by its eccentric path makes it
peculiarly useful for the determination of the planetary masses.
At perihelion it penetrates within the orbit of Mercury ; it

[1] Leverrier showed (*Comptes Rendus*, t. xxv. 1847, p. 564) that the problem
of the disturbances suffered by Lexell's comet was a far less determinate
one than it had been made to appear in the *Mécanique Céleste*. It is
possible that this body may, in 1779, have been finally thrust out of our
system ; it is also possible (as Laplace concluded) that it may be revolving
too far from the sun to be accessible to our view ; but it is much more
probable that its orbit still retains a family likeness to the one temporarily
assigned to it by Jovian influence in 1767, in which case Leverrier's calculations afford criteria for its eventual re-identification.

considerably transcends at aphelion the farthest excursion of Pallas. Its vicinity to the first-named planet in August 1835 offered the first convenient opportunity of placing that body in the astronomical balance. Its weight or mass had previously been assumed, not ascertained; and the comparatively slight deviation from its regular course impressed upon the comet by its attractive power, showed that it had been assumed nearly twice too great.[1] That fundamental datum of planetary astronomy—the mass of Jupiter—was corrected by similar means; and it was reassuring to find the correction in satisfactory accord with that already introduced from observation of the asteroidal movements.

The fact that comets contract in approaching the sun had been noticed by Hevelius; Pingré admitted it with hesitating perplexity;[2] the example of Encke's comet rendered it conspicuous and undeniable. On the 28th of October 1828, the diameter of the nebulous matter composing this body was estimated at 312,000 miles. It was then about one and a half times as remote from the sun as the earth is at the time of the equinox. On the 24th of December following, its distance being reduced by nearly two-thirds, it was found to be only 14,000 miles across.[3] That is to say, it had shrunk during those two months of approach to $\frac{1}{11,000}$th part of its original volume! Yet it had still seventeen days' journey to make before reaching perihelion. The same curious circumstance was even more markedly apparent at its return in 1838. Its bulk, or the actual space occupied by it, was reduced, as it drew near the hearth of our system (so far at least as could be inferred from optical evidence), in the enormous proportion of 800,000 to 1. A corresponding expansion on each occasion accompanied its retirement from the sphere of observation. Similar changes

---

[1] From the observed results of a second appulse in 1848, the Mercurian mass is now estimated at about $\frac{1}{4,860,000}$ that of the sun; while the inverse relation assumed by Lagrange to exist between distance from the sun and density brought it out $\frac{1}{2,025,810}$. Laplace, *Exposition du Système du Monde*, t. ii. p. 50 (5th ed. 1824).

[2] Arago, *Annuaire*, 1832, p. 218.   [3] Hind, *The Comets*, p. 20.

of volume, though rarely to the same astounding extent, have
been perceived in other comets.   They still remain unexplained ; but it can scarcely be doubted that they are due to
the action of the same energetic internal forces which reveal
themselves in so many splendid and surprising cometary phenomena.

Another question of singular interest was raised by Encke's
acute inquiries into the movements and disturbances of the
first known "comet of short period."   He found from the
first that its revolutions were subject to some influence besides that of gravity.   After every possible allowance had
been made for the pulls, now backward, now forward, exerted
upon it by the several planets, there was still a surplus of
acceleration left unaccounted for.   Each return to perihelion
took place about two and a half hours sooner than received
theories warranted.   Here then was a " residual phenomenon "
of the utmost promise for the disclosure of novel truths.
Encke (in accordance with the opinion of Olbers) explained
it as due to the presence in space of some such "subtle
matter " as was long ago invoked by Euler [1] to be the agent of
eventual destruction for the fair scheme of planetary creation.
The apparent anomaly of accounting for an accelerative effect
by a retarding cause disappears when it is considered that any
check to the motion of bodies revolving round a centre of
attraction causes them to draw closer to it, thus shortening
their periods and quickening their circulation.   If space were
filled with a resisting medium capable of impeding, even in
the most infinitesimal degree, the swift course of the planets,
their orbits should necessarily be, not ellipses, but very close
elliptical spirals, along which they would slowly, but inevitably,
descend into the burning lap of the sun.   The circumstance
that no such tendency can be traced in their revolutions by no
means sets the question at rest.   For it might well be that an
effect totally imperceptible until after the lapse of countless
ages, as regards the solid orbs of our system, might be obvious

[1] *Phil. Trans.*, vol. xlvi. p. 204.

in the movements of bodies like comets of small mass and great bulk ; just as a feather or a gauze veil at once yields its motion to the resistance of the air, while a cannon-ball cuts its way through with comparatively slight loss of velocity.

It will thus be seen that issues of the most momentous character hang on the *time-keeping* of comets ; for plainly all must in some degree suffer the same kind of hindrance as Encke's, if the cause of that hindrance be the one suggested. More than half a century, however, elapsed before the slightest trace of similar symptoms could be detected in any of its congeners. At length, in 1880, Professor Oppolzer announced[1] that a comet, first seen by Pons in 1819, and rediscovered by Winnecke in 1858, having a period of 2052 days (5.6 years), was accelerated at each revolution precisely in the manner required by Encke's theory. The "resisting medium" was thereby generally admitted to have made good its footing. But Backlund's latest researches[2] (in continuation of those of Von Asten, cut short by his premature death) into the movements of Encke's comet have revealed a perplexing circumstance. They confirm Encke's results for the period covered by them, but exhibit the acceleration as *progressively diminishing* from 1865 to 1881. Uniformity of action, however, would seem to be an indispensable attribute of a true ethereal resistance.

The question is thus reopened, and with a renewal of interest ; for although we have to wait for a definitive answer, there is much to be learned from even the unsuccessful testing of various hypotheses. There seems, in the first place, no reason to suspect any physical change in the comet itself, such as would render its motion less sensitive to opposition. A diminution of bulk would have this effect, but the telescope reports its aspect unaltered. Can the change, we then ask, be in the condition of inter-planetary space? The character of the supposed resistance, it may be remarked, has been

[1] *Astr. Nach.*, No. 2314.
[2] *Mém. de St. Pétersbourg*, t. xxxii. (7th series), 1884. For a *précis* of results, see *Bulletin Astronomique*, t. i. p. 239.

often misapprehended. What Encke stipulated for was not a
medium equally diffused throughout the visible universe, such
as the ethereal vehicle of the vibrations of light, but a rare
fluid, rapidly increasing in density towards the sun.[1] This
cannot be a solar atmosphere, since it is mathematically certain,
as Laplace has shown,[2] that no envelope partaking of the sun's
axial rotation can extend farther from his surface than nine-
tenths of the mean distance of Mercury. Within such an
envelope Encke's comet can never penetrate. There is, besides, strong evidence of a physical kind that the *actual* depth
of the solar atmosphere bears a very minute proportion to
the *possible* depth theoretically assigned to it. That matter,
however, not atmospheric in its nature — that is, neither
forming one body with the sun nor altogether aëriform—exists
in its neighbourhood, can admit of no reasonable doubt. The
great lens-shaped mass of the zodiacal light, reaching out at
times far beyond the earth's orbit, may be regarded as an
extension of the corona, and, like the corona, is probably composed of matter in very various forms—cosmical dust, planetary
refuse, cometary débris, vaporous ejections. Now the changes
in shape and brightness visible in this singular feature of our
system may well be accompanied by changes in the power of
impeding motion of its constituting substances; and we may
say with confidence that they are intimately connected with
variations in solar activity. The state of the sun and his
appendages at the times of the successive approaches to
perihelion of Encke's comet should thus be taken into account
in studying the problem of its acceleration, evidently a more
intricate one than had been supposed. The comparison may
yet be the means of bringing to light hitherto unsuspected
relations.

The history of the next known "planetary" comet has proved
of even more curious interest than that of the first. It was
discovered by an Austrian officer named Wilhelm von Biela
at Josephstadt in Bohemia, February 27, 1826, and ten days

---

[1] *Month. Not.*, vol. xix. p. 72.   [2] *Mécanique Céleste*, t. ii. p. 197.

later by the French astronomer Gambart at Marseilles. Both observers computed its orbit, showed its remarkable similarity to that traversed by comets visible in 1772 and 1805, and connected them together as previous appearances of the body just detected, by assigning to its revolutions a period of between six and seven years. The two brief letters conveying these strikingly similar inferences were printed side by side in the same number of the *Astronomische Nachrichten* (No. 94); but Biela's priority in the discovery of the comet was justly recognised by the bestowal upon it of his name.

The object in question was at no time (subsequently to its appearance in 1805) visible to the naked eye. Its aspect in Sir John Herschel's great reflector on the 23d of September 1832, was described by him as that of a "conspicuous nebula," about 2½ to 3 minutes in diameter. No trace of a tail was discernible. While he was engaged in watching it, a small knot of minute stars (16th or 17th magnitude) was directly traversed by it, "and when on the cluster," he tells us,[1] it "presented the appearance of a nebula resolvable, and partly resolved into stars, the stars of the cluster being visible through the comet." Yet the depth of cometary matter through which such faint stellar rays penetrated undimmed, was, near the central parts of the globe, not less than 50,000 miles.

It is curious to find that this seemingly harmless, and we may perhaps add, effete body, gave occasion to the first (and not the last) cometary "scare" of this enlightened century. Its orbit, at the descending node, may be said to have intersected that of the earth; since, according as it *bulged in or out* under the disturbing influence of the planets, the passage of the comet was effected *inside* or *outside* the terrestrial track. Now certain calculations published by Olbers in 1828[2] showed that, on October 29, 1832, a considerable portion of its nebulous surroundings would actually sweep over the spot which, a month later, would be occupied by our planet. It needed no more to set the popular imagination in a ferment.

[1] *Month. Not.*, vol. ii. p. 117.   [2] *Astr. Nach.*, No. 128.

Astronomers after all could not, by an alarmed public, be held to be infallible. Their computations, it was averred, which a trifling oversight would suffice to vitiate, exhibited clearly enough the danger, but afforded no guarantee of safety from a collision, with all the terrific consequences frigidly enumerated by Laplace. Nor did the panic subside until Arago formally demonstrated that the earth and comet could by no possibility approach within less than fifty millions of miles.[1]

The return of the same body in 1845–46, was marked by an extraordinary circumstance. When first seen, November 28, it wore its usual aspect of a faint round patch of cosmical *fog;* but on December 19, Mr. Hind noticed that it had become distorted somewhat into the form of a pear; and ten days later it had divided into two separate objects. This singular duplication was first perceived at New Haven in America, December 29,[2] by Messrs. Herrick and Bradley, and by Lieutenant Maury at Washington, January 13, 1846. The earliest British observer of the phenomenon was Professor Challis. "I see *two* comets!" he exclaimed, putting his eye to the great equatoreal of the Cambridge Observatory on the night of January 15; then, distrustful of what his senses had told him, he called in his judgment to correct their improbable report by resolving one of the dubious objects into a hazy star.[3] On the 23d, however, both were again seen by him in unmistakable cometary shape, and until far on in March (Otto Struve caught a final glimpse of the pair on the 16th of April),[4] continued to be watched with equal curiosity and amazement by astronomers in every part of the

---

[1] *Annuaire,* 1832, p. 186.

[2] *Am. Journal of Science,* vol. i. (2d series), p. 293. Prof. Hubbard's calculations indicated a probability that the definitive separation of the two nuclei occurred as early as Sept. 30, 1844. *Astronomical Journal* (Gould's), vol. iv. p. 5. See also, on the subject of this comet, W. T. Lynn, *Intellectual Observer,* vol. xi. p. 208, and the Rev. E. Ledger, *Observatory,* August 1883, p. 244. [3] *Month. Not.,* vol. vii. p. 73.

[4] *Bulletin Ac. Imp. de St. Pétersbourg,* t. vi. col. 77. The latest observation of the parent nucleus was that of Argelander, April 27, at Bonn.

northern hemisphere. What Seneca reproved Ephorus for supposing to have taken place in 373 B.C.—what Pingré blamed Kepler for conjecturing in 1618, had then actually occurred under the attentive eyes of science in the middle of the nineteenth century !

At a distance from each other of about two-thirds the distance of the moon from the earth, the twin comets meantime moved on tranquilly, so far, at least, as their course through the heavens was concerned. Their extreme *lightness*, or the small amount of matter contained in each, could not have received a more signal illustration than by the fact that their revolutions round the sun were performed independently ; that is to say, they travelled side by side without experiencing any appreciable' mutual disturbance, thus plainly showing that at an interval of only 157,250 miles, their attractive power was virtually inoperative. Signs of internal agitation, however, were not wanting. Each fragment threw out a short tail in a direction perpendicular to the line joining their centres, and each developed a bright nucleus, although the original comet had exhibited neither of these signs of cometary vitality. A singular interchange of brilliancy was, besides, observed to take place between these small objects, each of which alternately outshone and was outshone by the other, while an arc of light, apparently proceeding from the more lustrous, at times bridged the intervening space. Obviously, the gravitational tie, rendered powerless by exiguity of matter, was here replaced by some other form of mutual action, the nature of which can as yet be dealt with only by conjecture.

Once more, in August 1852, the double comet returned to the neighbourhood of the sun, but under circumstances not the most advantageous for observation. Indeed, the companion was not detected until September 16, by Father Secchi at Rome, and was then perceived to have increased its distance from the originating body to a million and a quarter of miles, or about eight times the average interval at the former appearance. Both vanished shortly afterwards, and have never since been seen, notwithstanding the eager watch kept for objects of such singular interest, and the accurate knowledge of their track supplied by Santini's investigations. We can scarcely doubt that the fate has overtaken them which Newton assigned as the end of all cometary existence. *Diffundi tandem et spargi per cœlos universos.*[1]

A telescopic comet with a period of $7\frac{1}{2}$ years, discovered November 22, 1843, by M. Faye of the Paris Observatory, formed the subject of a characteristically patient and profound inquiry on the part of Leverrier, designed to test its suggested identity with Lexell's lost comet. The result was decisive against the hypothesis of Valz, the divergences between the orbits of the two bodies being found to increase instead of to diminish, as the history of the newcomer was traced backwards into the last century.[2] Faye's comet pursues the most nearly circular path of any similar known object; even at its nearest approach to the sun it remains farther off than Mars when he is most distant from it; and it has been proved by the admirable researches of Professor Axel Möller,[3] director of the Swedish observatory of Lund, to exhibit no trace of the action of a resisting medium.

No *great* comet appeared between the " star " which presided at the birth of Napoleon and the " vintage " comet of 1811. The latter was first descried by Flaugergues at Viviers, March 26, 1811; Wisniewski, at Neu-Tscherkask in Southern Russia, caught the last glimpse of it August 17, 1812. Two disappearances in the solar rays as the earth moved round in its orbit, and two reappearances after conjunction, were included in this unprecedentedly long period of visibility of 510 days. This relative permanence (so far as the inhabitants of Europe were concerned) was due to the high northern latitude attained near perihelion, combined with a certain leisureliness of movement along a path everywhere external to that of the earth. The magnificent luminous train of this body, on

[1] D'Arrest, *Astr. Nach.*, No. 1624.
[2] *Comptes Rendus*, t. xxv. p. 570.   [3] *Month. Not.*, vol. xii. p. 248.

October 15, the day of its nearest terrestrial approach, covered an arc of the heavens 23½ degrees in length, corresponding to a real extension of one hundred millions of miles. Its form was described by Sir William Herschel as that of " an inverted hollow cone," and its colour as yellowish, strongly contrasting with the bluish-green tint of the " head," round which it was flung like a transparent veil. The planetary disc of the head, 127,000 miles across, appeared to be composed of strongly condensed nebulous matter; but somewhat eccentrically situated within it was a star-like nucleus of a reddish tinge, which Herschel presumed to be solid, and ascertained, with his usual care, to have a diameter of 428 miles. From the total absence of phases, as well as from the vivacity of its radiance, he confidently inferred that its light was not borrowed, but inherent.[1]

This remarkable apparition formed the subject of a memoir [2] by Olbers, the striking, yet steadily reasoned-out suggestions contained in which there was at that time no means of following up with profit. Only of late has the " electrical theory," of which Zöllner [3] regarded Olbers as the founder, assumed a definite and measurable form, capable of being tested by the touchstone of fact, as knowledge makes its slow inroads on the fundamental mystery of the physical universe.

The paraboloidal shape of the bright envelope separated by a dark interval from the head of the great comet of 1811, and constituting, as it were, the *root* of its tail, seemed to the astronomer of Bremen to reveal the presence of a double repulsion; the expelled vapours accumulating where the two forces, solar and cometary, balanced each other, and being then swept backwards in a huge train. He accordingly distinguished three classes of these bodies:—First, comets which develop *no* matter subject to solar repulsion. These have no

[1] *Phil. Trans.*, vol. cii. pp. 118–124.
[2] *Ueber den Schweif des grossen Cometen von* 1811, *Monat. Corr.*, vol. xxv. pp. 3–22. Reprinted by Zöllner, *Ueber die Natur der Cometen*, pp. 3–15.
[3] *Natur der Cometen*, p. 148.

I

tails, and are probably mere nebulosities, without solid nuclei. Secondly, comets which are acted upon by solar repulsion *only*, and consequently throw out no emanations *towards* the sun. Of this kind was a bright comet visible in 1807.[1] Thirdly, comets, like that of 1811, giving evidence of action of both kinds. These are distinguished by a dark *hoop* encompassing the head and dividing it from the luminous envelope, as well as by an obscure caudal axis, resulting from the hollow, cone-like structure of the tail.

Again, the ingenious view recently put forward by M. Bredichin of Moscow as to the connection between the *form* of these appendages and the *kind* of matter composing them, was very clearly anticipated by Olbers. The amount of tail-curvature, he pointed out, depends in each case upon the proportion borne by the velocity of the ascending particles to that of the comet in its orbit; the swifter the outrush, the straighter the resulting tail. But the velocity of the ascending particles varies with the energy of their repulsion by the sun, and this again, it may be presumed, with their quality. Thus multiple tails are developed when the same comet throws off, as it approaches perihelion, specifically distinct substances. The long, straight ray which proceeded from the comet of 1807, for example, was doubtless made up of particles subject to a much more vigorous solar repulsion than those formed into the shorter, curved emanation issuing from it nearly in the same direction. In the comet of 1811, he calculated that the particles expelled from the head travelled to the remote extremity of the tail in eleven minutes, indicating by this enormous rapidity of movement (comparable to that of the transmission of light) the action of a force greatly more powerful than the opposing one of gravity. The not uncommon phenomena of multiple envelopes, on the other hand, he explained as due to the varying amounts of repulsion exercised by the nucleus itself on the different kinds of matter developed from it.

The movements and perturbations of the comet of 1811

[1] The subject of a classical memoir by Bessel, published in 1810.

were no less profoundly studied by Argelander than its physical constitution by Olbers. The orbit which he assigned to it is of such vast dimensions as to require no less than 3065 years for the completion of its circuit ; and to carry the body describing it at each revolution to fourteen times the distance from the sun of the frigid Neptune. Thus, when it last visited our neighbourhood, Achilles may have gazed on its imposing train as he lay on the sands all night bewailing the loss of Patroclus ;[1] and when it returns, it will perhaps be to shine upon the ruins of empires and civilisations still deep buried among the secrets of the coming time.

On the 26th of June 1819, while the head of a comet passed across the face of the sun, the earth was (in all probability) involved in its tail. But of this remarkable double event nothing was known until more than a month later, when the fact of its past occurrence emerged from the calculations of Olbers.[2] Nor had the comet itself been generally visible previous to the first days of July. Several observers, however, on the publication of these results, brought forward accounts of singular spots perceived by them upon the sun at the time of the transit, and the original drawing of one of them, Pastorff of Buchholtz, has been preserved. This undoubtedly authentic delineation[3] represents a round nebulous object with a *bright* spot in the centre, of decidedly cometary aspect, and not in the least like an ordinary solar " macula." Mr. Hind,[4] nevertheless, has shown that its position on the sun is irreconcilable with that which the comet must have occupied ; and Mr. Ranyard's discovery of a similar smaller drawing by the same author, dated May 26, 1828,[5] reduces to evanescence the probability of its connection with that body. Indeed, recent experience renders very doubtful the possibility of such an observation.

---

[1] If we adopt the chronology of Mädler, *Reden und Abhandl.*, p. 118.
[2] *Astr. Jahrbuch* (Bode's), 1823, p. 134.
[3] Reproduced in Webb's *Celestial Objects*, 4th ed.
[4] *Month. Not.*, vol. xxxvi. p. 309.   [5] *Celestial Objects*, p. 40, *note*.

The return of Halley's comet in 1835 was looked forward to as an opportunity for testing the truth of floating cometary theories, and did not altogether disappoint expectation. As early as 1817, its movements and disturbances since 1759 were proposed by the Turin Academy of Sciences as the subject of a prize awarded to Baron Damoiseau. Pontécoulant was adjudged a similar distinction by the Paris Academy in 1829; while Rosenberger's calculations were rewarded with the gold medal of the Royal Astronomical Society.[1] The result entirely disproved the hypothesis (designed to explain the invariability of the planetary periods) of what may be described as a *vortex* of attenuated matter moving *with* the planets, and offering, consequently, no resistance to their motion. For since Halley's comet revolves in the opposite direction— in other words, has a " retrograde " movement—it is plain that if compelled to *make head* against an ethereal current, it would rapidly be deprived of the tangential velocity which enables it to keep at its proper distance from the sun, and would thus gradually but conspicuously approach, and eventually be precipitated upon it. No such effect, however, has in this crucial instance been detected.

On the 6th of August 1835, a nearly circular misty object was seen at Rome not far from the predicted place of the comet. It was not, however, until the middle of September that it began to throw out a tail, which by the 15th of October had attained a length of about 24 degrees (on the 19th, at Madras, it extended to fully 30),[2] the head showing to the naked eye as a reddish star rather brighter than Aldebaran or Antares.[3] Some curious phenomena accompanied the process of tail-formation. An outrush of luminous matter, resembling in shape a partially opened fan, issued from the nucleus *towards* the sun, and at a certain point, like smoke driven before a high wind, was vehemently swept backwards in a prolonged train. The appearance of the comet at

---

[1] See Airy's Address, *Mem. R. A. S.*, vol. x. p. 376.
[2] Hind, *The Comets*, p. 47.   [3] Arago, *Annuaire*, 1836, p. 228.

this time was compared by Bessel,[1] who watched it with minute attention, to that of a blazing rocket. He made the singular observation that this fan of light, which seemed the source of supply for the tail, oscillated like a pendulum to and fro across a line joining the sun and nucleus, in a period of $4\frac{3}{5}$ days; and he was unable to escape from the conclusion [2] that a repulsive force, about twice as powerful as the attractive force of gravity, was concerned in the production of these remarkable effects. Nor did he hesitate to recur to the analogy of magnetic polarity, or to declare, still more emphatically than Olbers, "the emission of the tail to be a purely electrical phenomenon." [3]

The transformations undergone by this body were almost as strange and complete as those which affected the brigands in Dante's "*Inferno*." When first seen it wore the aspect of a nebula; later it put on the distinctive garb of a comet; it next appeared as a star; finally it dilated, first in a spherical, then in a paraboloidal form, until May 5, 1836, when it vanished, as if by melting into adjacent space from the excessive diffusion of its light. A very uncommon circumstance in its development was that it lost (it would appear) all trace of tail *previous* to its arrival at perihelion on the 16th of November. Nor did it begin to recover its elongated shape for more than two months afterwards. On the 23d of January Boguslawski perceived it as a star of the sixth magnitude, *without measurable disc.*[4] Only two nights later, Maclear, director of the Cape Observatory, found the head to be 131 seconds across.[5] And so rapidly did the augmentation of size progress, that Sir John Herschel, who was then observing at Feldhausen, estimated the actual bulk of this singular object to have increased forty-fold in the ensuing week. " I can hardly

---

[1] *Astr. Nach.*, No. 300.

[2] It deserves to be recorded that Robert Hooke drew a very similar inference from his observations of the comets of 1680 and 1682. *Month. Not.*, vol. xiv. pp. 77–83.

[3] *Briefwechsel zwischen Olbers und Bessel*, Bd. ii. p. 390.

[4] Herschel, *Results*, p. 405.   [5] *Mem. R. A. S.*, vol. x. p. 92.

doubt," he remarks, "that the comet was fairly evaporated in perihelio by the heat, and resolved into transparent vapour, and is now in process of rapid condensation and re-precipitation on the nucleus."[1] A plausible, but no longer admissible interpretation of this still unexplained phenomenon.

By means of an instrument devised by himself for testing the quality of light, Arago obtained decisive evidence that some at least of the radiance proceeding from Halley's comet was derived by reflection from the sun.[2] Indications of the same kind had been afforded[3] by the comet which suddenly appeared above the north-western horizon of Paris, July 3, 1819, after having enveloped (as already stated) our terrestrial abode in its filmy appendages; but the "polariscope" had not then reached the perfection subsequently given to it, and its testimony was accordingly far less reliable than in 1835. Such experiments, however, are in reality more beautiful and ingenious than instructive, since incandescent as well as obscure bodies possess the power of throwing back light incident upon them, and will consequently transmit to us from the neighbourhood of the sun rays partly direct, partly reflected, of which a certain proportion will exhibit the peculiarity known as polarisation.

The most brilliant comets of the century were suddenly rivalled if not surpassed by the extraordinary object which blazed out beside the sun, February 28, 1843. It was simultaneously perceived in Mexico and the United States, in Southern Europe, and at sea off the Cape of Good Hope, where the passengers on board the *Owen Glendower* were amazed by the sight of a "short, dagger-like object," closely following the sun towards the western horizon.[4] At Florence Amici found its distance from the sun's centre at noon to be only $1°23'$; and spectators at Parma were able, when sheltered from the direct glare of midday, to trace the tail to a length of

---

[1] *Results*, p. 401.   [2] *Annuaire*, 1836, p. 233.
[3] *Cosmos*, vol. i. p. 90, *note* (Otte's trans.)
[4] Herschel, *Outlines*, p. 399 (9th ed.)

four or five degrees. The full dimensions of this astonishing appurtenance began to be disclosed a few days later. On the 3d of March it measured 25°, and on the 11th, at Calcutta, Mr. Clerihew observed a second streamer, nearly twice as long as the first, and making an angle with it of 18°, to have been emitted in a single day. This rapidity of projection, Sir John Herschel remarks, " conveys an astounding impression of the intensity of the forces at work." " It is clear," he continues, "that *if we have to deal here with matter, such as we conceive it—* viz., *possessing inertia—at all,* it must be under the dominion of forces incomparably more energetic than gravitation, and quite of a different nature." [1]

On the 17th of March a silvery ray, some 40 degrees long and slightly curved at its extremity, shone out above the sunset clouds in this country. No previous intimation had been received of the possibility of such an apparition, and even astronomers—no lightning messages across the seas being as yet possible—were perplexed. The nature of the phenomenon, indeed, soon became evident, but the wonder of it did not diminish with the study of its attendant circumstances. Never before, within astronomical memory, had our system been traversed by a body pursuing such an adventurous career. The closest analogy was offered by the great comet of 1680 (Newton's), which rushed past the sun at a distance of only 144,000 miles ; but even this—on the cosmical scale—scarcely perceptible interval was reduced nearly one-half in the case we are now concerned with. The centre of the comet of 1843 approached the formidable luminary within 78,000 miles, leaving, it is estimated, a clear space of not more than 32,000 between the surfaces of the bodies thus brought into such perilous proximity. The escape of the wanderer was, however, secured by the extraordinary rapidity of its flight. It swept past perihelion at a rate—366 miles a second—which, if continued, would have carried it right round the sun in *two hours ;* and in only eleven minutes more than that short period

[1] *Outlines,* p. 398.

it actually described half the *curvature* of its orbit—an arc of
180°—although in travelling over the remaining half many
hundreds of sluggish years will doubtless be consumed.

The behaviour of this comet may be regarded as an *experimentum crucis* as to the nature of tails. For clearly no fixed
appendage many millions of miles in length could be whirled
like a brandished sabre from one side of the sun to the other
in 131 minutes. Cometary trains are then, as Olbers rightly
conceived them to be, emanations, not appendages—inconceivably rapid outflows of highly rarefied matter, the greater
part, if not all of which becomes permanently detached from
the nucleus.

That of the comet of 1843 reached, about the time that it
became visible in this country, the extravagant length of 200
millions of miles.[1] It was narrow, and bounded by nearly
parallel and nearly rectilinear lines, resembling—to borrow a
comparison of Aristotle's—a "road" through the constellations; and after the 3d of March showed no trace of hollow-
ness, the axis being, in fact, rather brighter than the edges.
Distinctly perceptible in it were those singular aurora-like
coruscations which gave to the "tresses" of Charles V.'s
comet the appearance—as Cardan described it—of "a torch
agitated by the wind," and have not unfrequently been observed to characterise other similar objects. A consideration
first adverted to by Olbers proves these to originate in our
own atmosphere. For owing to the great differences in the
distances from the earth of the origin and extremity of such
vast effluxes, the light proceeding from their various parts is
transmitted to our eyes in notably different intervals of time.
Consequently a luminous undulation, even though propagated
instantaneously from end to end of a comet's tail, would
appear to us to occupy many minutes in its progress. But the
coruscations in question pass as swiftly as a falling star. They
are, then, of terrestrial production.

[1] Boguslawski calculated that it extended on the 21st of March to
581 millions. *Report Brit. Ass.*, 1845, p. 89.

Periods of the utmost variety were by different computators assigned to the body, which arrived at perihelion, February 27, 1843, at 9.47 p.m. Professor Hubbard of Washington found that it required 533 years to complete a revolution; MM. Laugier and Mauvais of Paris considered the true term to be 35;[1] Clausen looked for its return at the end of between six and seven. All these estimates were indeed admittedly uncertain, the available data affording no sure means of determining the value of this element; yet there seems no doubt that they fitted in more naturally with a period counted by centuries than with one reckoned by decades. Nor could any previous appearance be satisfactorily made out, although the similarity of the course pursued by a brilliant comet in 1668, known as the "Spina" of Cassini, made an identification not impossible. This would imply a period of 175 years, and it was somewhat hastily assumed that a number of earlier celestial visitants might thus be connected as returns of the same body.

It may now be asked what were the conclusions regarding the nature of comets drawn by astronomers from the considerable mass of novel experience accumulated during the first half of this century? The first and best assured was that the matter composing them is in a state of extreme tenuity. Numerous and trustworthy observations showed that the feeblest rays of light might traverse some hundreds of thousands of miles of their substance, even where it was apparently most condensed, without being perceptibly weakened. Nay, instances were recorded in which stars were said to have gained in brightness from the process![2] On the 24th of June 1825, Olbers[3] saw the comet then visible all but obliterated by the central passage of a star too small to be distinguished with the naked eye, its own light remaining wholly unchanged. A similar effect was noted December 1,

---

[1] *Comptes Rendus,* t. xvi. p. 919.

[2] Piazzi noticed a considerable increase of lustre in a very faint star of the twelfth magnitude viewed through a comet. Mädler, *Reden, &c.,* p. 248, *note.*   [3] *Astr. Jahrbuch,* 1828, p. 151.

1811, when the great comet of that year approached so close to Atair, the *lucida* of the Eagle, that the star seemed to be transformed into the nucleus of the comet.[1]   Even the central blaze of Halley's comet in 1835 was powerless to impede the passage of stellar rays.   Struve[2] observed at Dorpat, on September 17, an all but central occultation; Glaisher[3] one (so far as he could ascertain) absolutely so eight days later at Cambridge.   In neither case was there any appreciable diminution of the star's light.   Again, on the 11th of October 1847, Mr. Dawes,[4] an exceptionally keen observer, distinctly saw a star of the tenth magnitude through the exact centre of a comet discovered on the 1st of that month by Maria Mitchell of Nantucket.

Examples, on the other hand, were not wanting of the diminution of stellar light under similar circumstances; but probably in general not more than would be accounted for by the illumination of the background with diffused nebulous radiance.[5]   In one solitary instance, however, on the 28th of November 1828, a star was alleged to have actually vanished behind a comet.[6]   The observer of this unique phenomenon was Wartmann of Geneva ; but his instrument was so defective as to leave its reality open to grave doubt, especially when it is considered that the eclipsing body was Encke's comet, which better equipped astronomers have, on various occasions, found to be perfectly translucent.

From the failure to detect any effects of refraction in the light of stars occulted by comets, it was inferred (though, as we now know, erroneously) that their composition is rather that

---

[1] Mädler, *Gesch. d. Astr.*, Bd. ii. p. 412.

[2] *Recueil de l'Ac. Imp. de St. Pétersbourg*, 1835, p. 143.

[3] Guillemin's *World of Comets*, trans. by J. Glaisher, p. 294, *note*.

[4] *Month. Not.*, vol. viii. p. 9.

[5] A real, though only partial stoppage of light seems indicated by Herschel's observations on the comet of 1807.   Stars seen through the tail, October 18, lost much of their lustre.   One near the head was only faintly visible by glimpses.   *Phil. Trans.*, vol. xcvii. p. 153.

[6] Arago, *Annuaire*, 1832, p. 205.

of dust than that of vapour ; that they consist not of any continuous substance, but of discrete solid particles, very finely divided and widely scattered. In conformity with this view was the known smallness of their masses. Laplace had shown that if the amount of matter forming Lexell's comet had been as much as $\frac{1}{5000}$ of that contained in our globe, the effect of its attraction, on the occasion of its approach within 1,438,000 miles of the earth, July 1, 1770, must have been apparent in the lengthening of the year. And that some comets, at any rate, possess masses immeasurably below this maximum value, was clearly proved by the undisturbed parallel march of the two fragments of Biela in 1846.

But the discovery in this branch most distinctive of the period under review, is that of "short period" comets, of which four [1] were known in 1850. These, by the character of their movements, serve as a link between the planetary and cometary worlds, and by the nature of their construction, seem to mark a stage in cometary decay. For that comets are rather transitory agglomerations, than permanent products of cosmical manufacture, appeared to be demonstrated by the division and disappearance of one amongst their number, as well as by the singular and rapid changes in appearance undergone by many, and the (seemingly) irrevocable diffusion of their substance visible in nearly all. They might then be defined, according to the ideas respecting them prevalent thirty-five or forty years ago, as bodies unconnected by origin with the solar system, but encountered, and to some extent appropriated by it in its progress through space, owing their visibility in great part, if not altogether, to light reflected from the sun, and their singular and striking forms to the action of repulsive forces emanating from him, the penalty of their evanescent splendour being paid in gradual waste and final dissipation and extinction.

[1] Viz., Encke's, Biela's, Faye's, and Brorsen's. A comet with a supposed period of $5\frac{1}{2}$ years, detected by De Vico at Rome, August 22, 1844, has, it would appear, made no subsequent return to perihelion.

CHAPTER VI.

INSTRUMENTAL ADVANCES.

It is impossible to follow with intelligent interest the course of astronomical discovery without feeling some curiosity as to the means by which such surprising results have been secured. Indeed, the bare acquaintance with *what* has been achieved, without any corresponding knowledge of *how* it has been achieved, supplies food for barren wonder rather than for fruitful and profitable thought. Ideas advance most readily along the solid ground of practical reality, and often find true sublimity while laying aside empty marvels. Progress is the result, not so much of sudden flights of genius, as of sustained, patient, often commonplace endeavour; and the true lesson of scientific history lies in the close connection which it discloses between the most brilliant developments of knowledge and the faithful accomplishment of his daily task by each individual thinker and worker.

It would be easy to fill a volume with the detailed account of the long succession of optical and mechanical improvements by means of which the observation of the heavens has been brought to its present degree of perfection; but we must here content ourselves with a summary sketch of the chief amongst them. The first place in our consideration is naturally claimed by the telescope.

This marvellous instrument, we need hardly remind our readers, is of two distinct kinds—that in which light is gathered together into a focus by *refraction*, and that in which the same end is attained by *reflection*. The image formed is in each case viewed through a magnifying lens, or combination of

lenses, called the eye-piece. Not for above a century after the
" optic glasses " invented or stumbled upon by the spectacle-
maker of Middleburg (1608) had become diffused over Europe,
did the reflecting telescope come, even in England, the place
of its birth, into general use. Its principle (a sufficiently
obvious one) had indeed been suggested by Mersenne as early
as 1639 ;[1] James Gregory in 1663 [2] described in detail a mode
of embodying that principle in a practical shape; and Newton,
adopting an original system of construction, actually produced
in 1668 a tiny speculum, one inch across, by means of which
the apparent distance of objects was reduced thirty-nine times.
Nevertheless, the exorbitantly long tubeless refractors, introduced by Huygens, maintained their reputation until Hadley
exhibited to the Royal Society in 1723 [3] a reflector sixty-two
inches in focal length which rivalled in performance, and of
course indefinitely surpassed in manageability, one of the
" aerial " kind of 123 feet.

The concave mirror system now gained a decided ascendant,
and was brought to unexampled perfection by James Short
of Edinburgh during the years 1732–68. Its capabilities were,
however, first fully developed by William Herschel. The
energy and inventiveness of this extraordinary man marked an
epoch wherever they were applied. His ardent desire to measure and gauge the stupendous array of worlds which his specula
revealed to him, made him continually intent upon adding
to their " space-penetrating power " by increasing their light-
gathering surface. These, as he was the first to explain,[4] are
in a constant proportion one to the other. For a telescope with
twice the linear aperture of another will collect four times as
much light, and will consequently disclose an object four times
as faint as could be seen with the first, or, what comes to the
same, an object equally bright at twice the distance. In other
words, it will possess double the space-penetrating power
of the smaller instrument. Herschel's great mirrors—the

[1] Grant, *Hist. Astr.*, p. 527.   [2] *Optica Promota*, p. 93.
[3] *Phil. Trans.*, vol. xxxii. p. 383.   [4] *Ibid.*, vol. xc. p. 65.

first examples of the giant telescopes of modern times—were then primarily engines for extending the bounds of the visible universe ; and from the sublimity of this "final cause" was derived the vivid enthusiasm which animated his efforts to success.

It seems probable that the seven-foot telescope constructed by him in 1775—that is, within little more than a year after his experiments in shaping and polishing metal had begun—already exceeded in effective power any work by an earlier optician ; and both his skill and his ambition rapidly developed. His efforts culminated, after mirrors of ten, twenty, and thirty feet focal length had successively left his hands, in the gigantic forty-foot, completed August 28, 1789. It was the first reflector in which only a single mirror was employed. In the "Gregorian" form, the focussed rays are, by a second reflection from a small concave[1] mirror, thrown *straight back* through a central aperture in the larger one, behind which the eye-piece is fixed. The object under examination is thus seen in the natural direction. The "Newtonian," on the other hand, shows the object in a line of sight at right angles to the true one, the light collected by the speculum being diverted to one side of the tube by the interposition of a small plane mirror situated at an angle of 45° to the axis of the instrument. Upon these two systems Herschel worked until 1787, when, becoming convinced of the supreme importance of economising light (necessarily wasted by the second reflection), he laid aside the small mirror of his forty-foot then in course of construction, and turned it into a "front-view" reflector. This was done—according to the plan proposed by Lemaire in 1732—by slightly inclining the speculum so as to enable the image formed by it to be viewed with an eye-glass fixed at the upper margin of the tube. The observer thus stood with his back turned to the object he was engaged in scrutinising.

---

[1] Cassegrain, a Frenchman, substituted in 1672 a *convex* for a *concave* secondary speculum. The tube was thereby enabled to be shortened by twice the focal length of the mirror in question. The great Melbourne reflector (four feet aperture, by Grubb) is constructed upon this plan.

The advantages of the increased brilliancy afforded by this modification were strikingly illustrated by the discovery, August 28 and September 17, 1789, of the two Saturnian satellites nearest the ring. Nevertheless, the monster telescope of Slough cannot be said to have realised the sanguine expectations of its constructor. The occasions on which it could be usefully employed were found to be extremely rare. It was injuriously affected by every change of temperature. The great weight (25 cwt.) of a speculum *four feet* in diameter rendered it peculiarly liable to distortion. With all imaginable care, the delicate lustre of its surface could not be preserved longer than two years,[1] when the difficult process of repolishing had to be undertaken. It was never used after 1811, when, having *gone blind* from damp, it lapsed by degrees into the condition of a museum inmate.

The extraordinarily high magnifying powers employed by Herschel constituted a novelty in optical astronomy scarcely less striking than the gigantic size of his specula. They had never previously been approached ; they have never since been surpassed ; and they seem to mark, for these latitudes at least, the very outside limit of practicability. The attempt to increase in this manner the efficacy of the telescope is speedily checked by atmospheric, to say nothing of other difficulties. Precisely in the same proportion as an object is magnified, the disturbances of the medium through which it is seen are magnified also. Even on the clearest and most tranquil nights, the air is never for a moment really still. The rays of light traversing it are continually broken by minute fluctuations of refractive power caused by changes of temperature and pressure, and the currents which these engender. With such luminous quiverings and waverings the astronomer has always more or less to reckon ; their absence is simply a question of degree ; if sufficiently magnified, they are at all times capable of rendering observation impossible.

[1] *Phil. Trans.*, vol. civ. p. 275, *note.*

Thus, such vast powers as 3000, 4000, 5000, even 6652,[1] which Herschel now and again applied to his great telescopes, must, save on the rarest occasions, prove an impediment rather than an aid to vision.   They were, however, used by him only for special purposes, and with the clearest discrimination of their advantages and drawbacks.   It is obvious that perfectly different ends are subserved by increasing the *aperture* and by increasing the *power* of a telescope.   In the one case, a larger quantity of light is captured and concentrated; in the other, the same amount is distributed over a wider area.   A diminution of brilliancy accordingly attends upon each augmentation of apparent size.   For this reason, such faint objects as nebulæ are most successfully observed with moderate powers applied to instruments of a great capacity for light, the details of their structure actually disappearing when highly magnified.   With stellar groups the reverse is the case.   Stars cannot be magnified, simply because they are too remote to have any sensible dimensions ; but the space between them can.   It was thus for the purpose of dividing very close double stars that Herschel increased to such an unprecedented extent the magnifying capabilities of his instruments; and to this improvement incidentally the discovery of Uranus, March 13, 1781,[2] was due.   For by the examination with strong lenses of an object which, even with a power of 227, presented a suspicious appearance, he was able at once to pronounce its disc to be real, and not merely "spurious," and so to distinguish it unerringly from the crowd of stars amidst which it was moving.

While the reflecting telescope was astonishing the world by its rapid development in the hands of Herschel, its unpretending rival was slowly making its way towards the position which the future had in store for it.   The great obstacle which long

---

[1] *Phil. Trans.*, vol. xc. p. 70. With the 40-foot, however, only very moderate powers seem to have been employed, whence Dr. Robinson argued a deficiency of defining power. *Proc. Roy. Irish Ac.*, vol. ii. p. 11.

[2] *Phil. Trans.*, vol. lxxi. p. 492.

stood in the way of the improvement of refractors was the defect known as "chromatic aberration." This is due to no other cause than that which produces the rainbow and the spectrum—the separation, or "dispersion" in their passage through a refracting medium, of the variously coloured rays composing a beam of white light. In an ordinary lens there is no common point of concentration; each colour has its own separate focus; and the resulting image, formed by the superposition of as many images as there are hues in the spectrum, is indistinctly terminated with a tinted border, eminently baffling to exactness of observation.

The extravagantly long telescopes of the seventeenth century were designed to *avoid* this evil (as well as another source of indistinct vision in the spherical shape of lenses); but no attempt to *remedy* it was made until an Essex gentleman succeeded, in 1733, in so combining lenses of flint and crown glass as to produce refraction without colour.[1] Mr. Chester More Hall was, however, equally indifferent to fame and profit, and took no pains to make his invention public. The *effective* discovery of the achromatic telescope was, accordingly, reserved for John Dollond, whose method of correcting at the same time chromatic and spherical aberration was laid before the Royal Society in 1758. Modern astronomy may be said to have been thereby rendered possible. Refractors have always been found better suited than reflectors to the ordinary work of observatories. They are, so to speak, of a more robust, as well as of a more plastic nature. They suffer less from vicissitudes of temperature and climate. They retain their efficiency with fewer precautions and under more trying circumstances. Above all, they co-operate more readily with mechanical appliances, and lend themselves with far greater facility to purposes of exact measurement.

A practical difficulty, however, impeded the realisation of

[1] It is remarkable that, as early as 1695, the possibility of an achromatic combination was inferred by David Gregory from the structure of the human eye. See his *Catoptricæ et Dioptricæ Sphericæ Elementa*, p. 98.

K

the brilliant prospects held out by Dollond's invention. It was found impossible to procure flint-glass, such as was needed for optical use—that is, of perfectly homogeneous quality—except in fragments of insignificant size. Discs of more than two or three inches in diameter were of extreme rarity; and the crushing excise duty imposed upon the article by the financial unwisdom of the Government, both limited its production, and, by rendering experiments too costly for repetition, barred its improvement.

Up to this time, Great Britain had left foreign competitors far behind in the instrumental department of astronomy. The quadrants and circles of Bird, Cary, and Ramsden were unapproached abroad. The reflecting telescope came into existence and reached maturity on British soil. The refracting telescope was cured of its inherent vices by British ingenuity. But with the opening of the nineteenth century, the almost unbroken monopoly of skill and contrivance which our countrymen had succeeded in establishing was invaded, and British workmen had to be content to exchange a position of supremacy for one of at least partial and temporary inferiority.

Somewhere about the time that Herschel set about polishing his first speculum, Pierre Louis Guinand, a Swiss artisan, living near Chaux-de-Fonds, in the canton of Neuchâtel, began to grind spectacles for his own use, and was thence led on to the rude construction of telescopes by fixing lenses in pasteboard tubes. The sight of an English achromatic, however, stirred a higher ambition, and he took the first opportunity of procuring some flint-glass from England (then the only source of supply), with the design of imitating an instrument the full capabilities of which he was destined to be the humble means of developing. The English glass proving of inferior quality, he conceived the possibility, unaided and ignorant of the art as he was, of himself making better, and spent seven years (1784–90) in fruitless experiments directed to that end. Failure only stimulated him to enlarge their scale. He bought some land near Les Brenets, constructed upon it a furnace capable of melting two quintals of glass, and reducing himself and his family to the barest necessaries of life, he poured his earnings (he at this time made bells for repeaters) unstintingly into his crucibles.[1] His undaunted resolution triumphed. In 1799 he carried to Paris and there showed to Lalande several discs of flawless crystal four to six inches in diameter. Lalande advised him to keep his secret, but in 1805 he was induced to remove to Munich, where he became the instructor of the immortal Fraunhofer. His return to Les Brenets in 1814 was signalised by the discovery of an ingenious mode of removing striated portions of glass by breaking and re-soldering the product of each melting, and he eventually attained to the manufacture of perfect discs up to 18 inches in diameter. An object-glass for which he had furnished the material to Cauchoix, procured him, in 1823, a royal invitation to settle in Paris; but he was no longer equal to the change, and died at the scene of his labours, February 13 following.

This same lens (12 inches across) was afterwards purchased by Sir James South, and the first observation made with it, February 13, 1830, disclosed to Sir John Herschel the sixth minute star in the central group of the Orion nebula, known as the "trapezium."[2] A still larger objective (of nearly 14 inches) made of Guinand's glass was secured about the same time in Paris, by Mr. Edward Cooper of Markree Castle, Ireland. The peculiarity of the method discovered at Les Brenets resided in the manipulation, not in the quality of the ingredients; the secret, that is to say, was not chemical, but mechanical.[3] It was communicated by Henry Guinand (a son of the inventor) to Bontemps, one of the directors of the glassworks at Choisy-le-Roi, and by him transmitted to Messrs. Chance of Birmingham, with whom he entered into partnership when the revolutionary troubles of

[1] Wolf, *Biographien*, Bd. ii. p. 301.   [2] *Month. Not.*, vol. i. p. 153, *note*.
[3] Henrivaux, *Encyclopédie Chimique*, t. v. fasc. 5, p. 363.

1848 obliged him to quit his native country. The celebrated American opticians, Alvan Clark & Sons, have derived from the Birmingham firm the materials for some of their finest telescopes, notably the 19-inch Chicago and 26-inch Washington equatoreals.

Two distinguished amateurs, meanwhile, were preparing to reassert on behalf of reflecting instruments their claim to the place of honour in the van of astronomical discovery. Of Mr. Lassell's specula something has already been said.[1] They were composed of an alloy of copper and tin, with a minute proportion of arsenic (after the example of Newton[2]), and were remarkable for perfection of figure and brilliancy of surface.

The resources of the Newtonian system were developed still more fully—it might almost be said to the uttermost—by the enterprise of an Irish nobleman. William Parsons, known as Lord Oxmantown until 1841, when, on his father's death, he succeeded to the title of Earl of Rosse, was born at York, June 17, 1800. His public duties began before his education was completed. He was returned to Parliament as member for King's County while still an undergraduate at Oxford, and continued to represent the same constituency for thirteen years (1821–34). From 1845 until his death, which took place at Birr Castle, Parsonstown, October 31, 1867, he sat, silent but assiduous, in the House of Lords as an Irish representative peer; he held the not unlaborious post of President of the Royal Society from 1849 to 1854; presided over the meeting of the British Association at Cork in 1843, and was elected Vice-Chancellor of Dublin University in 1862. In addition to these extensive demands upon his time and thoughts, were those derived from his position as (practically) the feudal chief of a large body of tenantry in times of great and anxious responsibility, to say nothing of the more genial claims of an unstinted hospitality. Yet, while neglecting no public or private duty, this model nobleman found leisure to render to

---

[1] See *ante,* p. 109.   [2] *Phil. Trans.,* vol. vii. p. 407.

science services so conspicuous as to entitle his name to a lasting place in its annals.

He early formed the design of reaching the limits of the attainable in enlarging the powers of the telescope, and the qualities of his mind conspired with the circumstances of his fortune to render the design a feasible one. From refractors, it was obvious that no such vast and rapid advance could be expected. English glass-manufacture was still in a backward state. So late as 1839, Simms (successor to the distinguished instrumentalist Edward Troughton) reported a specimen of crystal scarcely $7\frac{1}{2}$ inches in diameter, and perfect only over six, to be unique in the history of English glass-making.[1] Yet at that time the 15-inch achromatic of Pulkowa had already left the workshop of Fraunhofer's successors at Munich. It was not indeed until 1845, when the impost which had so long hampered their efforts was removed, that the optical artists of these islands were able to compete on equal terms with their rivals on the Continent. In the case of reflectors, however, there seemed no insurmountable obstacle to an almost unlimited increase of light-gathering capacity ; and it was here, after some unproductive experiments with fluid lenses, that Lord Oxmantown concentrated his energies.

He had to rely entirely on his own invention, and to earn his own experience. James Short had solved the problem of giving to metallic surfaces a perfect parabolic figure (the only one by which parallel incident rays can be brought to an exact focus) ; but so jealous was he of his secret, that he caused all his tools to be burnt before his death ;[2] nor was anything known of the processes by which Herschel had achieved his astonishing results. Moreover, Lord Oxmantown had no skilled workmen to assist him. His implements, both animate and inanimate, had to be formed by himself. Peasants taken from the plough were educated by him into efficient mechanics and engineers. The delicate and complex machinery needed in operations of such hairbreadth nicety as his enterprise

[1] J. Herschel, *The Telescope*, p. 39.   [2] *Month. Not.*, vol. xxix. p. 125.

involved, the steam-engine which was to set it in motion, at times the very crucibles in which his specula were cast, issued from his own workshops.[1]

In 1827 experiments on the composition of speculum-metal were set on foot, and the first polishing-machine ever driven by steam-power was contrived. But twelve arduous years of struggle with recurring difficulties passed before success began to dawn. A material less tractable than the alloy selected of four chemical equivalents of copper to one of tin [2] can scarcely be conceived. It is harder than steel, yet brittle as glass, crumbling into fragments with the slightest inadvertence of handling or treatment;[3] and the precision of figure requisite to secure good definition is almost beyond the power of language to convey. The quantities involved are so small as not alone to elude sight, but to confound imagination. Sir John Herschel tells us that "the *total* thickness to be abraded from the edge of a spherical speculum 48 inches in diameter and 40 feet focus, to convert it into a paraboloid, is only $\frac{1}{21.333}$ of an inch;"[4] yet upon this minute difference of form depends the clearness of the image, and, as a consequence, the entire efficiency of the instrument. "Almost infinite," indeed (in the phrase of the late Dr. Robinson), must be the exactitude of the operation adapted to bring about so delicate a result.

At length, in 1840, two specula, each three feet in diameter, were turned out in such perfection as to prompt a still bolder experiment. The various processes needed to ensure success were now ascertained and under control; all that was necessary was to repeat them on a larger scale. A gigantic mirror, six feet across and fifty-four in focal length, was accordingly cast on the 13th of April 1842; in two months it was ground down to figure by abrasion with emery and water, and daintily

---

[1] *Month. Not.*, vol. xxix. p. 129.

[2] A slight excess of copper renders the metal easier to work, but liable to tarnish. Robinson, *Proc. Roy. Irish Ac.*, vol. ii. p. 4.

[3] *Brit. Ass.*, 1843, Dr. Robinson's closing Address. *Athenæum*, Sept. 23, p. 866.   [4] *The Telescope*, p. 82.

polished with rouge; and by the month of February 1845 the "leviathan of Parsonstown" was available for the examination of the heavens.

The suitable mounting of this vast machine was a problem scarcely less difficult than its construction. The shape of a speculum needs to be maintained with an elaborate care equal to that used in imparting it. In fact, one of the most formidable obstacles to increasing the size of such reflecting surfaces consists in their liability to bend under their own weight. That of the great Rosse speculum was no less than *four tons*. Yet, although six inches in thickness, and composed of a material only a degree inferior in rigidity to wrought iron, the strong pressure of a man's hand at its back produced sufficient flexure to distort perceptibly the image of a star reflected in it.[1] Thus the delicacy of its form was perishable equally by the stress of its own gravity, and by the slightest irregularity in the means taken to counteract that stress. The problem of affording a perfectly equable support in all possible positions was solved by resting the speculum upon twenty-seven platforms of cast iron, felt-covered, and carefully fitted to the shape of the areas they were to carry, which platforms were themselves borne by a complex system of triangles and levers, ingeniously adapted to distribute the weight with complete uniformity.[2]

A tube which resembled, when erect, one of the ancient round towers of Ireland,[3] served as the habitation of the great mirror. It was constructed of deal staves bound together with iron hoops, was fifty-eight feet long (including the speculum-box), and seven in diameter. We are assured that the late Dean of Ely walked through it with umbrella uplifted.[4] Two piers of solid masonry, about fifty feet high, seventy long, and

---

[1] Lord Rosse in *Phil. Trans.*, vol. cxl. p. 302.

[2] This method is the same in principle with that applied by Grubb in 1834 to a 15-inch speculum for the observatory of Armagh. *Phil. Trans.*, vol. clix. p. 145.   [3] Robinson, *Proc. Roy. Ir. Ac.*, vol. iii. p. 120.

[4] Brewster, *North British Review*, vol. ii. p. 207.

twenty-three apart, flanked the huge engine on either side. Its lower extremity rested on an universal joint of cast iron ; above, it was slung in chains, and even in a gale of wind remained perfectly steady.   The weight of the entire, although amounting to fifteen tons, was so skilfully counterpoised, that the tube could with ease be raised or depressed by two men working a windlass.   Its horizontal range was limited by the lofty walls erected for its support to about ten degrees on each side of the meridian; but it moved vertically from near the horizon through the zenith as far as the pole.   Its construction was of the Newtonian kind, the observer looking into the side of the tube near its upper end, which a series of galleries and sliding stages enabled him to reach in any position.   It has also, though rarely, been used without a second mirror, as a "Herschelian" reflector.

The splendour of the celestial objects as viewed with this vast "light-grasper" surpassed all expectation.   "Never in my life," exclaims Sir James South, "did I see such glorious sidereal pictures!"[1]   The orb of Jupiter produced an effect compared to that of the introduction of a coach-lamp into the telescope ;[2] and certain star-clusters exhibited an appearance (we again quote Sir James South) "such as man before had never seen, and which for its magnificence baffles all description."   But it was in the examination of the nebulæ that the superiority of the new instrument was most strikingly displayed. A large number of these misty objects, which the utmost powers of Herschel's specula had failed to resolve into stars, yielded at once to the Parsonstown reflector ; while many others showed under entirely changed forms through the dis- closure of previously unseen details of structure.

One extremely curious result of the increase of light was the abolition of the distinction between the two classes of "annular" and "planetary" nebulæ. Up to that time, only four ring- shaped systems—two in the northern and two in the southern hemisphere—were known to astronomers ; they were now

[1] *Astr. Nach.*, No. 536.   [2] Airy, *Month. Not.*, vol. ix. p. 120.

reinforced by five of the planetary kind, the discs of which were observed to be centrally perforated; while the sharp marginal definition visible in weaker instruments was replaced by ragged edges or filamentous fringes.

Still more striking was the discovery of an entirely new and highly remarkable species of nebulæ. These were termed "spiral," from the more or less regular convolutions, re-sembling the whorls of a shell, in which the matter composing them appeared to be distributed. The first and most con-spicuous specimen of this class was met with in April 1845 ; it is situated in Canes Venatici, close to the tail of the Great Bear, and wore, in Sir J. Herschel's instruments, the aspect of a split ring encompassing a bright nucleus, thus presenting, as he supposed, a complete analogue to the system of the Milky Way. In the Rosse mirror it shone out as a vast whirlpool of light—a stupendous witness to the presence of cosmical activities on the grandest scale, yet regulated by laws as to the nature of which we are profoundly ignorant. Professor Stephen Alexander of New Jersey, however, concluded, from an in-vestigation (necessarily founded on highly precarious data) of the mechanical condition of these extraordinary agglomerations, that we see in them "the partially scattered fragments of enor-mous masses once rotating in a state of dynamical equilibrium." He further suggested "that the separation of these fragments may still be in progress,"[1] and traced back their origin to the disruption, through its own continually accelerated rotation, of a "primitive spheroid" of inconceivably vast dimensions. Such also, it was added (the curvilinear form of certain out-liers of the Milky Way, giving evidence of a spiral structure), is probably the history of our own cluster; the stars composing which, no longer held together in a delicately adjusted system like that of the sun and planets, are advancing through a period of seeming confusion towards an appointed goal of higher order and more perfect and harmonious adaptation.[2]

---

[1] *Astronomical Journal* (Gould's), vol. ii. p. 97.

[2] *Ibid.*, vol. ii. p. 160.

The class of spiral nebulæ included, in 1850, fourteen members, besides several in which the characteristic arrangement seemed partial or dubious.[1]   A tendency in the exterior stars of other clusters to gather into curved branches (as in our Galaxy) was likewise noted ; and the existence of unsuspected analogies was proclaimed by the significant combination in the "Owl" nebula (a large planetary in Ursa Major)[2] of the twisted forms of a spiral, with the perforations distinctive of an annular nebula.

Once more, by the achievements of the Parsonstown reflector, the supposition of a "shining fluid" filling vast regions of space was brought into (as it has since proved) undeserved discredit.   Although Lord Rosse himself rejected the inference that because many nebulæ had been resolved, all were resolvable, very few imitated his truly scientific caution ; and the results of Bond's investigations[3] with the Harvard College refractor quickened and strengthened the current of prevalent opinion.   It is now certain that the evidence furnished on both sides of the Atlantic as to the stellar composition of some conspicuous objects of this class (notably the Orion and "Dumb-bell" nebulæ) was delusive ; but the spectroscope alone was capable of meeting it with a categorical denial.   Meanwhile there seemed good ground for the persuasion, which now, for the last time, gained the upperhand, that nebulæ are, without exception, true "island-universes," or assemblages of distant suns.

Lord Rosse's telescope possesses a nominal power of 6000— that is, it shows the moon as if viewed with the naked eye at a distance of forty miles.   But this seeming advantage is neutralised by the weakening of the available light through excessive diffusion, as well as by the troubles of the surging sea of air through which the observation must necessarily be made.

[1] Lord Rosse in *Phil. Trans.*, vol. cxl. p. 505.

[2] No. 2343 of Herschel's (1864) Catalogue.   Before 1850 a star was visible in each of the two larger openings by which it is pierced ; since then one only.   Webb, *Celestial Objects* (4th ed.), p. 409.

[3] *Mem. Am. Ac.*, vol. iii. p. 87 ; and *Astr. Nach.*, No. 611.

Professor Newcomb, in fact, doubts whether with *any* telescope our satellite has ever been seen to such advantage as it would be if brought within 500 miles of the unarmed eye.[1]

The French opticians' rule of doubling the number of milli-metres contained in the aperture of an instrument to find the highest magnifying power *usefully* applicable to it, would give 3600 as the maximum for the leviathan of Birr Castle ; but in a climate like that of Ireland the occasions must be rare when even that limit can be reached. Indeed, the experience acquired by its use plainly shows that atmospheric, rather than mechanical difficulties impede a still further increase of tele-scopic power. Its construction may accordingly be said to mark the *ne plus ultra* of effort in one direction, and the beginning of its conversion towards another. It became thenceforward more and more obvious that the *conditions* of observation must be ameliorated before any added efficacy could be given to it. The full effect of an uncertain climate in nullifying optical improvements was recognised, and the attention of astronomers began to be turned towards the advantages offered by more tranquil and more translucent skies.

Even more important for the practical uses of astronomy than the optical qualities of the telescope is the manner of its mounting. There is a far greater likelihood of getting good work done with an imperfect instrument skilfully mounted, than with the most admirable performance of the optician of which the mechanical accessories are ill-arranged or incon-venient. Thus the astronomer is ultimately dependent upon the mechanician ; and so excellently have his needs been served, that the history of the ingenious contrivances by which discoveries have been prepared would supply a subject scarcely inferior in extent and instruction to the history of those dis-coveries themselves. But the limits of the present work barely admit of a passing glance at the subject.

There are two chief modes of using the telescope, to which

[1] *Pop. Astr.*, p. 145.

all others may be considered subordinate.[1] Either it may be
immovably directed towards the south, in other words, fixed
in the *plane of the meridian*, so as to intercept the heavenly
bodies at the moment of transit across that plane ; or it may
be arranged so as to follow the daily revolution of the sky,
thus keeping the object viewed permanently in sight, instead
of simply noting the instant of its flitting across the telescopic
field.   The first plan is that of the " transit instrument," the
second that of the " equatoreal."   Both were, by a remarkable
coincidence, introduced about 1690[2] by Olaus Römer, the
brilliant Danish astronomer who first measured the velocity of
light.

The uses of each are entirely different.   With the transit,
the really fundamental task of astronomy—the determination
of the movements of the heavenly bodies—is mainly accomplished ; while the investigation of their nature and peculiarities
is best conducted with the equatoreal.   One is the instrument
of mathematical, the other of descriptive astronomy.   One
furnishes the materials with which theories are constructed, and
the tests by which they are corrected ; the other registers new
facts, takes note of new appearances, sounds the depths, and
pries into every nook of the heavens.

The great improvement of giving to a telescope equatoreally
mounted an automatic movement by connecting it with clockwork, was proposed in 1674 by Robert Hooke.   Bradley in
1721 actually observed Mars with a telescope " moved by a
machine that made it keep pace with the stars ;"[3] and Von
Zach relates[4] that he had once followed Sirius for twelve hours

---

[1] This statement must be taken in the most general sense. Supple-
mentary observations of great value are now made at Greenwich with the
altitude and azimuth instrument, which likewise served Piazzi to determine
the places of his stars ; while a " prime vertical instrument " is prominent
at Pulkowa.

[2] As early as 1620, according to R. Wolf (*Gesch. der Astr.*, p. 587),
Father Scheiner made the experiment of connecting a telescope with an
axis directed to the pole.   [3] *Bradley's Miscellaneous Works*, p. 350.

[4] *Astr. Jahrbuch*, 1799 (published 1796), p. 115.

with a "heliostat" of Ramsden's construction. But these eighteenth-century attempts were of no practical effect. Movement by clockwork was virtually a complete novelty when it was adapted by Fraunhofer in 1824 to the Dorpat refractor. By simply giving to an axis unvaryingly directed towards the celestial pole an equable rotation with a period of twenty-four hours, a telescope attached to it, and pointed in *any* direction, will trace out on the sky a parallel of declination, thus necessarily accompanying the movement of any star upon which it may be fixed. It thus forms part of the large sum of Fraunhofer's merits to have secured this inestimable advantage to observers.

It was considered by Sir John Herschel that Lassell's application of equatoreal mounting to a nine-inch Newtonian in 1840 made an epoch in the history of "that eminently British instrument, the reflecting telescope." [1] Nearly a century earlier,[2] it is true, Short had fitted one of his Gregorians to a complicated system of circles in such a manner that, by moving a handle, it could be made to follow the revolution of the sky; but the arrangement did not obtain, nor did it deserve, general adoption. Lassell's plan was a totally different one; he employed the crossed axes of the true equatoreal, and his success removed, to a great extent, the fatal objection of inconvenience in use, until then unanswerably urged against reflectors. The very largest of these can now be mounted equatoreally; even the Rosse within its limited range has been for some years provided with a movement by clockwork along declination-parallels.

The art of accurately dividing circular arcs into the minute equal parts which serve as the units of astronomical measurement, remained, during the whole of the eighteenth century, almost exclusively in English hands. It was brought to a high degree of perfection by Graham, Bird, and Ramsden, all of whom, however, gave the preference to the old-fashioned mural quadrant and zenith sector over the entire circle, which

[1] *Month. Not.*, vol. xli. p. 189. [2] *Phil. Trans.*, vol. xlvi. p. 242.

Römer had already found the advantage of employing.　The five-foot vertical circle, which Piazzi with some difficulty induced Ramsden to complete for him in 1789, was the first divided instrument constructed in what may be called the modern style.　It was provided with magnifiers for reading off the divisions (one of the neglected improvements of Römer), and was set up above a smaller horizontal circle, forming an "altitude and azimuth" combination (again Römer's invention), by which both the elevation of a celestial object above the horizon and its position as referred to the horizon could be measured.　In the same year Borda invented the "repeating circle" (the principle of which had been suggested by Tobias Mayer in 1756[1]), a device for exterminating, so far as possible, errors of graduation by *repeating* an observation with different parts of the limb.　This was perhaps the earliest systematic effort to correct the imperfections of instruments by the manner of their use.

The manufacture of astronomical circles was brought to a very refined state of excellence early in the present century by Reichenbach at Munich, and (after 1818) by Repsold at Hamburg.　Bessel states[2] that the "reading-off" on an instrument of the kind by the latter artist was accurate to about $\frac{1}{80}$th of a human hair.　Meanwhile the traditional reputation of the English school was fully sustained; and Sir George Airy did not hesitate to express his opinion that the new method of graduating circles, published by Troughton in 1809,[3] was the "greatest improvement ever made in the art of instrument-making."[4]　But a more secure road to improvement than that of mere mechanical exactness was pointed out by Bessel.　His introduction of a regular theory of instrumental errors might almost be said to have created a new art of observation.　Every instrument, he declared in memorable words,[5] must be twice made—once by the artist, and again by

[1] Grant, *Hist. of Astr.*, p. 487.　　　[2] *Pop. Vorl.*, p. 546.
[3] *Phil. Trans.*, vol. xcix. p. 105.
[4] *Report Brit. Ass.*, 1832, p. 132.　　　[5] *Pop. Vorl.*, p. 432.

the observer. Knowledge is power. Defects that are ascertained and can be allowed for are as good as non-existent. Thus the truism that the best instrument is worthless in the hands of a careless or clumsy observer, became supplemented by the converse maxim that defective appliances may, by skilful use, be made to yield valuable results. The Königsberg observations—of which the first instalment was published in 1815—set the example of regular "reduction" for instrumental errors. Since then, it has become an elementary part of an astronomer's duty to study the *idiosyncrasy* of each one of the mechanical contrivances at his disposal, in order that its inevitable, 'but now certified deviations from ideal accuracy may be included amongst the numerous corrections by which the pure essence of (even approximate) truth is distilled from the rude impressions of sense.

Nor is this enough; for the casual circumstances attending each observation have to be taken into account with no less care than the inherent, or *constitutional* peculiarities of the instrument with which it is made. There is no "once for all" in astronomy. Vigilance can never sleep; patience can never tire. Variable as well as constant sources of error must be anxiously heeded; one infinitesimal inaccuracy must be weighed against another; all the forces and vicissitudes of nature—frosts, dews, winds, the interchanges of heat, the disturbing effects of gravity, the shiverings of the air, the tremors of the earth, the weight and vital warmth of the observer's own body, nay, the rate at which his brain receives and transmits its impressions, must all enter into his calculations, and be sifted out from his results.

It was in 1823 that Bessel drew attention to discrepancies in the times of transits given by different astronomers.[1] The quantities involved were far from insignificant. He was himself nearly a second in advance of all his contemporaries, Argelander lagging behind him as much as a second and a

[1] C. T. Anger, *Grundzüge der neueren astronomischen Beobachtungs-Kunst*, p. 3.

quarter.   Each individual, in fact, was found to have a certain definite *rate of perception*, which, under the name of "personal equation," now forms an important element in the correction of observations.

Such are the refinements upon which modern astronomy depends for its progress.   It is a science of hairbreadths and fractions of a second.   It exists only by the rigid enforcement of arduous accuracy and unwearying diligence.   Whatever secrets the universe still has in store for man will only be communicated on these terms.   They are, it must be acknowledged, difficult to comply with.   They involve an unceasing struggle against the infirmities of his nature and the instabilities of his position.   But the end is not unworthy the sacrifices demanded.   One additional ray of light thrown on the marvels of creation—a single, minutest encroachment upon the strongholds of ignorance, is recompense enough for a lifetime of toil. Or rather, the toil is its own reward, if pursued in the lofty spirit which alone becomes it.   For it leads through the abysses of space and the unending vistas of time to the very threshold of that infinity and eternity of which the disclosure is reserved for a life to come.

PART II.

RECENT PROGRESS OF ASTRONOMY.

—•—

CHAPTER I.

FOUNDATION OF ASTRONOMICAL PHYSICS.

In the year 1826, Heinrich Schwabe of Dessau, elated with the hope of speedily delivering himself from his hereditary incubus of an apothecary's shop,[1] obtained from Munich a small telescope and began to observe the sun. His choice of an object for his researches was instigated by his friend Harding of Göttingen. It was a peculiarly happy one. The changes visible in the solar surface were then generally regarded as no less capricious than the changes in the skies of our temperate regions. Consequently, the reckoning and registering of sun-spots was a task hardly more inviting to an astronomer than the reckoning and registering of summer clouds. Cassini, Keill, Lemonnier, Lalande, were unanimous in declaring that no trace of regularity could be detected in their appearances or effacements.[2] Delambre pronounced them "more curious than really useful."[3] Even Herschel, profoundly as he studied them, and intimately as he was convinced of their importance as symptoms of solar activity, saw no reason to suspect that their abundance and scarcity were subject to orderly alternation. One man alone in the eighteenth century, Christian

[1] Wolf, *Gesch. der Astr.*, p. 655.
[2] Manuel Johnson, *Mem. R. A. Soc.*, vol. xxvi. p. 197.
[3] *Astronomie Théorique et Pratique*, t. iii. p. 20.

L

Horrebow of Copenhagen, divined their periodical character, and foresaw the time when the effects of the sun's vicissitudes upon the globes revolving round him might be investigated with success ; but this prophetic utterance was of the nature of a soliloquy rather than of a communication, and remained hidden away in an unpublished journal until 1859, when it was brought to light in a general ransacking of archives.[1]

Indeed, Schwabe himself was far from anticipating the discovery which fell to his share. He compared his fortune to that of Saul, who, seeking his father's asses, found a kingdom.[2] For the hope which inspired his early resolution lay in quite another direction. His patient ambush was laid for a possible intra-mercurial planet, which, he thought, must sooner or later betray its existence in crossing the face of the sun. He took, however, the most effectual measures to secure whatever new knowledge might be accessible. During forty-three years his " imperturbable telescope "[3] never failed (weather and health permitting), to bring in its daily report as to how many, or if any, spots were visible on the sun's disc, the information obtained being day by day recorded on a simple and unvarying system. In 1843 he made his first announcement of a probable decennial period,[4] but it met with no general attention ; although Julius Schmidt of Bonn (afterwards director of the Athens Observatory) and Gautier of Geneva were impressed with his figures, and Littrow had himself, in 1836,[5] hinted at the likelihood of some kind of regular recurrence. Schwabe, however, worked on, gathering each year fresh evidence of a law such as he had indicated ; and when Humboldt published in 1851, in the third volume of his *Kosmos*,[6] a table of the sun-spot statistics collected by him from 1826 downwards, the strength of his case was perceived with, so to speak, a start of surprise ; the reality and importance of the

---

[1] Wolf, *Gesch. der Astr.*, p. 654.   [2] *Month. Not.*, vol. xvii. p. 241.
[3] *Mem. R. A. Soc.*, vol. xxvi. p. 200.   [4] *Astr. Nach.*, No. 495.
[5] Gehler's *Physikalisches Wörterbuch*, art. *Sonnenflecken*, p. 851.
[6] *Zweite Abth.*, p. 401.

discovery were simultaneously recognised, and the persevering Hofrath of Dessau found himself famous among astronomers. His merit—recognised by the bestowal of the Astronomical Society's Gold Medal in 1857—consisted in his choice of an original and appropriate line of work, and in the admirable tenacity of purpose with which he pursued it. His resources and acquirements were those of an ordinary amateur; he was distinguished solely by the (unfortunately rare) power of turning both to the best account. He died where he was born and had lived, April 11, 1875, at the ripe age of eighty-six.

Meanwhile an investigation of a totally different character, and conducted by totally different means, had been prosecuted to a very similar conclusion. Two years after Schwabe began his solitary observations, Humboldt gave the first impulse, at the Scientific Congress of Berlin in 1828, to a great international movement for attacking simultaneously, in various parts of the globe, the complex problem of terrestrial magnetism. Through the genius and energy of Gauss, Göttingen became its centre. Thence new apparatus and a new system for its employment issued; there, in 1833, the first regular magnetic observatory was founded, while Göttingen mean time was made the universal standard for magnetic observations. The letter addressed by Humboldt in April 1836 to the Duke of Sussex as President of the Royal Society, enlisted the co-operation of England. A network of magnetic stations was spread all over the British dominions, from Canada to Van Diemen's Land; measures were concerted with foreign authorities, and an expedition was fitted out, under the able command of Captain (afterwards Sir James) Clark Ross, for the special purpose of bringing intelligence on the subject from the dismal neighbourhood of the South Pole. In 1841, the elaborate organisation created by the disinterested efforts of scientific "agitators" was complete; Gauss's "magnetometers" were vibrating, under the view of attentive observers, in five continents, and simultaneous results began to be recorded.

Ten years later, in September 1851, Dr. John Lamont, the

Scotch director of the Munich Observatory, in reviewing the magnetic observations made at Göttingen and Munich from 1835 to 1850, perceived with some surprise that they gave unmistakable indications of a period which he estimated at $10\frac{1}{3}$ years.[1] The manner in which this periodicity manifested itself requires a word of explanation. The observations in question referred to what is called the "declination" of the magnetic needle—that is, to the position assumed by it with reference to the points of the compass when moving freely in a horizontal plane. Now this position—as was discovered by Graham in 1722—is subject to a small daily fluctuation, attaining its maximum towards the east about 8 A.M., and its maximum towards the west shortly before 2 P.M. In other words, the direction of the needle approaches (in these countries at the present time) nearest to the true north some four hours before noon, and departs farthest from it between one and two hours after noon. It was the *range* of this daily variation that Lamont found to increase and diminish once in every $10\frac{1}{3}$ years.

In the following winter, Sir Edward Sabine, ignorant as yet of Lamont's conclusion, undertook to examine a totally different set of observations. The materials in his hands had been collected at the British colonial stations of Toronto and Hobarton from 1843 to 1848, and had reference, not to the regular diurnal swing of the needle, but to those curious spasmodic vibrations, the inquiry into the laws of which was the primary object of the vast organisation set on foot by Humboldt and Gauss. Yet the upshot was practically the same. Once in about ten years magnetic disturbances (termed by Humboldt "storms") were perceived to reach a maximum of violence and frequency. Sabine was the first to note the coincidence between this unlooked-for result and Schwabe's sun-spot period. He showed that, so far as observation had yet gone, the two cycles of change agreed perfectly both in duration and phase, maximum corresponding to maximum, minimum to minimum. What the nature of the connection could be that bound to-

[1] *Annalen der Physik* (Poggendorff's), Bd. lxxxiv. p. 580.

gether by a common law effects so dissimilar as the rents in the luminous garment of the sun, and the swayings to and fro of the magnetic needle, was, and still remains beyond the reach of well-founded conjecture; but the fact was from the first undeniable.

The memoir containing this remarkable disclosure was presented to the Royal Society, March 18, and read May 6, 1852.[1] On the 31st of July following, Rudolf Wolf at Berne,[2] and on the 18th of August Alfred Gautier at Sion,[3] announced, separately and independently, perfectly similar conclusions. This triple event is perhaps the most striking instance of the successful employment of the Baconian method of co-operation in discovery, by which "particulars" are amassed by one set of investigators—corresponding to the "Depredators" and "Inoculators" of Solomon's House—while inductions are drawn from them by another and a higher class—the "Interpreters of Nature." Yet even here the convergence of two distinct lines of research was wholly fortuitous, and skilful combination owed the most brilliant part of its success to the unsought bounty of what we call Fortune.

The exactness of the coincidence thus brought to light was fully confirmed by further inquiries. A diligent search through the scattered records of sun-spot observations, from the time of Galileo and Scheiner onwards, put Wolf[4] in possession of materials by which he was enabled to correct Schwabe's loosely indicated decennial period to one of slightly over eleven (11.11) years; and he further showed that this fell in with the ebb and flow of magnetic change even better than Lamont's $10\frac{1}{3}$ year cycle. For the first time, also, the analogy was pointed out between the "light-curve," or zigzagged line representing on paper the varying intensity in the lustre of certain stars, and the similar delineation of spot-frequency; the ascent from minimum to

[1] *Phil. Trans.*, vol. cxlii. p. 103.
[2] *Mittheilungen der Naturforschenden Gesellschaft*, 1852, p. 183.
[3] *Archives des Sciences*, t. xxi. p. 194.
[4] *Neue Untersuchungen, Mitth. Naturf. Ges.*, 1852, p. 249.

maximum being, in both cases, usually steeper than the descent from maximum to minimum ; while an additional point of resemblance was furnished by the irregularities in height of the various maxima.   In other words, both the number of spots on the sun and the brightness of variable stars increase, as a rule, more rapidly than they decrease ; nor does the amount of that increase, in either instance, show any approach to uniformity.

The endeavour, suggested by the very nature of the phenomenon, to connect sun-spots with weather was less successful. The first attempt of the kind was made by Sir William Herschel in the first year of the present century, and a very notable one it was.   Meteorological statistics, save of the scantiest and most casual kind, did not then exist; but the price of corn from year to year was on record, and this, with full recognition of its inadequacy, he adopted as his criterion.   Nor was he much better off for information respecting the solar condition. What little he could obtain, however, served, as he believed, to confirm his surmise that a copious emission of light and heat accompanies an abundant formation of " openings " in the dazzling substance whence our supply of those indispensable commodities is derived.[1]   He gathered, in short, from his inquiries very much what he had expected to gather, namely, that the price of wheat was high when the sun showed an unsullied surface, and that food and spots became plentiful together.

This plausible inference, however, was scarcely borne out by a more exact collocation of facts.   Schwabe failed to detect any reflection of the sun-spot period in his meteorological register.   Gautier[2] reached a provisional conclusion the reverse —though not markedly the reverse—of Herschel's.   Wolf, in 1852, derived from an examination of Vogel's collection of Zürich Chronicles (1000–1800 A.D.) evidence showing (as he thought) that minimum years were usually wet and stormy,

---

[1] *Phil. Trans.*, vol. xci. p. 316.
[2] *Bibliothèque Universelle de Genève*, t. li. p. 336.

maximum years dry and genial;[1] but a subsequent review of
the subject in 1859 convinced him that no relation of any
kind between the two kinds of effects was traceable.[2] With
the singular affection of our atmosphere known as the Aurora
Borealis (more properly Aurora Polaris) the case was different.
Here the Zürich Chronicles set Wolf on the right track in
leading him to associate such luminous manifestations with
a disturbed condition of the sun; since subsequent detailed
observation has exhibited the curve of auroral frequency as
following with such fidelity the jagged lines figuring to the eye
the fluctuations of solar and magnetic activity, as to leave no
reasonable doubt that all three rise and sink together under
the influence of a common cause. As long ago as 1714,[3]
Halley had conjectured that the Northern Lights were due to
magnetic "effluvia," but there was no evidence on the subject
forthcoming until Hiorter observed at Upsala in 1741 their
agitating influence upon the magnetic needle. That the effect
was no casual one was made superabundantly clear by Arago's
researches in 1819 and subsequent years. Now both were
perceived to be swayed by the same obscure power of cosmical
disturbance.

The sun is not the only one of the heavenly bodies by which
the magnetism of the earth is affected. Proofs of a similar
kind of lunar action were laid by Kreil in 1841 before the
Bohemian Society of Sciences, and were fully substantiated,
though with minor corrections, by Sabine's more extended
researches. It has thus been ascertained that each lunar day,
or the interval of twenty-four hours and about fifty-four minutes
between two successive meridian passages of our satellite, is
marked by a perceptible, though very small double oscillation

---

[1] *Neue Untersuchungen*, p. 269.

[2] *Die Sonne und ihre Flecken*, p. 30. Arago was the first who attempted
to decide the question by keeping, through a series of years, a parallel
register of sun-spots and weather; but the data regarding the solar condition
collected at the Paris Observatory from 1822 to 1830 were not sufficiently
precise to found any inference upon.

[3] *Phil. Trans.*, vol. xxix. p. 421.

of the needle—two progressive movements from east to west, and two returns from west to east.[1]   Moreover, the lunar, like the solar influence (as was proved in each case by Sabine's analysis of the Hobarton and Toronto observations), extends to all three "magnetic elements," affecting not only the position of the horizontal or *declination*-needle, but also the dip and intensity.   It seems not unreasonable to attribute some portion of the same subtle power to the planets, and even to the stars, though with effects rendered imperceptible by distance.

We have now to speak of the discovery and application to the heavenly bodies of a totally new method of investigation. Spectrum analysis may be shortly described as a mode of distinguishing the various species of matter by the kind of light proceeding from each.   This definition at once explains how it is that, unlike every other system of chemical analysis, it has proved available in astronomy.   Light, so far as *quality* is concerned, ignores distance.   No intrinsic change (that we yet know of) is produced in it by a journey from the farthest bounds of the visible universe; so that, provided only that in *quantity* it remain sufficient for the purpose, its peculiarities can be equally well studied whether the source of its vibrations be one foot or a hundred billion miles distant.   Now the most obvious distinction between one kind of light and another resides in colour.   But of this distinction the eye takes cognisance in an æsthetic, not in a scientific sense.   It finds gladness in the "thousand tints" of nature, but can neither analyse nor define them.   Here the refracting prism—or the combination of prisms known as the "spectroscope"—comes to its aid, teaching it to measure as well as to perceive.   It furnishes, in a word, an accurate scale of colour.   The various rays which, entering the eye together in a confused crowd, produce a compound impression made up of undistinguishable elements, are, by the mere passage through a triangular piece of glass, separated one from the other, and ranged side by side in orderly succession, so that it becomes possible to tell at a

[1] *Phil. Trans.*, vol. cxliii. p. 558, and vol. cxlvi. p. 505.

glance what kinds of light are present, and what absent. Thus, if we could only be assured that the various chemical substances, when made to glow by heat, emit characteristic rays—rays, that is, occupying a place in the spectrum reserved for them, and for them *only*—we should at once be in possession of a mode of identifying such substances with the utmost readiness and certainty. This assurance, which forms the solid basis of spectrum analysis, was obtained slowly and with difficulty.

The first to employ the prism in the examination of various flames (for it is only in a state of vapour that matter emits distinctive light) was a young Scotchman named Thomas Melvill, who died in 1753, at the age of twenty-seven. He studied the spectrum of burning spirits, into which were introduced successively sal ammoniac, potash, alum, nitre, and sea-salt, and observed the singular predominance, under almost all circumstances, of a particular shade of yellow light, perfectly definite in its degree of refrangibility [1]—in other words, taking up a perfectly definite position in the spectrum. His experiments were repeated by Morgan,[2] Wollaston, and—with far superior precision and diligence — by Fraunhofer.[3] The great Munich optician, whose work was completely original, rediscovered Melvill's deep yellow ray and measured its place in the colour-scale. It has since become well known as the "sodium line," and has played a very important part in the history of spectrum analysis. Nevertheless, its ubiquity and conspicuousness long impeded progress. It was elicited by the combustion of a surprising variety of substances—sulphur, alcohol, ivory, wood, paper; its persistent visibility suggesting the accomplishment of some universal process of nature rather than the presence of one individual kind of matter. But if spectrum analysis were to exist as a science at all, it could only be by attaining certainty as to the *unvarying* association of one special substance with each special quality of light.

---

[1] *Observations on Light and Colours,* p. 35.
[2] *Phil. Trans.,* vol. lxxv. p. 190.
[3] *Denkschriften* (Munich Ac. of Sc.), 1814-15, Bd. v. p. 197.

Thus perplexed, Fox Talbot [1] hesitated in 1826 to enounce this fundamental principle.   He was inclined to believe that the presence in the spectrum of any individual ray told unerringly of the volatilisation in the flame under scrutiny of some body as whose badge or distinctive symbol that ray might be regarded ; but the continual prominence of the yellow beam staggered him.   It appeared, indeed, without fail where sodium *was ;* but it also appeared where it might be thought only reasonable to conclude that sodium *was not.*   Nor was it until thirty years later that William Swan,[2] by pointing out the extreme delicacy of the spectral test, and the singularly wide dispersion of sodium, made it appear probable (but even then only probable) that the questionable yellow line was really due invariably to that substance.   Common salt (chloride of sodium) is, in fact, the most diffusive of solids.   It floats in the air ; it flows with water ; every grain of dust has its attendant particle ; its absolute exclusion approaches the impossible.   And withal the light that it gives in burning is so intense and concentrated, that if a single grain be divided into 180 million parts, and one alone of such inconceivably minute fragments be present in a source of light, the spectroscope will show unmistakably its characteristic beam.

Amongst the pioneers of knowledge in this direction were Sir John Herschel [3]—who, however, applied himself to the subject in the interests of optics, not of chemistry—W. A. Miller,[4] and Wheatstone.   The last especially made a notable advance when, in the course of his studies on the " prismatic decomposition " of the electric light, he reached the significant conclusion that the rays visible in its spectrum were different for each kind of metal employed as " electrodes."[5]   Thus

---

[1] *Edinburgh Journal of Science,* vol. v. p. 77.   See also *Phil. Mag.,* Feb. 1834, vol. iv. p. 112.   [2] *Ed. Phil. Trans.,* vol. xxi. p. 411.

[3] *On the Absorption of Light by Coloured Media, Ed. Phil. Trans.,* vol. ix. p. 445 (1823).   [4] *Phil. Mag.,* vol. xxvii. (ser. iii.), p. 81.

[5] *Report Brit. Ass.,* 1835, p. 11 (pt. ii.)   *Electrodes* are the terminals from one to the other of which the electric spark passes, volatilising and

indications of a wider principle were to be found in several quarters, but no positive certainty on any single point was obtained until, in 1859, Gustav Kirchhoff, professor of physics in the University of Heidelberg, and his colleague, the eminent chemist Robert Bunsen, took the matter in hand. By them the general question as to the necessary and invariable connection of certain rays in the spectrum with certain kinds of matter, was first resolutely confronted, and first definitively answered. It was answered affirmatively—else there could have been no science of spectrum analysis—as the result of experiments more numerous, more stringent, and more precise than had previously been undertaken.[1] And the assurance of their conclusion was rendered doubly sure by the discovery, through the peculiarities of their light alone, of two new metals, named, from the blue and red rays by which they were respectively distinguished, " Cæsium," and " Rubidium."[2] Both were immediately afterwards actually obtained in small quantities by evaporation of the Dürckheim mineral waters.

The link connecting this important result with astronomy may now be indicated. In the year 1802 it occurred to William Hyde Wollaston to substitute for the round hole used by Newton and his successors for the admittance of light to be examined with the prism, an elongated "crevice" $\frac{1}{20}$th of an inch in width. He thereupon perceived that the spectrum, thus formed of light, as it were, *purified* by the abolition of overlapping images, was traversed by seven dark lines. These he took to be natural boundaries of the various colours,[3] and, satisfied with this quasi-explanation, allowed the subject to drop. It was independently taken up after twelve years by a man of higher genius. In the course of experiments on light, directed towards the perfecting of his achromatic lenses,

rendering incandescent in its transit some particles of their substance, the characteristic light of which accordingly flashes out in the spectrum.

[1] *Phil. Mag.*, vol. xx. p. 93.
[2] *Annalen der Physik*, Bd. cxiii. p. 357.
[3] *Phil. Trans.*, vol. xcii. p. 378.

Fraunhofer, by means of a slit and a telescope, made the surprising discovery that the solar spectrum is crossed, not by seven, but by thousands of obscure transverse streaks.[1] Of these he counted some 600, and carefully mapped 324; while a few of the most conspicuous he set up (if we may be permitted the expression) as landmarks, measuring their distances apart with a theodolite, and affixing to them the letters of the alphabet by which they are still universally known. He went further. The same system of examination applied to the rest of the heavenly bodies showed the mild effulgence of the moon and planets to be deficient in precisely the same rays as sunlight; while in the stars it disclosed the differences in likeness which are always an earnest of increased knowledge. The spectra of Sirius and Castor, instead of being delicately ruled crosswise throughout, like that of the sun, were seen to be interrupted by three massive bars of darkness—two in the blue and one in the green;[2] the light of Pollux, on the other hand, seemed precisely similar to sunlight attenuated by distance or reflection, and that of Capella, Betelgeux, and Procyon to share some of its peculiarities. One solar line especially—that marked in his map with the letter D—proved common to all the four last-mentioned stars; and it was remarkable that it exactly coincided in position with the conspicuous yellow beam (afterwards, as we have said, identified with the light of glowing sodium) which he had already found to accompany most kinds of combustion. Moreover, both the *dark* solar and the *bright* terrestrial "D lines" were displayed by the refined Munich appliances as double.

In this striking correspondence, discovered by Fraunhofer in 1815, was contained the very essence of solar chemistry; but its true significance did not become apparent until long afterwards. Fraunhofer was by profession, not a physicist, but a practical optician. Time pressed; he could not and would not deviate from his appointed track; all that was

[1] *Denkschriften*, Bd. v. p. 202.
[2] *Ibid.*, p. 220; *Edin. Jour. of Science*, vol. viii. p. 9.

possible to him was to indicate the road to discovery, and exhort others to follow it.[1]

Partially and inconclusively at first this was done. The "fixed lines" (as they were called) of the solar spectrum took up the position of a standing problem, to the solution of which no approach seemed possible. Conjectures as to their origin were indeed rife. An explanation put forward by Zantedeschi [2] and others, and dubiously favoured by Sir David Brewster and Dr. J. H. Gladstone,[3] was that they resulted from "interference"—that is, a destruction of the motion producing in our eyes the sensation of light, by the superposition of two light-waves in such a manner that the crests of one exactly fill up the hollows of the other. This effect was supposed to be brought about by imperfections in the optical apparatus employed.

A more plausible view was that the atmosphere of the earth was the agent by which sunlight was deprived of its missing beams. For a few of them this is actually the case. Brewster found in 1832 that certain dark lines, which were invisible when the sun stood high in the heavens, became increasingly conspicuous as he approached the horizon.[4] These are the well-known "atmospheric lines;" but the immense majority of their companions in the spectrum remain quite unaffected by the thickness of the stratum of air traversed by the sunlight containing them. They are then obviously due to another cause.

There remained the true interpretation—absorption in the *sun's* atmosphere; and this, too, was extensively canvassed. But a remarkable observation made by Professor Forbes of Edinburgh [5] on the occasion of the annular eclipse of May 15,

---

[1] *Denkschriften*, Bd. v. p. 222.   [2] *Arch. des Sciences*, 1849, p. 43.
[3] *Phil. Trans.*, vol. cl. p. 159, *note.*
[4] *Ed. Phil. Trans.*, vol. xii. p. 528.
[5] *Phil. Trans.*, vol. cxxvi. p. 453. "I conceive," he says, "that this result proves decisively that the sun's atmosphere has nothing to do with the production of this singular phenomenon," p. 455. And Brewster's

1836, appeared to throw discredit upon it. If the problematical dark lines were really occasioned by the stoppage of certain rays through the action of a vaporous envelope surrounding the sun, they ought, it seemed, to be strongest in light proceeding from his edges, which, cutting that envelope obliquely, passed through a much greater depth of it. But the circle of light left by the interposing moon, and of course derived entirely from the rim of the solar disc, yielded to Forbes's examination precisely the same spectrum as light coming from its central parts. This circumstance helped to baffle inquirers, already sufficiently perplexed. It still remains an anomaly, of which no completely satisfactory explanation has been offered.

Convincing evidence as to the true nature of the solar lines was however at length, in the autumn of 1859, brought forward at Heidelberg. Kirchhoff's *experimentum crucis* in the matter was a very simple one. He threw bright sunshine across a space occupied by vapour of sodium, and perceived with astonishment that the dark Fraunhofer line D, instead of being effaced by flame giving a *luminous* ray of the same refrangibility, was deepened and thickened by the superposition. He tried the same experiment, substituting for sunbeams light from a Drummond lamp, and with a similar result. A dark furrow, corresponding in every respect to the solar D line, was instantly seen to interrupt the otherwise unbroken radiance of its spectrum. The inference was irresistible, that the effect thus produced artificially was brought about naturally in the same way, and that sodium formed an ingredient in the glowing atmosphere of the sun.[1] This first discovery was quickly followed up by the identification of numerous bright rays in the spectra of other metallic bodies with others of the hitherto mysterious Fraunhofer lines. Kirchhoff was thus led to the conclusion that (besides sodium) iron, magnesium, calcium, and chromium are certainly solar constituents, and that copper, zinc, barium, and nickel are also

well-founded opinion that it had much to do with it was thereby in fact, overthrown.

[1] *Monatsberichte*, Berlin, 1859, p. 664.

present, though in smaller quantities.[1] As to cobalt, he hesitated to pronounce, but its existence in the sun has since been established.

These memorable results were founded upon a general principle first enunciated by Kirchhoff in a communication to the Berlin Academy, December 15, 1859, and afterwards more fully developed by him.[2] It may be expressed as follows : Substances of every kind are opaque to the precise rays which they emit at the same temperature ; that is to say, they stop the kinds of light or heat which they are then actually in a condition to radiate. But it does not follow that *cool* bodies absorb the rays which they would give out if sufficiently heated. Hydrogen at ordinary temperatures, for instance, is almost perfectly transparent, but if raised to the glowing point—as by the passage of electricity—it *then* becomes capable of arresting, and at the same time of displaying in its own spectrum light of four distinct colours.

This principle, well understood, discloses the whole secret of solar chemistry. It gives the key to the hieroglyphics of the Fraunhofer lines. The same characters which are written *bright* in terrestrial spectra are written *dark* in the unrolled sheaf of sun-rays ; the meaning remains unchanged. It must, however, be remembered that they are only *relatively* dark. The substances stopping those particular tints in the neighbourhood of the sun are at the same time vividly glowing with the very same. Remove the dazzling solar background, by contrast with which they show as obscure, and they will be seen, and have, under certain circumstances, actually been seen, in all their native splendour. It is because the atmosphere of the

---

[1] *Abhandlungen*, Berlin, 1861, pp, 80, 81.

[2] *Ibid.*, 1861, p. 77 ; *Annalen der Physik*, Bd. cxix. p. 275. A similar conclusion, reached by Balfour Stewart in 1858 for heat-rays (*Ed. Phil. Trans.*, vol. xxii. p. 13), was, in 1860, without previous knowledge of Kirchhoff's work, extended to light (*Phil. Mag.*, vol. xx. p. 534) ; but his experiments wanted the precision of those executed at Heidelberg. Ångström, too, had foreshadowed it in 1853 (*Phil. Mag.*, vol. ix. p. 328), as indeed Euler had done nearly a century earlier.

sun is cooler than the globe it envelopes that the different
kinds of vapour constituting that atmosphere take more than
they give, absorb more light than they are capable of emitting ;
raise them to the same temperature as the sun itself, and their
powers of emission and absorption being brought exactly to
the same level, the thousands of dusky rays in the solar spectrum
will be at once obliterated.

The establishment of the *terrestrial* science of spectrum
analysis was due, as we have seen, equally to Kirchhoff and
Bunsen, but its *celestial* application to Kirchhoff alone. He
effected this object of the aspirations, more or less dim, of
many other thinkers and workers, by the union of two separate
though closely related lines of research—the study of the
different kinds of light *emitted* by various bodies, and the study
of the different kinds of light *absorbed* by them.   The latter
branch appears to have been first entered upon by Dr. Thomas
Young in 1803 ;[1] it was pursued by the younger Herschel,[2]
by William Allen Miller, Brewster, and Gladstone.   Brewster
indeed made, in 1833,[3] a formal attempt to found what might
be called an *inverse* system of analysis with the prism, based
upon absorption ; and his efforts were repeated, just a quarter
of a century later, by Gladstone.[4] But no general point of
view was attained ; nor, it may be added, was it by this path
attainable.

Kirchhoff's map of the solar spectrum, drawn to scale with
exquisite accuracy, and printed in three shades of ink to
convey the graduated obscurity of the lines, was published in
the Transactions of the Berlin Academy for 1861 and 1862.[5]
Representations of the principal lines belonging to various
elementary bodies formed, as it were, a series of marginal notes
accompanying the great solar scroll, and enabling the veriest

[1] *Miscellaneous Works*, vol. i. p. 189.
[2] *Ed. Phil. Trans.*, vol. ix. p. 458.   [3] *Ibid.*, vol. xii. p. 519.
[4] *Quart. Jour. Chem. Soc.*, vol. x. p. 79.
[5] A facsimile accompanied Professor Roscoe's translation of Kirchhoff's
"Researches on the Solar Spectrum" (London, 1862-63).

tyro in the new science to decipher its meaning at a glance. Where the dark solar and bright metallic rays agreed in position, it might safely be inferred that the metal emitting them was a solar constituent; and such coincidences were numerous. In the case of iron alone, no less than sixty occurred in one-half of the spectral area, rendering the chances [1] absolutely overwhelming against mere casual conjunction. The preparation of this elaborate picture proved so trying to the eyes that Kirchhoff was compelled by failing vision to resign the latter half of the task to his pupil Hofmann. The complete map measured nearly eight feet in length.

The conclusions reached by Kirchhoff were no sooner announced than they took their place, with scarcely a dissenting voice, among the established truths of science. The broad result, that the dark lines in the spectrum of the sun afford an index to its chemical composition no less reliable than any of the tests used in the laboratory, was equally captivating to the imagination of the vulgar, and authentic in the judgment of the learned; and, like all genuine advances in the knowledge of Nature, it stimulated curiosity far more than it gratified it. Now the history of how discoveries were missed is often quite as instructive as the history of how they were made; it may then be worth while to expend a few words on the thoughts and trials by which, in the present case, the actual event was heralded.

Three times it seemed on the verge of being anticipated. The experiment, which in Kirchhoff's hands proved decisive, of passing sunlight through glowing vapours and examining the superposed spectra, was performed by Professor W. A. Miller of King's College in 1845.[2] Nay, more, it was performed with express reference to the question, then already (as has been noted) in debate, of the possible production of Fraunhofer's lines by absorption in a solar atmosphere. Yet it led to nothing.

[1] Estimated by Kirchhoff at *a trillion to one. Abhandl.*, 1861, p. 79.
[2] *Phil. Mag.*, vol. xxvii. (3d series), p. 90.

M

178   *HISTORY OF ASTRONOMY.*

Again, at Paris in 1849, with a view to testing the asserted coincidence between the solar D line and the bright yellow beam in the spectrum of the electric arc (really due to the un-suspected presence of sodium), Léon Foucault threw a ray of sunshine across the arc and observed its spectrum.[1] He was surprised to see that the D line was rendered more intensely dark by the combination of lights. To assure himself still further, he substituted a reflected image of one of the white-hot carbon-points for the sunbeam, with an identical result. *The same ray was missing.* It needed but another step to have generalised this result, and thus laid hold of a natural truth of the highest importance ; but that step was not taken. Foucault, keen and brilliant though he was, rested satisfied with the information that the *voltaic arc* had the power of stopping the kind of light emitted by it ; he asked no further question, and was consequently the bearer of no further intelli-gence on the subject.

The truth conveyed by this remarkable experiment was, however, divined by one eminent man. Professor Stokes of Cambridge stated to Sir William Thomson, shortly after it had been made, his conviction that an absorbing atmosphere of sodium surrounded the sun. And so forcibly was his hearer impressed with the weight of the arguments based upon the absolute agreement of the D line in the solar spectrum with the yellow ray of burning sodium (then freshly certified by W. H. Miller), combined with Foucault's "reversal" of that ray, that he regularly inculcated, in his public lectures on natural philosophy at Glasgow, five or six years before Kirch-hoff's discovery, not only the *fact* of the presence of sodium in the solar neighbourhood, but also the *principle* of the study of solar and stellar chemistry in the spectra of flames.[2] Yet it does not appear to have occurred to either of these two dis-tinguished professors—themselves amongst the foremost of their time in the successful search for new truths—to verify

[1] *L'Institut,* Feb. 7, 1849, p. 45 ; *Phil. Mag.,* vol. xix. (4th series), p. 193.
[2] *Ann. d. Phys.,* vol. cxviii. p. 110.

practically a sagacious conjecture in which was contained the possibility of a scientific revolution. It is just to add, that Kirchhoff was unacquainted, when he undertook his investigation, either with the experiment of Foucault or the speculation of Stokes.

It may here be useful—since without some clear ideas on the subject no proper understanding of recent astronomical progress is possible—to take a cursory view of the elementary principles of spectrum analysis. To many of our readers they are doubtless already familiar; but it is better to be trite in repetition than obscure from lack of explanation.

The spectrum, then, of a body is simply the light proceeding from it *spread out* by refraction [1] into a brilliant variegated band, passing from brownish-red through crimson, orange, yellow, green, and azure into dusky violet. The reason of this spreading-out or "dispersion" is that the various colours have different wave-lengths, and consequently meet with different degrees of retardation in traversing the denser medium of the prism. The shortest and quickest vibrations (producing the sensation we call "violet") are thrown farthest away from their original path—in other words, suffer the widest "deviation;" the longest and slowest (the red) travel much nearer to it. Thus the sheaf of rays which would otherwise combine into a patch of white light are separated through the divergence of their tracks after refraction by a prism, so as to form a tinted riband. This *visible* spectrum is prolonged *invisibly* at both ends by a long range of vibrations, either too rapid or too sluggish to affect the eye as light, but recognisable through their chemical and heating effects.

Now all incandescent solid or liquid substances, and even gases ignited under great pressure, give what is called a "continuous spectrum;" that is to say, the light derived from them is of every conceivable hue. Sorted out with the prism, its tints merge imperceptibly one into the other, uninterrupted

---

[1] Spectra may be produced by *diffraction* as well as by *refraction;* but we are here only concerned with the matter in its simplest aspect.

by any dark spaces. No colours, in short, are missing. But gases and vapours rendered luminous by heat emit rays of only a few tints—sometimes of one only—which accordingly form an interrupted spectrum, usually designated as one of lines or bands. And since these rays are perfectly definite and characteristic—not being the same for any two substances —it is easy to tell what kind of matter is concerned in producing them. We may suppose that the inconceivably minute particles which by their rapid thrillings agitate the ethereal medium so as to produce light, are free to give out their peculiar tone of vibration only when floating apart from each other in gaseous form; but when crowded together into a condensed mass, the clear ring of the distinctive note is drowned, so to speak, in an universal molecular clang. Thus prismatic analysis has no power to identify individual kinds of matter except when they present themselves as glowing vapours.

A spectrum is said to be "reversed" when lines previously seen bright on a dark background appear dark on a bright background. In this form it is equally characteristic of chemical composition with the "direct" spectrum, being due to *absorption*, as the latter is to *emission*. And absorption and emission are, by Kirchhoff's law, strictly correlative. This is easily understood by the analogy of sound. For, just as a tuning-fork responds to sound-waves of its own pitch, but remains indifferent to those of any other, so those particles of matter whose nature it is, when set swinging by heat, to vibrate a certain number of times in a second, thus giving rise to light of a particular shade of colour, appropriate those same vibrations, and those only, when transmitted past them,—or, phrasing it otherwise, are *opaque* to them, and *transparent* to all others.

It should further be explained that the *shape* of the bright or dark spaces in the spectrum has nothing whatever to do with the nature of the phenomena. The "lines" and "bands" so frequently spoken of are seen as such for no other reason than because the light forming them is admitted through a narrow, straight opening. Change that opening into a fine crescent or a sinuous curve, and the "lines" will at once appear as crescents or curves.

Resuming in a sentence what has been already explained, we find that the prismatic analysis of the heavenly bodies was founded upon three classes of facts : First, the unmistakable character of the light given by each different kind of glowing vapour ; secondly, the identity of the light absorbed with the light emitted by each ; thirdly, the coincidences observed between rays missing from the solar spectrum and rays absorbed by various terrestrial substances. Thus, a realm of knowledge, pronounced by Morinus[1] in the seventeenth century, and no less dogmatically by Auguste Comte[2] in the nineteenth, hopelessly out of reach of the human intellect, was thrown freely open, and the chemistry of the sun and stars took its place amongst the foremost of the experimental sciences.

The immediate increase of knowledge was not the chief result of Kirchhoff's labours; still more important was the change in the scope and methods of astronomy, which, set on foot in 1852 by the detection of a common period affecting at once the spots on the sun and the magnetism of the earth, was extended and accelerated by the discovery of spectrum analysis. The nature of that change is concisely indicated by the heading of the present chapter; we would now ask our readers to endeavour to realise somewhat distinctly what is implied by the "foundation of astronomical physics."

Some two centuries and fourscore years ago, Kepler drew a forecast of what he called a "physical astronomy"—a science treating of the efficient causes of planetary motion, and holding the "key to the inner astronomy."[3] What Kepler dreamed of and groped after, Newton realised. He showed the beautiful and symmetrical revolutions of the solar system to be governed by a uniformly acting cause, and that cause no other than the familiar force of gravity which gives stability to all our

---

[1] *Astrologia Gallica* (1661), p. 189.

[2] *Pos. Phil.*, vol. i. pp. 114–115 (Martineau's trans.)

[3] *Proem. Astronomiæ Pars Optica* (1604), *Op.*, t. ii.

terrestrial surroundings. The world under our feet was thus for the first time brought into physical connection with the worlds peopling space, and a very tangible relationship was demonstrated as existing between what used to be called the "corruptible" matter of the earth and the "incorruptible" matter of the heavens.

This process of unification of the cosmos—this levelling of the celestial with the sublunary—was carried no farther until the fact unexpectedly emerged from a vast and complicated mass of observations, that the magnetism of the earth is subject to subtle influences emanating, certainly from some, and presumably (were their amount sufficient to be perceptible) from all of the heavenly bodies; the inference being thus rendered at least plausible, that a force not less universal than gravity itself, but with whose modes of action we are as yet unacquainted, pervades the universe, and forms, it might be said, an intangible bond of sympathy between its parts. Now for the investigation of this influence two roads are open. It may be pursued by observation either of the bodies from which it emanates, or of the effects which it produces—that is to say, either by the astronomer or by the physicist, or, better still, by both concurrently. ` Their acquisitions are mutually profitable; nor can either be considered as independent of the other. Any important accession to knowledge respecting the sun, for example, may be expected to cast a reflected light on the still obscure subject of terrestrial magnetism; while discoveries in magnetism or its *alter ego* electricity must profoundly affect solar inquiries.

The establishment of the new method of spectrum analysis drew far closer this alliance between celestial and terrestrial science. Indeed, they have come to merge so intimately one into the other, that it is no easier to trace their respective boundaries than it is to draw a clear dividing-line between the animal and vegetable kingdoms. Yet up to the middle of the present century, astronomy, while maintaining her strict union with mathematics, looked with indifference on the rest of the sciences; it was enough that she possessed the telescope and the calculus. Now the materials for her inductions are supplied by the chemist, the electrician, the inquirer into the most recondite mysteries of light and the molecular constitution of matter. She is concerned with what the geologist, the meteorologist, even the biologist, has to say; she can afford to close her ears to no new truth of the physical order. Her position of lofty isolation has been exchanged for one of community and mutual aid. The astronomer has become, in the highest sense of the term, a physicist; while the physicist is bound to be something of an astronomer.

This, then, is what is designed to be conveyed by the "foundation of astronomical or cosmical physics." It means the establishment of a science of Nature whose conclusions are not only presumed by analogy, but are ascertained by observation, to be valid wherever light can travel and gravity is obeyed—a science by which the nature of the stars can be studied upon the earth, and the nature of the earth can be better known by study of the stars—a science, in a word, which is, or aims at being, one and universal, even as Nature—the visible reflection of the invisible highest Unity—is one and universal.

It is not too much to say that a new birth of knowledge has ensued. The astronomy so signally promoted by Bessel[1]—the astronomy placed by Comte[2] at the head of the hierarchy of the sciences—was the science of the *movements* of the heavenly bodies. And there were those who began to regard it as a science which, from its very perfection, had ceased to be interesting—whose tale of discoveries was told, and whose farther advance must be in the line of minute technical improvements, not of novel and stirring disclosures. But the science of the *nature* of the heavenly bodies is one only in the beginning of its career. It is full of the audacities, the inconsistencies, the imperfections, the possibilities of youth. It promises everything; it has already performed much; it will

[1] *Pop. Vorl.*, pp. 14, 19, 408.   [2] *Pos. Phil.*, p. 115.

doubtless perform much more.   The means at its disposal are
vast, and are being daily augmented.   What has so far been
secured by them it must now be our task to extricate from
more doubtful surroundings and place in due order before our
readers.

CHAPTER II.

SOLAR OBSERVATIONS AND THEORIES.

THE zeal with which solar studies have been pursued during the last quarter of a century has already gone far to redeem the neglect of the two preceding ones. Since Schwabe's discovery was published in 1851, observers have multiplied, new facts have been rapidly accumulated, and the previous comparative quiescence of thought on the great subject of the constitution of the sun, has been replaced by a bewildering variety of speculations, conjectures, and more or less justifiable inferences. It is satisfactory to find this novel impulse not only shared, but to a large extent guided, by our countrymen.

William Rutter Dawes, one of many clergymen eminent in astronomy, observed, in 1852, with the help of a solar eye-piece of his own devising, some curious details of spot-structure.[1] The umbra—heretofore taken for the darkest part of the spot—was seen to be suffused with a mottled, nebulous illumination, in marked contrast with the striated appearance of the penumbra; while through this "cloudy stratum" a "black opening" permitted the eye to divine further unfathomable depths beyond. The *hole* thus disclosed—evidently the true nucleus—was found to be present in all considerable, as well as in many small maculæ.

Again, the whirling motions of some of these objects were noticed by him. The remarkable form of one sketched at Wateringbury, in Kent, January 17, 1852, gave him the means

[1] *Mem. R. A. S.*, vol. xxi. p. 157.

of detecting and measuring a rotatory movement of the whole spot round the black nucleus at the rate of 100 degrees in six days. "It appeared," he said, "as if some prodigious ascending force of a whirlwind character, in bursting through the cloudy stratum and the two higher and luminous strata, had given to the whole a movement resembling its own."[1] An interpretation founded, as is easily seen, on the Herschelian theory, then still in full credit.

An instance of the same kind was observed by the late Mr. W. R. Birt in 1860,[2] and cyclonic movements are now a well-recognised feature of sun-spots. They are, however, as Father Secchi[3] concluded from his long experience, but temporary and casual. Scarcely three per cent. of all spots visible exhibit the spiral structure which should invariably result if a conflict of opposing, or the friction of unequal, currents were essential, and not merely incidental to their origin. A whirlpool phase not unfrequently accompanies their formation, and may be renewed at periods of recrudescence or dissolution; but it is both partial and inconstant, sometimes affecting only one side of a spot, sometimes slackening gradually its movement in one direction, to resume it, after a brief pause, in the opposite. Persistent and uniform motions, such as the analogy of terrestrial storms would absolutely require, are not to be found. So that the "cyclonic theory" of sun-spots, suggested by Herschel in 1847,[4] and urged, from a different point of view, by Faye in 1872, may be said to have completely broken down.

The *drift* of spots over the sun's surface was first systematically investigated by Carrington. Before narrating what he *did*, it is worth while to pause for a moment to consider who he *was*. Nor will it take long to tell. Richard Christopher Carrington was a self-constituted astronomer, with the will and the courage and the instinct of thoughtful labour in him.

Born at Chelsea in May 1826, he entered Trinity College,

---

[1] *Mem. R. A. S.*, vol. xxi. p. 160.　　[2] *Month. Not.*, vol. xxi. p. 144.
[3] *Le Soleil*, t. i. pp. 87–90 (2d ed. 1875).　　[4] See *ante*, p. 75.

Cambridge, in 1844. He was intended for the Church, but Professor Challis's lectures diverted him to astronomy, and he resolved, as soon as he had taken his degree, to prepare, with all possible diligence, to follow his new vocation. His father, who was a brewer on a large scale at Brentford, offered no opposition; ample means were at his disposal; nevertheless, he chose to serve an apprenticeship of three years as observer in the University of Durham, as though his sole object had been to earn a livelihood. He quitted the post only when he found that its restricted opportunities offered no further prospect of self-improvement.

He now built an observatory of his own at Redhill in Surrey, with the design of completing Bessel's and Argelander's survey of the northern heavens by adding to it the circumpolar stars omitted from their view. This project, successfully carried out between 1854 and 1857, had another and still larger one superposed upon it before it had even begun to be executed. In 1852, while the Redhill Observatory was in course of erection, the discovery of the coincidence between the sun-spot and magnetic periods was announced. Carrington was profoundly interested, and devoted his enforced leisure to the examination of records, both written and depicted, of past solar observations. Struck with their fragmentary and inconsistent character, he resolved to "appropriate," as he said, by "close and methodical research," the eleven-year period next ensuing.[1] He calculated rightly that he should have the field pretty nearly to himself; for many reasons conspire to make public observatories slow in taking up new subjects, and amateurs with freedom to choose, and means to treat them effectually, were even scarcer then than they are now.

The execution of this laborious task was commenced November 9, 1853. It was intended to be merely a *parergon*—a "second subject," upon which daylight energies might be spent, while the hours of night were reserved for cataloguing those stars that "are bereft of the baths of ocean." Its results, how-

[1] *Observations at Redhill* (1863), Introduction.

ever, proved of the highest interest, although the vicissitudes
of life barred the completion, in its full integrity, of the original
design.   By the death, in 1858, of the elder Carrington, the
charge of the brewery devolved upon his son; and eventually
absorbed so much of his care that it was found advisable to
bring the solar observations to a premature close, March 24,
1861.

His scientific life may be said to have closed with them.
Attacked four years later with severe, and, in its results, permanent illness, he disposed of the Brentford business, and withdrew to Churt, near Farnham, in Surrey.   There, in a lonely
spot, on the top of a detached conical hill known as the
" Devil's Jump," he built a second observatory, and erected
an instrument which he was no longer able to use with pristine
effectiveness; and there, November 27, 1875, he died of the
rupture of a blood-vessel on the brain, before he had completed
his fiftieth year.[1]

His observations of sun-spots were of a *geometrical* character.
They concerned positions and movements, leaving out of sight
physical peculiarities.   Indeed, the prudence with which he
limited his task to what came strictly within the range of his
powers to accomplish, was one of Carrington's most valuable
qualities.   The method of his observations, moreover, was
chosen with the same practical sagacity as their objects.   As
early as 1847, Sir John Herschel had recommended the daily
self-registration of sun-spots,[2] and he enforced the suggestion,
with more immediate prospect of success, in 1854.[3]   The art
of celestial photography, however, was even then in a purely
tentative stage, and Carrington wisely resolved to waste no time
on dubious experiments, but employ the means of registration
and measurement actually at his command.   These were very
simple, yet very effective.   To the " helioscope " employed by
Father Scheiner [4] two centuries and a quarter earlier a species
of micrometer was added.   The image of the sun was pro-

[1] *Month. Not.*, vol. xxxvi. p. 142.   [2] *Cape Observations*, p. 435, *note*.
[3] *Month. Not.*, vol. x. p. 158.   [4] *Rosa Ursina*, lib. iii. p. 348.

jected upon a screen by means of a firmly clamped telescope,
in the focus of which were placed two cross wires forming
angles of 45° with the meridian. The six instants were then
carefully noted at which these were met by the edges of the
disc as it traversed the screen and by the nucleus of the spot
to be measured.[1] A short process of calculation then gave the
exact position of the spot as referred to the sun's centre.

From a series of 5290 observations made in this way, together
with a great number of accurate drawings, Carrington derived
conclusions of great importance on each of the three points
which he had proposed to himself to investigate. These were :
the law of the sun's rotation, the existence and direction of
systematic currents, and the distribution of spots on the solar
surface.

Grave discrepancies were early perceived to exist between
the determinations of the sun's rotation by different observers.
Galileo, with "comfortable generality," estimated the period at
"about a lunar month;"[2] Scheiner, at twenty-seven days.[3]
Cassini, in 1678, made it 25.58 ; Delambre, in 1775, no more
than twenty-five days. Later inquiries brought these divergences within no more tolerable limits. Laugier's result of
25.34 days—obtained in 1841—enjoyed the highest credit, yet
it differed widely in one direction from that of Böhm (1852),
giving 25.52 days, and in the other from that of Kysaeus (1846),
giving 25.09 days. Now the cause of these variations was
really obvious from the first, although for a long time strangely
overlooked. Father Scheiner pointed out in 1630 that different
spots gave different periods, adding the significant remark
that one at a distance from the solar equator revolved more
slowly than those nearer to it.[4] But the hint was wasted. For
upwards of two centuries ideas on the subject were either

[1] *Observations at Redhill*, p. 8.   [2] *Op.*, t. iii. p. 402.
[3] *Rosa Ursina*, lib. iv. p. 601. Both Galileo and Scheiner spoke of the
apparent or "synodical" period, which is about one and a third days
longer than the *true* or "sidereal" one. The difference is caused by the
revolution of the earth in its orbit in the same direction with the sun's
rotation on its axis.   [4] *Rosa Ursina*, lib. iii. p. 260.

retrograde or stationary. What were called the "proper motions" of spots were, however, recognised by Schröter,[1] and utterly baffled Laugier,[2] who despaired of obtaining any concordant result as to the sun's rotation except by taking the mean of a number of discordant ones. At last, in 1855, a valuable course of observations made at Capo di Monte, Naples, in 1845–46, enabled C. H. F. Peters[3] (now of Hamilton College, Clinton, N.Y.) to set in the clearest light the insecurity of determinations based on the assumption of fixity in objects visibly affected by movements uncertain both in amount and direction.

Such was the state of affairs when Carrington entered upon his task. Everything was in confusion; the most that could be said was that the confusion had come to be distinctly admitted and referred to its true source. What he discovered was this: that the sun, or at least the outer shell of the sun visible to us, has *no single period of rotation*, but drifts round, carrying the spots with it, at a rate continually accelerated from the poles to the equator. In other words, the time of axial revolution is *shortest at the equator* and lengthens with increase of latitude. Carrington devised a mathematical formula by which the rate or "law" of this lengthening was conveniently expressed; but it was a purely empirical one. It was a concise statement, but implied no physical interpretation. It summarised, but did not explain the facts. An assumed "mean period" for the solar rotation of 25.38 days (twenty-five days nine hours, very nearly), was thus found to be *actually* conformed to only in two parallels of solar latitude (14° north and south), while the equatorial period was slightly less than twenty-five, and that of latitudes 50° rose to twenty-seven days and a half.[4] These curious results gave quite a new direction to ideas on solar physics.

The other two "elements" of the sun's rotation were also

[1] Faye, *Comptes Rendus*, t. lx. p. 818.   [2] *Ibid.*, t. xii. p. 648.
[3] *Proc. Am. Ass. Adv. of Science*, 1855, p. 85.
[4] *Observations at Redhill*, p. 221.

ascertained by Carrington with hitherto unattained precision. He fixed the inclination of its axis to the ecliptic at 82° 45'; the longitude of the ascending node at 73° 40' (both for the epoch 1850 A.D.) These data—which have not yet been improved upon—suffice to determine the position in space of the sun's equator. Its north pole is directed towards a star in the coils of the Dragon, midway between Vega and the Pole-star; its plane intersects that of the earth's orbit in such a way that our planet finds itself in the same level on or about the 3d of June and the 5th of December, when any spots visible on the disc cross it in apparently straight lines. At other times, the paths pursued by them seem curved—downwards between June and December, upwards between December and June.

A singular peculiarity in the distribution of sun-spots emerged from Carrington's studies at the time of the minimum of 1856. Two broad belts of the solar surface, as we have seen, are frequented by them, of which the limits may be put at 6° and 35° of latitude, one zone lying so far north, the other as much south of the solar equator. Individual equatorial spots are not uncommon, but nearer to the poles than 35° they are a rare exception. Carrington observed—as an extreme instance —in July 1858, one in south latitude 44°; and Peters, in June 1846, watched, during several days, a spot in 50° 24' north latitude. But beyond this no true macula has ever been seen; for Lahire's reported observation of one in latitude 70° is now believed to have had its place on the solar globe erroneously assigned; and the "veiled spots" described by Trouvelot in 1875[1] as occurring within 10° of the pole can only be regarded as, at the most, the same kind of disturbance in an undeveloped form.

But the novelty of Carrington's observations consisted in the detection of certain changes in distribution concurrent with the progress of the eleven-year period. As the minimum approached, the spot-zones contracted towards the equator, and

[1] *Am. Jour. of Science*, vol. xi. p. 169.

there finally vanished; then, as if by a fresh impulse, spots suddenly reappeared in high latitudes, and spread downwards with the development of the new phase of activity. Scarcely had this remark been made public,[1] when Wolf[2] found a confirmation of its general truth in Böhm's observations during the years 1833–36; and a perfectly similar behaviour was noted both by Spörer and Secchi at the minimum epoch of 1867. The ensuing period gave less marked indications; but it may be looked upon as established that the activity manifested in sun-spots widens its range with the growth of its intensity, and becomes reduced in space and strength simultaneously—a feature of which no theory has yet given any tolerable account.

Gustav Spörer, born at Berlin in 1822, began to observe sun-spots with the view of assigning the law of solar rotation in December 1860. His means were at first very limited, but his assiduity and success attracted attention, and a Government endowment was procured for the little solar observatory organised by him at Anclam, in Pomerania. Unaware of Carrington's discovery (first made known in January 1859), he arrived at and published, in June 1861,[3] a similar conclusion as to the equatorial quickening of the sun's movement on its axis. His sun-spot observations were continued until 1867, and he afterwards became one of the most zealous students of solar prominences, upon which subject he is at present employed at the " astro-physical " observatory of Potsdam.

The time had now evidently come for a fundamental revision of current notions respecting the nature of the sun. Herschel's theory of a cool, dark, habitable globe, surrounded by, and protected against the radiations of a luminous and heat-giving envelope, was shattered by the first *dicta* of spectrum analysis. Traces of it may be found for a few years subsequent to 1859,[4]

[1] *Month. Not.*, vol. xix. p. 1.
[2] *Vierteljahrsschrift der Naturfors. Gesellschaft* (Zürich), 1859, p. 252.
[3] *Astr. Nach.*, No. 1315.
[4] As late as 1866 an elaborate treatise in its support was written by M. F. Coyteux, entitled *Qu'est ce que le Soleil? Peut-il être habité?* and answering the question in the affirmative.

but they are obviously survivals from an earlier order of ideas, doomed to speedy extinction. It needs only a moment's consideration of what was implied in the discovery of the origin of the Fraunhofer lines to see the incompatibility of the new facts with the old conceptions. It implied, not only the presence near the sun, as glowing vapours, of bodies highly refractory to heat, but that these glowing vapours formed the *relatively cool envelope* of a still hotter internal mass. Kirchhoff, accordingly, included in his great memoir " On the Solar Spectrum," read before the Berlin Academy of Sciences, July 11, 1861, an exposition of the views on the subject to which his memorable investigations had led him. They may be briefly summarised as follows.

Since the body of the sun gives a continuous spectrum, it must be either solid or liquid,[1] while the interruptions in its light prove it to be surrounded by a complex atmosphere of metallic vapours, somewhat cooler than itself. Spots are simply clouds due to local depressions of temperature, differing in no respect from terrestrial clouds except as regards the kinds of matter composing them. These *sun-clouds* take their origin in the zones of encounter between polar and equatorial currents in the solar atmosphere.

This explanation was liable to all the objections urged against the "cumulus theory" on the one hand, and the "trade-wind theory" on the other. Setting aside its propounder, it was consistently upheld perhaps by no man eminent in science except Spörer; and his advocacy of it tended rather to delay the recognition of his own merits than to promote its general adoption.

M. Faye, of the Paris Academy of Sciences, was the first to propose a coherent scheme of the solar constitution covering the whole range of new discovery. The fundamental ideas on the subject now in vogue here made their first connected

[1] The subsequent researches of Plücker, Frankland, Wüllner, and others showed that gases strongly compressed give an absolutely unbroken spectrum.

N

appearance. Much, indeed, remained to be modified and corrected; but the transition was finally made from the old to the new order of conceptions. The essence of the change of view thus effected may be conveyed in a single sentence. The sun was thenceforth regarded, not as a mere heated body, or—still more remotely from the truth—as a cool body unaccountably spun round with a cocoon of fire, but as a vast *heat-radiating machine.* The terrestrial analogy was abandoned in one more particular besides that of temperature. The solar system of circulation, instead of being adapted, like that of the earth, to the distribution of heat received from without, was seen to be directed towards the transportation towards the surface of the heat contained within. Polar and equatorial currents, tending to a purely superficial equalisation of temperature, were replaced by vertical currents bringing up successive portions of the intensely heated interior mass, to contribute their share in turns to the radiation into space which might be called the proper function of a sun.

Faye's views, which were communicated to the Academy of Sciences, January 16, 1865,[1] were avowedly based on the anomalous mode of solar rotation discovered by Carrington. This may be regarded either as an acceleration increasing from the poles to the equator, or as a retardation increasing from the equator to the poles, according to the rate of revolution we choose to assume for the unseen nucleus. Faye preferred to consider it as a retardation produced by ascending currents continually left behind as the sphere widened in which the matter composing them was forced to travel. He further supposed that the depth from which these vertical currents started, and consequently the amount of retardation effected by their ascent to the surface, became progressively greater as the poles were approached; but this was plainly an arbitrary expedient of theory, confronted with inconvenient and uncompromising facts.

The extreme internal mobility betrayed by Carrington's and

---

[1] *Comptes Rendus,* t. lx. pp. 89, 138.

Spörer's observations led to the inference that the matter composing the sun was mainly or wholly gaseous. This had already been suggested by Father Secchi[1] a year earlier, and by Sir John Herschel in April 1864;[2] but it first obtained general currency through Faye's more elaborate presentation. A physical basis was afforded for the view by Cagniard de la Tour's experiments in 1822,[3] proving that, under conditions of great heat and pressure, the vaporous state was compatible with a very considerable density. The position was strengthened when Andrews showed, in 1869,[4] that above a fixed limit of temperature, varying for different bodies, true liquefaction is impossible, even though the pressure be so tremendous as to retain the gas within the same space that enclosed the liquid. The opinion that the mass of the sun is gaseous now commands a very general assent; although the gaseity admitted is of such a nature as to afford the consistence rather of honey or pitch than of the aeriform fluids with which we are familiar.

On another important point the course of subsequent thought was powerfully influenced by Faye's conclusions in 1865. Arago somewhat hastily inferred from experiments with the polariscope the wholly gaseous nature of the visible disc of the sun. Kirchhoff, on the contrary, believed (erroneously, as we now know) that the brilliant continuous spectrum derived from it proved it to be a white-hot solid or liquid. Herschel and Secchi[5] indicated a cloud-like structure as that which would best harmonise the whole of the evidence at command. The novelty introduced by Faye consisted in regarding the photosphere no longer "as a defined surface, in the mathematical sense, but as a limit to which, in the general fluid mass, ascending currents carry the physical or chemical phenomena of incandescence."[6] Uprushing floods of mixed

[1] *Bull. Meteor. dell' Osservatorio del Coll. Rom.*, Jan. 1, 1864, p. 4.
[2] *Quart. Jour. of Science*, vol. i. p. 222.
[3] *Ann. de Chim. et de Phys.*, t. xxii. p. 127.
[4] *Phil. Trans.*, vol. clix. p. 575.
[5] *Les Mondes*, Dec. 22, 1864, p. 707.
[6] *Comptes Rendus*, t. lx. p. 147.

vapours with strong affinities—say of calcium or sodium and oxygen—at last attain a region cool enough to permit their combination ; a fine dust of solid or liquid compound particles (of lime or soda, for example) there collects into the photospheric clouds, and descending by its own weight in torrents of incandescent rain, is *dissociated* by the fierce heat below, and replaced by ascending and combining currents of similar constitution.

This first attempt to assign the part played in cosmical physics by chemical affinities, was marked by the importation into the theory of the sun of the now familiar phrase *dissociation*. It is indeed tolerably certain that no such combinations as those contemplated by Faye occur at the photospheric level, since the temperature there must be enormously higher than would be needed to reduce all metallic earths and oxides ; but molecular changes of some kind, dependent perhaps in part upon electrical conditions, in part upon the effects of radiation into space, most likely replace them. The conjecture (originally thrown out, it would seem, by Faye himself) was countenanced by Ångström,[1] and has recently been advocated by Professor Hastings of Baltimore,[2] that the photospheric clouds are composed of particles of some member of the carbon-triad [3] precipitated from its mounting vapour just where the temperature is lowered by expansion and radiation to the boiling-point of that substance. But the question is one which must for the present remain within the sphere of interesting and admissible speculation.

In Faye's theory, sun-spots were regarded as simply breaks in the photospheric clouds, where the rising currents had strength to tear them asunder. It followed that they were regions of increased heat—regions, in fact, where the temperature was too high to permit the occurrence of the precipitations to which the photosphere is due. Their obscurity was attributed to deficiency of emissive power. But here it was irresistibly

[1] *Recherches sur le Spectre Solaire,* p. 38.
[2] *Am. Jour. of Science,* 1881, vol. xxi. p. 41.
[3] Carbon, silicon, and boron.

objected by Professors Balfour Stewart and Kirchhoff that emissive and absorptive power being strictly correlative, the supposed defect of radiation would be exactly compensated by an increase of transparency. The light from the farther photosphere would then, *shining across the whole body of the sun,* completely fill up, to the eye, the gap in the hither photosphere, and no macula at all would remain visible. Besides, we now know that ignited gases under a pressure far less than that which must exist at even a small distance below the solar surface, give light equally brilliant and uninterrupted with that derived from solid bodies.

After every deduction, however, has been made, we still find that several ideas of permanent value were embodied in this comprehensive sketch of the solar constitution. The principal of these were : first, that the sun is a mainly gaseous body ; secondly, that its stores of heat are rendered available at the surface by means of vertical convection-currents—by the bodily transport, that is to say, of intensely hot matter upwards, and of comparatively cool matter downwards; thirdly, that the photosphere is a surface of condensation, forming the limit set by the cold of space to this circulating process, and that a similar formation must attend, at a certain stage, the cooling of every cosmical body.

To Mr. Warren De la Rue belongs the honour of having obtained the earliest results of substantial value in celestial photography. What had been done previously was interesting in the way of promise, but much could not be claimed for it as actual performance. Some " pioneering experiments " were made by Dr. J. W. Draper of New York in 1840, resulting in the production of a few " moon-pictures " one-inch in diameter ;[1] but slight encouragement was derived from them, either to himself or others. Bond of Cambridge (U.S), however, got impressions of Vega and Castor in 1845,[2] and in 1850 secured with the Harvard 15-inch refractor that daguerreotype of the

[1] H. Draper, *Quart. Jour. of Sc.*, vol. i. p. 381 ; also *Phil. Mag.*, vol. xvii. 1840, p. 222.   [2] *Proc. Roy. Soc.*, vol. xiii. p. 511.

moon with which the career of extra-terrestrial photography
may be said to have formally opened.   It was shown in London at the Great Exhibition of 1851, and determined the
direction of De la Rue's efforts.   Yet it did little more than
prove the art to be a possible one.

Warren De la Rue was born in Guernsey in 1815, was
educated at the École Sainte-Barbe in Paris, and made a large
fortune as a paper manufacturer in England.   The material
supplies for his scientific campaign were thus amply and early
provided.   Towards the end of 1853 he took some successful
lunar photographs.   They were remarkable as the first examples
of the application to astronomical light-painting of the collodion
process, invented by Archer in 1851; and also of the use of
reflectors (Mr. De la Rue's was one of thirteen inches, constructed by himself) for that kind of work.   The absence of
a driving apparatus was, however, very sensibly felt; the
difficulty of moving the instrument by hand so as accurately
to follow the moon's apparent motion being such as to cause
the discontinuance of the experiments until 1857, when the
want was supplied.   Mr. De la Rue's new observatory, built
in that year at Cranford, twelve miles west of Hyde Park, was
specially dedicated to celestial photography; and there he
immediately applied to the heavenly bodies the stereoscopic
method of obtaining relief, and turned his attention to the
delicate business of photographing the sun.

A solar daguerreotype [1] was taken at Paris, April 2, 1845, by
MM. Foucault and Fizeau, acting on a suggestion from Arago.
But the attempt, though far from being unsuccessful, does not,
at that time, seem to have been repeated.   Its great difficulty
consisted in the enormous light-power of the object to be
represented, rendering an inconceivably short period of exposure indispensable, under pain of getting completely " burnt-
up" plates.   In 1857 Mr. De la Rue was commissioned by
the Royal Society to construct an instrument specially adapted
to the purpose for the Kew observatory.   The resulting "photo-

[1] Reproduced in Arago's *Popular Astronomy*, plate xii. vol. 1.

heliograph" may be described as a small telescope (of $3\frac{1}{2}$ inches aperture and 50 focus), with a plate-holder at the eye-end, guarded in front by a spring-slide, the rapid movement of which across the field of view secured for the sensitive plate a virtually instantaneous exposure. By its means the first solar light-pictures of real value were taken, and the autographic record of the solar condition recommended by Sir John Herschel was commenced and continued at Kew during fourteen years—1858–72. The work of photographing the sun is now carried on in every quarter of the globe, from the Mauritius to Massachusetts, and the days are few indeed on which the self-betrayal of the camera can be evaded by our chief luminary. In the year 1883, the incorporation of Indian with Greenwich pictures afforded a record of the state of the solar surface on 340 days; and the missing twenty-five were doubtless provided for elsewhere.

The conclusions arrived at by photographic means at Kew were communicated to the Royal Society in a series of papers drawn up jointly by Messrs. De la Rue, Balfour Stewart, and Benjamin Loewy, in 1865 and subsequent years. They influenced materially the progress of thought on the subject they were concerned with.

By its rotation the sun itself offers opportunities for bringing the stereoscope to bear upon it. Two pictures, taken at an interval of twenty-six minutes, show just the amount of difference needed to give, by their combination, the maximum effect of solidity.[1] Mr. De la Rue thus obtained, in 1861, a stereoscopic view of a sun-spot and surrounding faculæ, representing the various parts in their true mutual relations. " I have ascertained in this way," he wrote,[2] "that the faculæ occupy the highest portions of the sun's photosphere, the spots appearing like holes in the penumbræ, which appeared lower than the regions surrounding them ; in one case, parts of the faculæ were discovered to be sailing over a spot apparently at some considerable height above it." Thus Wilson's inference as to the

[1] *Report Brit. Ass.*, 1859, p. 148.   [2] *Phil. Trans.*, vol. clii. p. 407.

depressed nature of spots received, after the lapse of not far
from a century, proof of the most simple, direct, and convincing kind.   A careful application of Wilson's own geometrical
test gave results only a trifle less decisive.   Of 694 spots
observed, 78 per cent. showed, as they traversed the disc, the
expected effects of perspective;[1] and their absence in the
remaining 22 per cent. might be easily explained by internal
commotions producing irregularities of structure.   The absolute
depth of spot-cavities—at least of their sloping sides—was
determined by Father Secchi through measurement of the
"parallax of profundity"[2]—that is, of apparent displacements
attendant on the sun's rotation, due to depression below the
sun's surface.   He found that it in every case fell short of
4000 miles, and averaged not more than 1321, corresponding,
on the terrestrial scale, to an excavation in the earth's crust
of $1\frac{1}{5}$ miles.   There may be, however, and probably are, depths
below this depth, of which the eye takes not even indirect
cognisance; so that it would be hasty to pronounce spots to
be a merely superficial phenomenon.

The opinion of the Kew observers as to the nature of such
disturbances was strongly swayed by another curious result of
the "statistical method" of inquiry.   They found that of 1137
instances of spots accompanied by faculæ, 584 had those
faculæ chiefly or entirely on the left, 508 showed a nearly
equal distribution, while 45 only had faculous appendages
mainly on the right side.[3]   Now, the rotation of the sun, as we
see it, is performed from left to right; so that the marked
tendency of the faculæ was a lagging one.   This was easily
accounted for by supposing the matter composing them to have
been flung upwards from a considerable depth, whence it would
reach the surface with the lesser *absolute* velocity belonging to
a smaller circle of revolution, and would consequently fall
behind the cavities or "spots" formed by its abstraction.

[1] *Researches in Solar Physics*, part i. p. 20.
[2] Both the phrase and the method were suggested by Faye.   *Comptes
Rendus*, t. lxi. p. 1082.   [3] *Proc. Roy. Soc.*, vol. xiv. p. 39.

The ideas of M. Faye were, on two fundamental points, contradicted by the Kew investigators. He held spots to be regions of *uprush* and of heightened temperature; they believed their obscurity to be due to a *downrush* of comparatively cool vapours. On which side does the truth lie?

Observing at Ville-Urbanne, March 6, 1865, M. Chacornac saw floods of photospheric matter visibly precipitating themselves into the abyss opened by a great spot, and carrying with them small neighbouring maculæ.[1] Similar instances were repeatedly noted by Father Secchi, who considered the existence of a kind of *suction* in spots to be quite beyond question.[2] The tendency in their vicinity, to put it otherwise, is *centripetal,* not *centrifugal;* and this alone seems to negative the supposition of a central uprush.

A fresh witness was now at hand. The application of the spectroscope to the direct examination of the sun's surface dates from March 4, 1866, when Mr. Norman Lockyer began his inquiry into the cause of the darkening in spots.[3] The answer was prompt and unmistakable, and was again, in this case, adverse to the French theorist's view. The obscurations in question were found to be produced by no deficiency of emissive power, but by an increase of absorptive action. The background of variegated light remains unchanged, but more of it is stopped by the interposition of a dense mass of relatively cool vapours. The spectrum of a sun-spot is crossed by the same set of multitudinous dark lines, with some minor differences, visible in the ordinary solar spectrum. We must then conclude that the same vapours which are dispersed over the unbroken solar surface are accumulated in the umbral cavity, the compression incident to such accumulation being betrayed by the thickening of certain lines of absorption. But there is also a general absorption, extending almost continuously from one end of the spot-spectrum to the other. And this is explained in Professor Hastings's ingenious speculation by a

[1] Lockyer, *Contributions to Solar Physics,* p. 70.
[2] *Le Soleil,* p. 87. [3] *Proc. Roy. Soc.,* vol. xv. p. 256.

deposition of *soot*, or something analogous—in other words, by the presence, as a slowly settling fine dust, of cold, dark particles of carbon or silicon.[1]

An inquiry, however, prosecuted by Professor Young of Princeton, New Jersey, during the latter half of 1883, has set the matter in a new light. Using a spectroscope of exceptionally high dispersive power, he succeeded to a considerable extent in "resolving" the supposed continuous obscurity of spot-spectra into a countless army of fine dark lines set very close together.[2] The substances producing this darkening or absorption are then in a gaseous state, and the "soot" theory collapses. We may add, with some confidence, that their temperature, although affected by great irregularities, is in general lower than that of the encircling photosphere. Professor Langley in 1875 [3] fully confirmed Professor Henry's discovery in 1845, that the nuclei of spots radiate far less heat than equal areas of the unbroken disc; but this tells us little or nothing as to their real thermal condition. The character of their spectra, however, makes it extremely probable that it is one of comparative coolness.

As to the movements of the constipated vapours forming spots, the spectroscope is also competent to supply information. The principle of the method by which it is procured will be explained farther on. Suffice it here to say that the transport, at any considerable velocity, to or from the eye of the gaseous material giving bright or dark lines, can be measured by the displacement of such lines from their previously known normal positions. In this way movements have been detected in or above spots of enormous rapidity, ranging up to 320 *miles per second.*[4] But the result, so far, has been to negative conjectures either of uprushes or downrushes as part of the regular internal economy of spots.

A new theory of sun-spots, started by Faye in 1872, and still advocated by him, is sufficiently plausible to merit some brief

[1] *Am. Jour.*, vol. xxi. p. 42.   [2] *Phil. Mag.*, vol. xvi. p. 460.
[3] *Comptes Rendus*, t. lxxx. p. 746.   [4] Young, *The Sun*, p. 99.

attention. He had been foremost in pointing out that the observations of Carrington and Spörer absolutely forbade the supposition that any phenomenon at all resembling our trade-winds exists in the sun. The "proper movements" of spots give no evidence of regular currents either towards or from the poles. The systematic drift of the photosphere is strictly a drift in longitude ; its direction is everywhere parallel to the equator. This fact being once clearly recognised, the "solar tornado " hypothesis at once fell to pieces ; but M. Faye[1] perceived another source of vorticose motion in the unequal rotating velocities of contiguous portions of the photosphere. The "pores " with which the whole surface of the sun is studded he took to be the smaller eddies resulting from these inequalities ; the spots to be such eddies developed into whirlpools. It only needs to thrust a stick into a stream to produce the kind of effect designated. And it happens that the differences of angular movement adverted to attain a maximum just about the latitudes where spots are most frequent and conspicuous.

There are, however, two fatal objections. One (already mentioned) is the total absence of the regular swirling motion —in a direction contrary to that of the hands of a watch north of the solar equator, in the opposite sense south of it—which should impress itself upon every lineament of a sun-spot if the cause assigned were a primary producing, and not merely (as it possibly may be) a secondary determining one. The other, pointed out by Professor Young,[2] is that the cause is inadequate to the effect. The difference of movement, or *relative drift*, supposed to occasion such prodigious disturbances, amounts, at the utmost, for two portions of the photosphere 123 miles apart, to about five yards a minute. Thus the friction of contiguous sections must be quite insignificant.

One other view remains to be noticed. It is that urged by Father Secchi in and after the year 1872, and adopted with some useful modifications by Professor Young.[3] Spots are manifestly associated with violent eruptive action, giving rise

[1] *Comptes Rendus*, t. lxxv. p. 1664. [2] *The Sun*, p. 174. [3] *Ibid.*, p. 175.

to the faculæ and prominences which usually garnish their borders. It is accordingly contended that upon the withdrawal of matter from below by the flinging up of a prominence must ensue a sinking-in of the surface, into which the partially cooled erupted vapours rush and settle, producing just the kind of darkening by increased absorption told of by the spectroscope. Round the edges of the cavity the rupture of the photospheric shell will form lines of weakness provocative of further eruptions, which will, in their turn, deepen and enlarge the cavity. The phenomenon will thus tend to per-petuate itself, until equilibrium is at last restored by internal processes. A sun-spot might then be described as an inverted terrestrial volcano, in which the outbursts of heated matter take place on the borders instead of at the centre of the crater, while the cooled products gather in the centre instead of at the borders. A real analogy is, however, probably masked by superficial unlikeness. Both in earth and sun—but in the sun to an enormously greater extent—the same fundamental conditions of volcanic action are found. These are heat and pressure. Matter, in which inconceivable powers of expansion are lodged by virtue of the suppressed fury of its interstitial movements, is held down in the rigid grasp of its own weight. The slightest disturbance of this delicately adjusted balance of forces suffices to produce an outbreak. The gun is ready loaded; it only needs to pull the trigger. It is true that we cannot, in either case, tell exactly how the trigger is pulled— whether by local increase of heat or local relief of pressure, or by both in combination; but it is easy to see that the erup-tive capacities of our own quiescent little globe must, in the sun, be intensified to a degree beyond the reach even of imagination.

The "volcanic hypothesis" of sun-spots makes no attempt to explain their peculiarities of distribution either in space or time—their preference for two zones of the solar surface, or their marked periodicity. It is thus far indeed from being completely satisfactory; yet it seems the least misleading way of conceiving the facts that can be suggested in the present state of our knowledge.

A singular circumstance has now to be recounted. On the 1st of September 1859, while Carrington was engaged in his daily work of measuring the positions of sun-spots, he was startled by the sudden appearance of two patches of peculiarly intense light within the area of the largest group visible. His first idea was that a ray of unmitigated sunshine had penetrated the screen employed to reduce the brilliancy of the image; but, having quickly convinced himself to the contrary, he ran to summon an additional witness of an unmistakably remarkable occurrence. On his return he was disappointed to find the strange luminous outburst already on the wane; shortly afterwards the last trace vanished. Its entire duration was five minutes—from 11.18 to 11.23 A.M., Greenwich time; and during those five minutes it had traversed a space estimated at 35,000 miles! No perceptible change took place in the details of the group of spots visited by this transitory conflagration, which, it was accordingly inferred, took place at a considerable height above it.[1]

Carrington's account was precisely confirmed by an observation made at Highgate. Mr. R. Hodgson described the appearance seen by him as that "of a very brilliant star of light, much brighter than the sun's surface, most dazzling to the protected eye, illuminating the upper edges of the adjacent spots and streaks, not unlike in effect the edging of the clouds at sunset."[2]

This unique phenomenon seemed as if specially designed to accentuate the inference of a sympathetic relation between the earth and the sun. From the 28th of August to the 4th of September 1859, a magnetic storm of unparalleled intensity, extent, and duration, was in progress over the entire globe. Telegraphic communication was everywhere interrupted—except, indeed, that it was, in some cases, found practicable to work the lines *without batteries*, by the agency of the earth-

---

[1] *Month. Not.*, vol. xx. p. 13.　　　　[2] *Ibid.*, p. 15.

currents alone;[1] sparks issued from the wires; gorgeous
auroræ draped the skies in solemn crimson over both hemispheres, and even within the tropics; the magnetic needle lost
all trace of continuity in its movements, and darted to and fro
as if stricken with inexplicable panic. The coincidence was
drawn even closer. *At the very instant*[2] of the solar outburst
witnessed by Carrington and Hodgson, the photographic apparatus at Kew registered a marked disturbance of all the three
magnetic elements; while, shortly after the ensuing midnight,
the electric agitation culminated, thrilling the earth with subtle
vibrations, and lighting up the atmosphere from pole to pole
with the coruscating splendours which, perhaps, dimly recall
the times when our ancient planet itself shone as a star.

  Here then, at least, the sun was—in Professor Balfour
Stewart's phrase—"taken in the act"[3] of stirring up terrestrial
commotions. Nor have instances since been wanting of an
indubitable connection between outbreaks of individual spots
and magnetic disturbances—four such were recorded in 1882—
although the peculiar features of the event of September 1,
1859, have not recurred. An attempt was made to explain
them by Professor Piazzi Smyth,[4] who suggested that the flying
luminous objects seen on that occasion were nothing else than
a pair of unusually large meteors ignited by retardation in the
solar atmosphere. But the inadequacy of the conjecture hardly
needs to be pointed out. The sudden development of light
was certainly no accidental occurrence, but marked the climax
of some systematic commotion already for some days in progress. If we were to look for its terrestrial analogue, we should
rather find it in the "auroral beam" which traversed the
heavens during a vivid display of polar lights, November 17,
1882, and shared, there is every reason to believe, their
electrical origin and character.[5]

  [1] *Am. Jour.*, vol. xxix. (2d series), pp. 94–95.
  [2] The magnetic disturbance took place at 11.15 A.M., three minutes
before the solar blaze compelled the attention of Carrington.
    [3] *Phil. Trans.*, vol. cli. p. 428.   [4] *Month. Not.*, vol. xx. p. 88.
    [5] See J. Rand Capron, *Phil. Mag.*, May 1883.

Meantime M. Rudolf Wolf, transferred to the direction of the Zürich Observatory, had relaxed none of his zeal in the investigation of sun-spot periodicity. A laborious revision of the entire subject with the aid of fresh materials led him, in 1859,[1] to the conclusion that while the *mean* period differed little from that arrived at in 1852 of 11.11 years, very considerable fluctuations on either side of that mean were rather the rule than the exception. Indeed, the phrase " sun-spot period " must be understood as fitting very loosely the great fact it is taken to represent; so loosely, that the interval between two maxima may rise to sixteen and a half or sink below seven and a half years.[2] In 1861[3] Wolf showed, and the remark was fully confirmed by the Kew observations, that the shortest periods brought the most acute crises, and *vice versâ ;* as if for each wave of disturbance a strictly equal amount of energy were available, which might spend itself lavishly and rapidly, or slowly and parsimoniously, but could in no case be exceeded. The further inclusion of recurring solar commotions within a cycle of fifty-five and a half years was simultaneously pointed out; and Hermann Fritz showed soon after that the aurora borealis is subject to an identical double periodicity.[4] The same inquirer has more recently detected both for auroræ and sun-spots a " secular period " of 222 years,[5] and the Kew observations indicate for the latter, oscillations accomplished within twenty-six and twenty-four days.[6] The more closely spot-fluctuations are looked into, indeed, the more complex they prove. Maxima

[1] *Mittheilungen über die Sonnenflecken,* No. ix., *Vierteljahrsschrift der Naturforschenden Gesellschaft in Zürich,* Jahrgang 4.

[2] *Mitth.,* No. lii. p. 58 (1881).

[3] *Ibid.,* No. xii. p. 192. Mr. Joseph Baxendell, of Manchester, reached independently a similar conclusion. See *Month. Not.,* vol. xxi. p. 141.

[4] Wolf, *Mitth.,* No. xv. p. 107, &c. Olmsted, following Hansteen, had already, in 1856, sought to establish an auroral period of sixty-five years. *Smithsonian Contributions,* vol. viii. p. 37.

[5] Hahn, *Ueber die Beziehungen der Sonnenfleckenperiode zu meteorologischen Erscheinungen,* p. 99 (1877).

[6] *Report Brit. Ass.,* 1881, p. 518; 1883, p. 418.

of one order are superposed upon, or in part neutralised by, maxima of another order; originating causes are masked by modifying causes; the larger waves of the commotion are indented with minor undulations, and these again crisped with tiny ripples, while the whole rises and falls with the swell of the great secular wave, scarcely perceptible in its progress because so vast in scale.

The idea that solar maculation depends in some way upon the position of the planets occurred to Galileo in 1612.[1] It has been industriously sifted by a whole bevy of modern solar physicists. Wolf in 1859[2] found reason to believe that the eleven-year curve is determined by the action of Jupiter, modified by that of Saturn, and diversified by influences proceeding from the Earth and Venus. Its tempting approach to agreement with Jupiter's period of revolution round the sun, indeed, irresistibly suggested a causal connection; yet it does not seem that the most skilful " coaxing" of figures can bring about a fundamental harmony. Carrington pointed out in 1863 that while, during *eight successive periods*, from 1770 downwards, there were approximate coincidences between Jupiter's aphelion passages and sun-spot maxima, the relation had been almost exactly reversed in the two periods preceding that date;[3] and the latest conclusion of M. Wolf himself is that the Jovian origin must be abandoned.[4] Nevertheless it is still held by M. Duponchel[5] of Paris, who accommodates discrepancies with the help of perturbations by the large exterior planets; and it deserves notice that his prediction of an abnormal lengthening of the maximum due in 1882, through certain peculiarities in the positions of Uranus and Neptune about that time, has been remarkably verified by the event.

That outbreaks of solar activity are *modified* by influences depending upon planetary configuration has been tolerably well ascertained by the Kew observations. This no less signi-

[1] *Opere*, t. iii. p. 412.　　　　[2] *Mitth.*, Nos. viii. and xviii.
[3] *Observations at Redhill*, p. 248.　　[4] *Comptes Rendus*, t. xcv. p. 1249.
[5] *Ibid.*, t. xciii. p. 827; t. xcvi. p. 1418.

ficant than surprising result was imparted by Professor Balfour Stewart to the ΄Royal Society of Edinburgh, April 18, 1864.[1] The method of research by which it was arrived at (said to have been *privately* recommended by Galileo[2]) consisted in studying the " behaviour " of each spot as it crossed the disc. This, it was found, was almost always marked, about the same epochs, with a common character. If one rent in the photosphere widened as the central meridian of the sun was approached, those in its train were pretty sure to do likewise ; if it closed up, its successors followed suit. Moreover, the controlling power was perceived to travel onwards at a rate quicker than that of the earth's annual revolution. It followed, in short, with much fidelity, the orbital movement of Venus. Its nature is of such a kind as to assuage outbreaks on the side of the sun turned towards the planet, and to aggravate them on the opposite hemisphere.[3] The action both of Jupiter and Mercury is, it would seem, the same in kind though less in degree. That of the earth is more difficult to determine, but it can scarcely be doubted that it is similarly exercised. It has even been attempted to invert the process, and arrive at the period of an unknown planet through the observation of sun-spots. Professor Balfour Stewart has shown that inequalities in their development exist corresponding severally to the revolution of such a body round the sun in twenty-four days, and to its "synodical periods" or successive meetings with Jupiter, Venus, and Mercury.[4] But the prediction still awaits fulfilment.

The question so much discussed, as to the influence of sun-spots on weather, does not yet admit of a satisfactory answer. The facts of meteorology are too complex for easy or certain classification. Effects owning dependence on one

---

[1] *Ed. Phil. Trans.*, vol. xxiii. p. 499.

[2] *Researches in Solar Physics*, ser. ii. p. 46 (privately printed). The Rev. Mr. Selwyn is responsible for the statement, for which he gives no authority.   [3] *Proc. Roy. Soc.*, vols. xiv. p. 59, xx. p. 210.

[4] *Report Brit. Ass.*, 1881, p. 518.

cause often wear the livery of another; the meaning of observed particulars may be inverted by situation; and yet it is only by the collection and collocation of particulars that we can hope to reach any general law. There is, however, a good deal of evidence to support the opinion—the grounds for which were primarily derived from the labours of Mr. Meldrum at the Mauritius—that increased rainfall and atmospheric agitation attend spot-maxima; while Herschel's conjecture of a more copious emission of light and heat about the same epochs is so far from having been borne out by modern investigations, that the probabilities seem rather to lean the other way.

The examination of what we may call the *texture* of the sun's surface derived new interest from a remarkable announcement made by Mr. James Nasmyth in 1862.[1] He had made (as he supposed) the discovery that the entire luminous stratum of the sun is composed of a multitude of elongated shining objects on a darker background, shaped much like willow-leaves, of vast size, crossing each other in all possible directions, and endowed with unceasing relative motions. A lively controversy ensued. In England and abroad, the most powerful telescopes were directed to a scrutiny encompassed with varied difficulties. The results, on the whole, were such as to invalidate the precision of the disclosures made by the Hammerfield reflector. Mr. Dawes was especially emphatic in declaring that Nasmyth's "willow-leaves" were nothing more than the "nodules" of Sir William Herschel seen under a misleading aspect of uniformity; and there is little doubt that he was right. It is, however, admitted that something of the kind may be seen in the penumbræ and "bridges" of spots, presenting an appearance compared by Dawes himself in 1852 to that of a piece of coarse straw-thatching left untrimmed at the edges.[2]

The term "granulated," suggested by Dawes in 1864,[3] best describes the mottled aspect of the solar disc as shown by

Report Brit. Ass., 1862, p. 16 (pt. ii.)   [2] *Mem. R. A. Soc.,* vol. xxi. p. 161.
[3] *Month. Not.,* vol. xxiv. p. 162.

modern telescopes and cameras. The grains, or rather the "floccules," with which it is thickly strewn, have been resolved by Langley, under exceptionally favourable conditions, into "granules" not above 100 miles in diameter; and from these relatively minute elements, composing, jointly, about one-fifth of the visible photosphere,[1] he estimates that three-quarters of the entire light of the sun are derived.[2] Janssen goes so far as to say that if the whole surface were as bright as its brightest parts, its luminous emission would be ten to twenty times greater than it actually is.[3]

The rapid changes in the forms of these solar cloud-summits are beautifully shown in the marvellous photographs taken by Janssen at Meudon, with exposures reduced at times to $\frac{1}{100,000}$ of a second! By their means, also, the curious phenomenon known as the *réseau photosphérique* has been made evident.[4] This consists in the diffusion over the entire disc of fleeting blurred patches, as if of imperfect definition, due, doubtless, to agitations in the intervening solar atmosphere. The same cause may perhaps account for the evanescent obscurations described by Father Perry of Stonyhurst before the Royal Astronomical Society, May 9, 1884.[5]

The "grains," or more brilliant parts of the photosphere, are now generally held to represent the upper terminations of ascending and condensing currents, while the darker interstices (Herschel's "pores") mark the positions of descending cooler ones. In the penumbræ of spots, the glowing streams rushing up from the tremendous sub-solar furnace are bent sideways by the powerful indraught, so as to change their vertical for a nearly horizontal motion, and are thus taken, as it were, in flank by the eye, instead of being seen end-on in mamelon-form. This gives a plausible explanation of the channelled

[1] *Am. Jour. of Science*, vol. vii. 1874, p. 92.
[2] Young, *The Sun*, p. 103.   [3] *Ann. Bur. Long.*, 1879, p. 679.
[4] *Ibid.*, 1878, p. 689.
[5] *Observatory*, vol. vii. p. 154. Father Perry sought to identify the objects observed by him with Trouvelot's "veiled spots;" Mr. Ranyard suggested the more probable analogy of the *réseau photosphérique.*

structure of penumbræ which suggested the comparison to a rude thatch. Accepting this theory as in the main correct, we perceive that the very same circulatory process which, in its spasms of activity, gives rise to spots, produces in its regular course the singular " marbled " appearance, for the recording of which we are no longer at the mercy of the fugitive or delusive impressions of the human retina. And precisely this circulatory process it is which gives to our great luminary its permanence as a *sun*, or warming and illuminating body.

CHAPTER III.

RECENT SOLAR ECLIPSES.

BY observations made during a series of five remarkable eclipses, comprised within a period of eleven years, knowledge of the solar surroundings was advanced nearly to its present stage. Each of these events brought with it a fresh disclosure of a definite and unmistakable character. We will now briefly review this orderly sequence of discovery.

Photography was first systematically applied to solve the problems presented by the eclipsed sun, July 18, 1860. It is true that a very creditable daguerreotype, taken by Busch at Königsberg during the eclipse of 1851, is still valuable as a record of the corona of that year ; and some subsequent attempts were made to register partial phases of solar occultation ; but the ground remained practically unbroken until 1860.

In that year the track of totality crossed Spain, and thither, accordingly, Mr. Warren De la Rue transported his photo-helio-graph, and Father Secchi his six-inch Cauchoix refractor. The question then primarily at issue was that relating to the nature of the red protuberances. Although, as already stated, the evidence collected in 1851 gave a reasonable certainty of their connection with the sun, objectors were not silenced ; and when the side of incredulity was supported by so considerable an authority as M. Faye, it was impossible to treat it with contempt. Two crucial tests were available. If it could be shown that the fantastic shapes suspended above the edge of the dark moon were seen under an identical aspect from two distant stations, that fact alone would annihilate the theory of optical illusion or "mirage;" while the certainty that they were progressively concealed by the advancing moon on one side, and uncovered on the other, would effectually detach them from dependence on our satellite, and establish them as solar appendages.

Now both these tests were eminently capable of being applied by photography. But the difficulty arose that nothing was known as to the chemical power of the rosy prominence-light, while everything depended on its right estimation. A shot had to be fired, as it were, in the dark. It was a matter of some surprise, and of no small congratulation, that, in both cases, the shot took effect.

Mr. De la Rue occupied a station at Rivabellosa, in the Upper Ebro valley; Father Secchi set up his instrument at Desierto de las Palmas, about 250 miles to the south-east, overlooking the Mediterranean. From the totally eclipsed sun, with its strange garland of flames, each observer derived several perfectly successful impressions, which were found, on comparison, to agree in the most minute details. This at once settled the fundamental question as to the substantial reality of these objects; while their solar character was demonstrated by the passage of the moon *in front* of them, indisputably attested by pictures taken at successive stages of the eclipse. That forms seeming to defy all laws of equilibrium were, nevertheless, not wholly evanescent, appeared from their identity at an interval of seven minutes, during which the lunar shadow was in transit from one station to the other; and the singular energy of their "actinic" rays was shown by the record on the sensitive plates of some prominences invisible in the telescope. Moreover, photographic evidence strongly confirmed the inference—previously drawn by Grant and others, and now repeated with fuller assurance by F. Secchi—that an uninterrupted stratum of prominence-matter encompasses the sun on all sides, forming a reservoir from which gigantic jets issue, and into which they subside.

Thus a beginning of accurate knowledge regarding the surroundings of the sun was made, and the value of the brief moments of eclipse indefinitely increased by supplementing transient visual impressions with the faithful and lasting records of the camera.

In the year 1868 the history of eclipse spectroscopy virtually began, as that of eclipse photography in 1860; that is to say, the respective methods then first gave definite results. On the 18th of August 1868, the Indian and Malayan peninsulas were traversed by a lunar shadow producing total obscuration during five minutes and thirty-eight seconds. Two English and two French expeditions were despatched to the distant regions favoured by an event so propitious to the advance of knowledge, chiefly to obtain the verdict of the prism as to the composition of prominences. Nor were they despatched in vain. An identical discovery was made by nearly all the observers. At Jamkandi, in the Western Ghauts, where Lieutenant (now Colonel) Herschel was posted, unremitting bad weather threatened to baffle his eager expectations; but during the lapse of the critical five and a half minutes the clouds broke, and across the driving wrack a "long, finger-like projection" jutted out over the margin of the dark lunar globe. In another moment the spectroscope was pointed towards it; three bright lines—red, orange, and blue—flashed out, and the problem was solved.[1] The problem was solved in this general sense, that the composition out of glowing vapours of the objects infelicitously termed "protuberances" or "prominences" was no longer doubtful; although further inquiry was needed for the determination of the particular species to which those vapours belonged.

Similar, but more complete observations were made, with less atmospheric hindrance, by Tennant and Janssen at Guntoor, by Pogson at Masulipatam, and by Rayet at Wha-Tonne, on the coast of the Malay peninsula, the last observer counting as many as nine bright lines.[2] Amongst them it was not difficult

[1] *Proc. Roy. Soc.*, vol. xvii. p. 116.   [2] *Comptes Rendus*, t. lxvii. p. 757.

to recognise the characteristic light of hydrogen; and it was generally, though over-hastily, assumed that the orange ray matched the luminous emissions of sodium. But fuller opportunities were at hand.

The eclipse of 1868 is chiefly memorable for having taught astronomers to do without eclipses, so far, at least, as one particular branch of solar inquiry is concerned. Inspired by the beauty and brilliancy of the variously tinted prominence-lines revealed to him by his spectroscope, Janssen exclaimed to those about him, "Je verrai ces lignes-là en dehors des éclipses!" On the following morning he carried into execution the plan which formed itself in his brain while the phenomenon which suggested it was still before his eyes. It rests upon an easily intelligible principle.

The glare of our own atmosphere alone hides the appendages of the sun from our daily view. To a spectator on an airless planet, the central globe would appear attended by all its splendid retinue of crimson prominences, silvery corona, and far-spreading zodiacal light, projected on the star-spangled, black background of an absolutely unilluminated sky. Now the spectroscope offers the means of indefinitely weakening atmospheric glare by diffusing a constant amount of it over an indefinitely widened area. But monochromatic or "bright-line" light is, by its nature, incapable of being so diffused. It can, of course, be *deviated* by refraction to any extent desired; but it always remains equally concentrated, in whatever direction it may be thrown. Hence, when it is mixed up with continuous light—as in the case of the solar flames shining through our atmosphere—it derives a *relative* gain in intensity from every addition to the dispersive power of the spectroscope with which the heterogeneous mass of beams is analysed. Employ prisms enough, and eventually the undiminished rays of persistent colour will stand out from the continually fading rainbow-tinted band, by which they were at first effectually veiled.

This Janssen saw by a flash of intuition while the eclipse

was in progress ; and this he realised at 10 A.M. next morning, August 19, 1868—the date of the beginning of spectroscopic work at the margin of the unobscured sun. During the whole of that day and many subsequent ones, he enjoyed, as he said, the advantage of a prolonged eclipse. The intense interest with which he surveyed the region suddenly laid bare to his scrutiny, was heightened by evidences of rapid and violent change. On the 18th of August, during the eclipse, a vast spiral structure, *at least* 89,000 miles high, was perceived, planted in surprising splendour on the rim of the interposed moon. It was formed, as Major Tennant judged from its appearance in his photographs, by the encounter of two mounting torrents of flame, and was distinguished as the "Great Horn." Next day it was in ruins; hardly a trace remained to show where it had been.[1] Janssen's spectroscope furnished him besides with the strongest confirmation of what had already been reported by the telescope and the camera as to the continuous nature of the scarlet "sierras" lying at the base of the prominences. Everywhere at the sun's edge the same bright lines appeared.

It was not until the 4th of September that Janssen thought fit to send news of his discovery to Europe. He little dreamed of being anticipated ; nor did he indeed grudge that science should advance at the expense of his own undivided fame. A few minutes before his despatch was handed to the Secretary of the Paris Academy of Sciences, a communication similar in purport had been received from Mr. Norman Lockyer. There is no need to discuss the narrow and wearisome question of priority ; each of the competitors deserves, and has obtained, full credit for his invention. With noteworthy and confident prescience, Mr. Lockyer, in 1866, before anything was yet known regarding the constitution of the "red flames," ordered a strongly dispersive spectroscope for the express purpose of viewing, apart from eclipses, the bright-line spectrum which he expected them to give. Various delays, however,

---

[1] *Comptes Rendus*, t. lxvii. p. 839.

supervened, and the instrument was not in his hands until October 16, 1868. On the 20th he picked up the vivid rays, of which the presence and (approximately) the positions had in the meantime become known. But there is little doubt that, even without that previous knowledge, they would have been found; and that the eclipse of August 18 only accelerated a discovery already assured.

Mr. Huggins, too, had been groping for prominence-lines during two years and a half with the aid of various apparatus at his observatory of Tulse Hill;[1] but not until he knew where to look did he succeed in seeing them. It should be added that the *principle* of the method was suggested to Lieutenant Herschel by the phenomena of the eclipse, and was briefly described in his report.[2]

Astronomers, thus liberated, by the acquisition of power to view them at any time, from the necessity of studying prominences during eclipses, were able to concentrate the whole of their attention on the corona. The first thing to be done was to ascertain the character of its spectrum. This was seen in 1868 only as a faintly continuous one; for Rayet, who seems to have perceived its characteristic bright line far above the summits of the flames, connected it, nevertheless, with those objects. On the other hand, Lieutenant Campbell ascertained on the same occasion the polarisation of the coronal light in planes passing through the sun's centre,[3] thereby showing that light to be, in whole or in part, reflected sunshine. But if reflected sunshine, it was objected, the chief at least of the dark Fraunhofer lines should be visible in it, as they are visible in moonbeams, sky illumination, and all other sun-derived light. The objection was well founded, but was prematurely urged, as we shall see.

On the 7th of August 1869, a track of total eclipse crossed the continent of North America diagonally, entering at Behring's Straits, and issuing on the coast of North Carolina. It was

---

[1] *Month. Not.*, vol. xxviii. p. 88.
[2] *Proc. Roy. Soc.*, vol. xvii. p. 119.   [3] *Ibid.*, p. 123.

beset with observers; but the most effective work was done in Iowa. At Des Moines, Professor Harkness of the Naval Observatory, Washington, obtained from the corona an "absolutely continuous spectrum," slightly less bright than that of the full moon, but traversed by a single green ray.[1] The same green ray was seen at Burlington and its position measured by Professor Young of Dartmouth College.[2] It was found to coincide with that of a dark line of iron in the solar spectrum, numbered 1474 on Kirchhoff's scale. This was perplexing; since it seemed, at first sight, to compel the inference that the corona was actually composed of vapour of iron,[3] so attenuated as to give only one line of secondary importance out of the many hundreds belonging to it. But in 1876 Young was able, by the use of greatly increased dispersion, to resolve the Fraunhofer line "1474" into a pair, of which one component is due to iron, the other (the more refrangible) to the coronal gas.[4] This substance, of which nothing is known to terrestrial chemistry, is luminous at least half a million of miles above the sun's surface, and must be considerably lighter even than hydrogen.

A further trophy was carried off by American skill[5] sixteen months after the determination due to it of the distinctive spectrum of the corona. The eclipse of December 22, 1870, though lasting only two minutes and ten seconds, drew observers from the New, as well as from the Old World to the shores of the Mediterranean. Janssen issued from Paris in a balloon, carrying with him the *vital parts* of a reflector specially constructed to collect evidence about the corona. But he reached Oran only to find himself shut behind a cloud-curtain more impervious than the Prussian lines. Everywhere the sky was more or less overcast. Mr. Lockyer's journey from England to Sicily, and shipwreck in the *Psyche*, were

[1] *Washington Observations*, 1867, App. ii., Harkness's Report, p. 60.
[2] *Am. Jour.*, vol. xlviii. (2d series), p. 377.
[3] This view was never assented to by either Young or Lockyer.
[4] *Am. Jour.*, vol. xi. (3d series), p. 429.
[5] Everything in such observations depends upon the proper manipulation of the slit of the spectroscope.

recompensed with a glimpse of the solar aureola during *one second and a half!* Three parties stationed at various heights on Mount Etna saw absolutely nothing. Nevertheless, important information was snatched in despite of the elements.

The prominent event was Young's discovery of the "reversing layer." As the surviving solar crescent narrowed before the encroaching moon, "the dark lines of the spectrum," he tells us, "and the spectrum itself, gradually faded away, until all at once, as suddenly as a bursting rocket shoots out its stars, the whole field of view was filled with bright lines more numerous than one could count. The phenomenon was so sudden, so unexpected, and so wonderfully beautiful, as to force an involuntary exclamation." [1] Its duration was about two seconds, and the impression produced was that of a complete reversal of the Fraunhofer spectrum—that is, the substitution of a bright for every dark line.

Now something of the kind was theoretically necessary to account for the dusky rays in sunlight which have taught us so much, and have yet much more to teach us; so that, although surprising from its transitory splendour, the appearance could not strictly be called "unexpected." Moreover, its premonitory symptom in the fading out of those rays had been actually described by Father Secchi in 1868,[2] and looked for by Young as the moon covered the sun in August 1869. But with the slit of his spectroscope placed *normally* to the sun's limb, the bright lines gave a flash too thin to catch the eye. In 1870 the position of the slit was *tangential*—it ran along the shallow bed of incandescent vapours, instead of cutting across it: hence his success.

The same observation was made at Xerez de la Frontera by Mr. Pye, a member of Young's party; and, although an exceedingly delicate one, has since frequently been repeated. The whole Fraunhofer series appeared bright (omitting other instances) to Maclear, Herschel, and Fyers in 1871, at the beginning or end of totality; to Pogson during a period (perhaps

[1] *Mem. R. A. Soc.*, vol. xli. p. 435.   [2] *Comptes Rendus*, t. lxvii. p. 1019.

erroneously estimated) of from five to seven seconds, at the break up of an annular eclipse, June 6, 1872; to Stone at Klipfontein, April 16, 1874, when he saw "the field full of bright lines." [1] But between the picture presented by the " véritable pluie de lignes brillantes," [2] which descended into M. Trépied's spectroscope for three seconds after the disappearance of the sun, May 17, 1882, and the familiar one of the dark-line solar spectrum, certain differences were perceived, showing their relation to be not simply that of a positive to a negative impression.

A "reversing layer," or stratum of mixed vapours, glowing, but at a lower temperature than that of the actual solar surface, was an integral part of Kirchhoff's theory of the production of the Fraunhofer lines. Here it was assumed that the missing rays were stopped, and here also it was assumed that the missing rays would be seen bright, could they be isolated from the overpowering splendour of their background. This isolation is effected by eclipses, with the result—beautifully confirmatory of theory—of *reversing*, or turning from dark to bright, the Fraunhofer spectrum. But there is a difficulty. If absorption be in truth thus localised, it should appear greatly strengthened near the edges of the solar disc. This, however, is not the case. Kirchhoff met the objection by giving a great depth to the reversing stratum, whereby the difference in length of the paths across that stratum traversed by rays from the sun's limb and centre, became *relatively* insignificant. In other words, he supposed that the chief part of the light absent from the spectrum was arrested in the region of the corona. This view is rendered wholly untenable by the character of the coronal spectrum.

Faye, on the other hand, abolished the reversing layer altogether (there was at that time no ocular demonstration of its existence); or rather, sunk it out of sight below the visible level of the photosphere, and got the necessary absorption done in the interstices of the photospheric clouds by the vapours in

[1] *Mem. R. A. Soc.*, vol. xli. p. 43.   [2] *Comptes Rendus*, t. xciv. p. 1640.

which they float, and from which they condense. It was, however, at once seen that the lines thus produced would be bright,
not dark, since the brilliant cumuli would be cooled, by their
greater power of radiation, below the temperature of the
surrounding medium. A better explanation was offered by
Professor Hastings of Baltimore in 1881.[1] He maintains that
Young's stratum, of which the thickness is estimated at about
600 miles,[2] represents only the upper margin of a *reversing
ocean*, in which the granules of the photosphere float at various
depths. The necessary difference of temperature is derived
from the coolness of the descending vapours, which bathe the
radiating particles and rob them of certain characteristic beams.
We are thus driven to suppose that only a small part of the
absorption betrayed by the Fraunhofer lines takes place in the
complex layer disclosed by eclipses;[3] so that a strict correspondence between its bright rays and the solar dusky rays
is not to be expected, and would, in fact, prove somewhat
embarrassing. M. Trépied's detection of differences is, for
this reason, especially valuable, and we may hope that, before
long, an instantaneous photograph of the complete "rainbow-
flash" accompanying totality will afford a more stable support
to theory on the subject than it can yet claim.

The last of the five eclipses which we have grouped together
for separate consideration, was visible in Southern India and
Australia, December 12, 1871. Some splendid photographs
were secured by the English parties on the Malabar coast,
showing, for the first time, the remarkable branching forms of
the coronal emanations ; but the most conspicuous result was

[1] *Am. Jour. of Science*, vol. xxi. p. 33.
[2] Pulsifer's observations at Fort Worth in 1878 gave a *minimum* depth
of 524 miles (*Am. Jour. of Science*, vol. xvii. p. 495).
[3] This cannot be due to the shallowness of the layer, since a few feet (or
even, as in the case of sodium, a few millimetres) of glowing vapour can
be experimentally shown capable of producing the amount of absorption
present in the solar spectrum. We must then assume that its temperature
is so nearly on a level with that of the photosphere that it replaces almost
all the light it absorbs.

Janssen's detection of some of the dark Fraunhofer lines long vainly sought in the continuous spectrum of the corona. Chief amongst these was the D line of sodium, the original index, it might be said, to solar chemistry. No proof could be afforded more decisive than this faint *echoing back* of the distinctive notes of the Fraunhofer spectrum, that the polariscope had spoken the truth in asserting a large part of the coronal radiance to be reflected sunlight. But it is (especially at certain epochs) so drenched in original luminous emissions, that its characteristic features are almost obliterated. Janssen's success in seizing them was due in part to the extreme purity of the air at Sholoor, in the Neilgherries, where he was stationed; in part to the use of an instrument adapted by its large aperture and short focus to give an image of the utmost possible luminosity.

His observations further "peremptorily demonstrated" the presence of hydrogen far outside the region of prominences, and forming an integral constituent of the corona. This important fact was simultaneously attested by Lockyer at Baikul, and by Respighi at Poodacottah, each making separate trial of a "slitless spectroscope" devised for the occasion. This consists simply of a prism placed outside the object-glass of a telescope or the lens of a camera, whereby the radiance encompassing the eclipsed sun is separated into as many differently tinted rings as it contains different kinds of light. These tinted rings were viewed by Respighi through a telescope, and were photographed by Lockyer, with the same result of showing hydrogen to ascend uniformly from the sun's surface to a height of fully 200,000 miles. Another notable observation made by Herschel and Tennant at Dodabetta showed the green ray " 1474 " to be just as bright in a "rift" as in the adjacent streamer. The visible structure of the corona was thus seen to be independent of the distribution of the gases which enter into its composition.

By means, then, of the five great eclipses of 1860–71 it was ascertained: first, that the prominences, and at least the

lower part of the corona, are genuine solar appurtenances;
secondly, that the prominences are composed of hydrogen and
other gases in a state of incandescence, and rise, as irregular
outliers, from a continuous envelope of the same materials,
some thousands of miles in thickness; thirdly, that the corona
is of a highly complex constitution, being made up in part of
glowing vapours, in part of matter capable of reflecting sunlight. We may now proceed to consider the results of subsequent eclipses.

These have raised and have helped to solve some very
curious questions. Indeed, every carefully watched total
eclipse of the sun stimulates, as well as appeases curiosity,
and leaves a legacy of outstanding doubt, continually, as time
and inquiry go on, removed, but continually replaced. It
cannot be denied that the corona is a perplexing phenomenon,
and that it does not become less perplexing as we know more
about it. It presented itself under quite a new and strange
aspect on the occasion of the eclipse which visited the Western
States of North America July 29, 1878. The conditions of observation were peculiarly favourable. The weather was superb;
above the Rocky Mountains the sky was of such purity as to
permit the detection, with the naked eye, of Jupiter's satellites
on several successive nights. The opportunity of advancing
knowledge was made the most of. Nearly a hundred astronomers (including several Englishmen) occupied twelve separate
posts, and prepared for an attack in force.

The question had often suggested itself, and was a natural
one to ask, whether the corona sympathises with the general
condition of the sun? whether, either in shape or brilliancy,
it varies with the progress of the sun-spot period? A more
propitious moment for getting this question answered could
hardly have been chosen than that at which the eclipse occurred.
Solar disturbance was just then at its lowest ebb. The development of spots for the month of July 1878 was represented
on Wolf's system of "relative numbers" by the fraction 0.1,
as against 135.4 for December 1870, an epoch of maximum

activity. The "chromosphere"[1] was, for the most part, shallow and quiescent; its depth, above the spot-zones, had sunk from about 6000 to 2000 miles; prominences were few and faint. Obviously, if a type of corona corresponding to a minimum of sun-spots existed, it should be seen then or never. It *was* seen; but while, in some respects, it agreed with anticipation, in others it completely set it at naught.

The corona of 1878, as compared with those of 1869, 1870, and 1871, was generally admitted to be shrunken in its main outlines, and much reduced in brilliancy. Mr. Lockyer pronounced it ten times fainter than in 1871; Professor Harkness estimated its light at less than one-seventh that derived from the mist-blotted aureola of 1870.[2] In shape, too, it was markedly different. When sun-spots are numerous, the corona appears to be most fully developed above the spot-zones, thus offering to our eyes a rudely quadrilateral contour. The four great luminous sheaves forming the corners of the square are made up of rays curving together from each side into "synclinal" or ogival groups, each of which may be compared to the petal of a flower. To Janssen, in 1871, the eclipsing moon seemed like the dark heart of a gigantic dahlia, painted in light on the sky; and the similitude to the ornament on a compass-card used by Sir George Airy in 1851, well conveys the decorative effect of the beamy, radiated kind of aureola never, it would appear, absent when solar activity is at a tolerably high pitch. In his splendid volume on eclipses,[3] Mr. Ranyard first generalised the peculiarity of the synclinal structures by a comparison of records; but the symmetry of their arrangement, though frequently striking, is liable to be confused by secondary formations. Nothing of all this, however, was visible in 1878. Instead, there was seen, as the

[1] The rosy envelope of prominence-matter was so named by Lockyer in 1868 (*Phil. Trans.*, vol. clix. p. 430); and the appellation, its defiance of Greek grammar notwithstanding, has had vitality to survive and prevail.

[2] *Bull. Phil. Soc. Washington*, vol. iii. p. 118.

[3] *Mem. R. A. Soc.*, vol. xli. 1879.

groundwork of the corona, a ring of pearly light, nebulous to the eye, but shown by telescopes and in photographs to have a fibrous texture, as if made up of bundles of fine hairs.   North and south a series of short, vivid, electrical-looking flame-brushes diverged with conspicuous regularity from each of the solar poles.   Their direction was not towards the centre of the sun, but towards each summit of his axis, so that the farther rays on either side started almost tangentially to the surface.   It is difficult not to connect this unusual display of polar activity[1] with the great relative depth of the chromo-sphere in those regions, noticed by Trouvelot previous to the eclipse.[2]

But the leading, and a truly amazing, characteristic of the phenomenon was formed by two vast, faintly-luminous *wings* of light, expanded on either side of the sun in the direction of the ecliptic.   These were missed by very few careful onlookers ; but the extent assigned to them varied with skill in, and facilities for, seeing.   By far the most striking observations were made by Newcomb at Separation (Wyoming), by Cleveland Abbe from the shoulder of Pike's Peak, and by Langley at its summit, an elevation of 14,100 feet above the sea.   Never before had an eclipse been viewed from anything approaching that altitude, or under so translucent a sky.   A proof of the great reduction in atmospheric glare was afforded by the perceptibility of the corona for above four minutes after totality was over.   For the 165 seconds of its duration, the remarkable streamers above alluded to continued "persistently visible," stretching away right and left of the sun to a distance of at least ten million miles !   One branch was traced over an apparent extent of fully twelve lunar diameters, without sign of a definite termination having been reached ; and there

[1] Professor W. A. Norton observed a similar phenomenon in 1869, accompanied by some symptoms of equatorial emission.   This is the more remarkable as 1869 was a year of many sun-spots.   His evidence, though unsupported, and adverse to the theory of varying types, should not be overlooked.   See *Am. Jour. of Sc.*, vol. i. (3d ser.), p. 1.

[2] *Wash. Obs.*, 1876, App. iii. p. 80.

were no grounds for supposing the other more restricted. The axis of the longest ray was found to coincide exactly, so far as could be judged, with the ecliptic.[1] Pale cross-beams were seen by Young and Abbe.

The resemblance to the zodiacal light was striking; and a community of origin between that enigmatical member of our system and the corona was irresistibly suggested. We should, indeed, expect to see, under such exceptionally favourable atmospheric conditions as Professor Langley enjoyed on Pike's Peak, the *roots* of the zodiacal light presenting near the sun just such an appearance as he witnessed; but we can imagine no reason why their visibility should be associated with a low state of solar activity. Nevertheless this seems to be the case with the streamers which astonished astronomers in 1878. Once before, in August 1867, similar emanations had been described and depicted by Grosch [2] of the Santiago Observatory; and then, too, sun-spots were at a minimum. Moreover, they were seen combined with the same symptoms of polar excitement visible eleven years later. The reality of the presumed connection will be solidly established should the peculiar corona of 1867 and 1878 reappear in 1889.

An alternative explanation was offered by the meteoric hypothesis. Professor Cleveland Abbe was fully persuaded that the long rays carefully observed by him were nothing else than streams of meteorites rushing towards or from perihelion; and it is quite certain that the solar neighbourhood must be crowded with such bodies. But there are no grounds for supposing that they affect the ecliptic more than any other of the infinite number of planes passing through the sun's centre. On the contrary, everything we know leads us to believe that meteorites, like their cometary allies, yield no obedience to the *rules of the road* which bind the planets, but travel in either direction indifferently, and in paths inclined at any angle to the fundamental plane of our system. Besides, the peculiar

---

[1] *Wash. Obs.*, 1876, App. iii. p. 209.   [2] *Astr. Nach.*, No. 1737.

structure at the base of the streamers displayed in the photographs, the curved rays meeting in pointed arches like Gothic windows, the visible upspringing tendency, the filamentous texture, speak unmistakably of the action of forces proceeding *from* the sun, not of extraneous matter circling round him.

Again, it may be asked what possible relation can exist between the zodiacal-plane and the sun's internal activity? For it is a remakable fact that to this approximately, and not to the level of the solar equator, the streamers conformed. We are acquainted with no such relation; but it may be remarked that the coronal axis of symmetry has frequently been observed during eclipses to be inclined at an appreciable angle to the solar axis of rotation, and the corresponding "magnetic equator" might quite conceivably be the scene of emanations induced by some form of electrical repulsion.

The surest, though not the most striking, proof of sympathetic change in the corona is afforded by the analysis of its light. In 1878 the bright lines so conspicuous in the coronal spectrum in 1870 and 1871 were discovered to have faded to the very limits of visibility. Several skilled observers failed to see them at all; but Young and Eastman succeeded in tracing both the hydrogen and the green "1474" rays all round the sun, to a height estimated at 340,000 miles. The substances emitting them were thus present, though in a low state of incandescence. The continuous spectrum was relatively strong; a faint reflection of the Fraunhofer lines was traced in it; and polarisation was undoubted, increasing towards the limb, whereas in 1870 it reached a maximum at a considerable distance from it. Experiments with Edison's tasimeter showed that the corona radiates a sensible amount of heat.

The next promising eclipse occurred May 17, 1882. The concourse of astronomers which has become usual on such occasions assembled this time at Sohag, in Upper Egypt. Rarely have seventy-four seconds been turned to such account. To each observer a special task was assigned, and the advantages of a strict division of labour were visible in the variety and amount of the information gained.

The year 1882 was one of numerous sun-spots. On the eve of the eclipse twenty-three separate maculæ were counted. If there were any truth in the theory which connected coronal forms with fluctuations in solar activity, it might be anticipated that the vast ecliptical expansions and polar "brushes" of 1878 would be found replaced by the star-like structure of 1871. This expectation was literally fulfilled. No zodiacal streamers were to be seen. The universal failure to perceive them, after express search in a sky of the most transparent purity, justifies the emphatic assertion that *they were not there.* Instead, the type of corona observed in India eleven years earlier was reproduced, with its shining aigrettes, complex texture, and brilliant decorative effect.

Concordant testimony was given by the spectroscope. The reflected light derived from the corona was weaker than in 1878, while its original emissions were proportionately intensified. A number of new bright lines were discovered. Tacchini determined four in the red end of the spectrum; Thollon perceived several in the violet; and Dr. Schuster measured and photographed about thirty.[1] The Fraunhofer lines autographically recorded in the continuous spectrum were not less numerous. This was the first successful attempt to photograph the spectrum of the corona as seen with an ordinary slit-spectroscope. The slitless spectroscope, or " prismatic camera," although its statements are necessarily of a far looser character, was, however, also profitably employed. The uncommon strength in the chromospheric regions of the violet light concentrated in the two lines H and K, attributed to calcium, was strikingly brought out by it; and Dr. Schuster, using plates sensitised in the infrared by Captain Abney's newly invented process, obtained an annular impression of the solar nimbus, probably corresponding to an invisible red hydrocarbon band made known by Captain Abney's researches.[2]

[1] *Proc. Roy. Soc.*, vol. xxxv. p. 154.   [2] *Observatory*, vol. v. p. 209.

Dr. Schuster's photographs of the corona itself were the most extensive, as well as the most detailed, of any yet secured.   One rift imprinted itself on the plates to a distance of nearly a diameter and a half from the limb ; and the transparency of the streamers was shown by the delineation through them of the delicate tracery beyond.   The singular and picturesque feature was added of a bright comet, self-depicted in all the exquisite grace of swift movement betrayed by the fine curve of its tail, hurrying away from, possibly, its only visit to our sun, and rendered momentarily visible by the withdrawal of the splendour in which it had been, and was again quickly veiled.

From a careful study of these valuable records Dr. Huggins derived the idea of a possible mode of photographing the corona *without an eclipse.*[1]   As already stated, its ordinary invisibility is entirely due to the "glare" or reflected light diffused through our atmosphere.   But Dr. Huggins found, on examining Schuster's negatives, that a large proportion of the light in the coronal spectrum, both continuous and interrupted, is collected in the violet region between the Fraunhofer lines G and H.   There, then, he hoped that, all other rays being excluded, it might prove strong enough to vanquish inimical glare, and stamp on prepared plates, through *local* superiority in illuminative power, the forms of the appendage by which it is emitted.

His experiments were begun towards the end of May 1882, and by September 28 he had obtained a fair earnest of success. The exclusion of all other qualities of light save that with which he desired to operate, was at first effected by the interposition of screens of purple glass, or other similarly absorbing media ; later, however, his purpose was more simply and efficaciously realised by using chloride of silver as his sensitive material, that substance being chemically inert to all other but

[1] *Proc. Roy. Soc.,* vol. xxxiv. p. 409.   Experiments directed to the same end had been made by Dr. O. Lohse at Potsdam, 1878–80 ; not without some faint promise of ultimate success.   *Astr. Nach.,* No. 2486.

those precise rays in which the corona has the advantage.[1] Of the genuineness of the impressions left upon his plates there can be no question. Their satisfactory agreement with the Egyptian photographs fully attest the truth of their pretensions as coronal autographs. "Not only the general features," Captain Abney bore them witness,[2] "are the same, but details, such as rifts and streamers, have the same position and form." It was found, moreover, that the corona photographed during the total eclipse of May 6, 1883, was intermediate in shape between the coronas photographed by Dr. Huggins before and after that event, each picture taking its proper place in a series of progressive modifications highly interesting in themselves, and emphatic in their testimony to the value of the method employed to record them. In this climate, however, and near the sea-level, it can never be brought to the perfection of which it gives promise.

The prosperous result of the Sohag observations stimulated the desire to repeat them on the first favourable opportunity. This offered itself one year later, May 6, 1883, yet not without the drawbacks incident to terrestrial conditions. The eclipse promised was of rare length, giving no less than five minutes and twenty-three seconds of total obscurity, but its path was almost exclusively a "water-track." It touched land only on the outskirts of the Marquesas group in the Southern Pacific, and presented, as the one available foothold for observers, a coral reef named Caroline Island, seven and a half miles long by one and a half wide, unknown previous to 1874, and visited only for the sake of its stores of guano. Seldom has a more striking proof been given of the vividness of human curiosity as to the condition of the worlds outside our own, than in the assemblage of a group of distinguished men from the chief centres of civilisation, on a barren ridge, isolated in a vast and

---

[1] The sensitiveness of chloride of silver extends from h to H; that is, over the upper or more refrangible half of the space in which the main part of the coronal light is concentrated.

[2] *Proc. Roy. Soc.*, vol. xxxiv. p. 414.

tempestuous ocean, at a distance, in many cases, of 11,000 miles and upwards from the ordinary scene of their labours. And all these sacrifices—the cost and care of preparation, the transport and readjustment of delicate instruments, the contrivance of new and more subtle means of investigating phenomena—on the precarious chance of a clear sky during one particular five minutes! The event, though fortunate, emphasised the hazard of the venture. The observation of the eclipse was made possible only by the happy accident of a serene interval between two storms.

The American expedition was led by Professor Edward S. Holden, and to it were courteously permitted to be attached Messrs. Lawrance and Woods, photographers, sent out by the Royal Society of London.   M. Janssen was chief of the French Academy mission; he was accompanied from Meudon by Trouvelot, and joined from Vienna by Palisa, and from Rome by Tacchini.   A large share of the work done was directed to assuring or negativing previous results.   The circumstances of an eclipse favour illusion.   A single observation by a single observer, made under unfamiliar conditions, and at a moment of peculiar excitement, can scarcely be regarded as offering more than a suggestion for future inquiry.   But incredulity may be carried too far.   Janssen, for instance, felt compelled by the survival of unwise doubts, to devote some of the precious minutes of obscurity at Caroline Island to confirming what, in his own persuasion, needed no confirmation—that is, the presence of reflected Fraunhofer lines in the spectrum of the corona.   Trouvelot and Palisa, on the other hand, instituted an exhaustive, but fruitless search for the spurious "intra-mercurial" planet announced by Swift and Watson in 1878.

New information, however, was not deficient.   The corona proved identical in type with that of 1882, agreeably to what was expected at an epoch of protracted solar activity.   The characteristic aigrettes (of which five appeared in Mr. Dixon's sketch) were of even greater brilliancy than in the preceding year, and the chemical intensity of the coronal light—then first measured with some precision—was found to exceed that of full moonlight. Janssen's photographs, owing to the considerable apertures (six and eight inches) of his object-glasses, and the long exposures permitted by the duration of totality, were singularly perfect; they gave a greater extension to the corona than could be traced with the telescope,[1] and showed its forms as absolutely fixed and of remarkable complexity.

The English pictures, taken with exposures up to sixty seconds, were likewise of great value. They exhibited details of structure from the limb to the tips of the streamers, which terminated definitely, and as it seemed actually, where the impressions on the plates ceased. The coronal spectrum was also successfully photographed, with a number of bright and dark lines; and a print was caught of some of the more prominent rays of the reversing layer just before and after totality. The use of the prismatic camera was baffled by the anomalous scarcity of prominences.

A highly suggestive observation was made during this eclipse by Professor Tacchini. One of the aigrettes of the corona displayed in his spectroscope, on a feebly continuous background, two of the bright bands familiar in the hydrocarbon spectrum of comets.[2] This requires confirmation; nevertheless, the analogy which it hints at is a tempting one. The resemblance of the silvery sheaves of the corona to the tails of comets had already given rise to much fruitless speculation; and the exertion of a repulsive force, such as is obviously at work in comets, by the sun on his surroundings, has been considered, by some solar physicists, absolutely necessary to explain the lowness of atmospheric pressure at his surface. The presence of carbon in the sun's atmosphere was inferred by Mr. Lockyer in 1878 from a comparison of photographs of the solar and electric-arc spectra;[3] Dr. Schuster, as has been mentioned, obtained indications of the same kind in 1882; and Captain Abney finds hydrocarbon bands in the

---

[1] *Comptes Rendus*, t. xcvii. p. 592.   [2] *Ibid.*, p. 594.
[3] *Proc. Roy. Soc.*, vol. xxvii. p. 308.

invisible or infra-red part of the Fraunhofer spectrum.[1] But the subject needs to be further investigated.

Another of the observations made at Caroline Island, although probably through some unexplained cause delusive, merits some brief notice. Using an ingenious apparatus for viewing simultaneously the spectrum from both sides of the sun, Professor Hastings saw (as he supposed), certain alter-nations, with the advance of the moon, in the respective heights above the right and left solar limbs of the coronal line " 1474," which were thought to imply an unexpected strength of diffusive action in our atmosphere. If this were true, then spectroscopic evidence as to the *extent* of the sun's gaseous surroundings should at once be discarded as misleading; but the simple consideration that if diffusion caused the observed effect, it should extend bright lines *across* the disc of the moon no less than on either side of it, suffices to show the fallacy of the inference.

The controversy is an old one as to the part played by our air in producing the radiance visible round the eclipsed sun. In its original form, it is true, it came to an end when Pro-fessor Harkness, in 1869,[2] pointed out that the shadow of the moon falls equally over the air and on the earth, and that if the sun had no luminous appendages, a circular space of almost absolute darkness would consequently surround the apparent places of the superposed sun and moon. Mr. Proc-tor,[3] with his usual ability, impressed this mathematically cer-tain truth (the precise opposite of the popular notion) upon public attention; and Sir John Herschel calculated that the diameter of the " negative halo" thus produced would be, in general, no less than 23°.

But about the same time a noteworthy circumstance relat-ing to the state of things in the solar vicinity was brought into view. On February 11, 1869, Messrs. Frankland and Lockyer communicated to the Royal Society a series of experi-

---

[1] *Report Brit. Ass.*, 1881, p. 524.   [2] *Wash. Obs.*, 1867, App. ii. p. 64.
[3] *The Sun*, p. 357.

ments on gaseous spectra under varying conditions of heat and density, leading them to the conclusion that the higher solar prominences exist in a medium of excessive tenuity, and that even at the base of the chromosphere the pressure is far below that at the earth's surface.[1] This inference was fully borne out by the researches of Wüllner; and Janssen expressed the opinion that the chromospheric gases are rarefied almost to the degree of an air-pump vacuum.[2] Hence was derived a general and fully justified conviction that there could be outside, and incumbent upon the chromosphere no such vast atmosphere as the corona appeared to represent. Upon the strength of which conviction the "glare" theory entered, chiefly under the auspices of Mr. Lockyer, upon the second stage of its existence.

The genuineness of the "inner corona" to a height of 5' or 6' from the limb was admitted; but it was supposed that by the detailed reflection of its light in our air the far more extensive "outer corona" was optically created, the irregularities of the moon's edge being called in to account for the rays and rifts by which its structure was varied. This view received some countenance from Maclear's observation, during the eclipse of 1870, of bright lines "everywhere"—even at the centre of the lunar disc. Here, indeed, was an undoubted case of atmospheric diffusion; but here, also, was a safe index to the extent of its occurrence. Light scatters equally in all directions; so that when the moon's face at the time of an eclipse shows (as is the common case) a blank in the spectroscope, it is quite certain that the corona is not noticeably enlarged by atmospheric causes. A sky drifted over with thin cirrous clouds and air charged with aqueous vapour amply accounted for the abnormal amount of scattering in 1870.

But even in 1870 positive evidence was obtained of the substantial reality of the radiated outer corona, in the appearance on the photographic plates exposed by Willard in Spain and by Brothers in Sicily, of identical dark rifts. The truth is,

[1] *Proc. Roy. Soc.*, vol. xvii. p. 289.   [2] *Comptes Rendus*, t. lxxiii. p. 434.

that far from being developed by misty air, it is peculiarly liable
to be effaced by it.   The purer the sky, the more extensive,
brilliant, and intricate in the details of its structure the corona
appears.   Take as an example General Myer's description of
the eclipse of 1869, as seen from the summit of White Top
Mountain, Virginia, at an elevation above the sea of 5530 feet,
in an atmosphere of peculiar clearness.

"To the unaided eye," he wrote,[1] "the eclipse presented,
during the total obscuration, a vision magnificent beyond
description.   As a centre stood the full and intensely black
disc of the moon, surrounded by the aureola of a soft bright
light, through which shot out, as if from the circumference of
the moon, straight, massive, silvery rays, seeming distinct and
separate from each other, to a distance of two or three diameters of the solar disc; the whole spectacle showing as on a
background of diffused rose-coloured light."

On the same day, at Des Moines, Newcomb could perceive,
through somewhat hazy air, no long rays, and the four-pointed
outline of the corona reached at its farthest only *a single
semi-diameter* of the moon from the limb.   The plain fact,
that our atmosphere acts rather as a veil to hide the coronal
radiance than as the medium through which it is visually
formed, emerges from the records of innumerable other observations.

Summing up what we have learned about the corona during
some forty minutes of scrutiny in as many years, we may state,
to begin with, that it is *not* a solar atmosphere.   It does not
gravitate upon the sun's surface and share his rotation, as our
air gravitates upon and shares the rotation of the earth; and
this for the simple reason that there is no visible growth of
pressure downwards (such as the spectroscope would infallibly
give notice of) in its gaseous constituents; whereas under the
sole influence of the sun's attractive power, their density should
be multiplied many million times in the descent through a mere
fraction of their actual depth.

[1] *Wash. Obs.*, 1867, App. ii. p. 195.

The corona is properly described as a solar appendage ; and may be conjecturally defined as matter in a perpetual state of efflux from, and influx to our great luminary, under the stress of electrical repulsion in one direction and of gravity in the other.[1] Its constitution is of a composite character. It is partly made up of self-luminous gases, chiefly hydrogen, and the unknown substance giving the green ray " 1474 ; " partly of solid or liquid particles, seen by reflected sunlight. There is a strong probability that it is affected by the periodic ebb and flow of solar activity, the rays emitted by the gases contained in it fading, and the continuous spectrum brightening, at times of minimum sun-spots, as if by a fall of temperature producing, on the one hand, a decline in luminosity of the incandescent materials existing near the sun, and, on the other, a condensation of vapours previously invisible into compact particles of some reflective capacity.

The most important lesson, however, derived from eclipses is that of independence of them. Some of its fruits in the daily study of prominences the next chapter will collect ; while the attainment, through Dr. Huggins's photographic method, of a corresponding power as regards the corona, may be expected to mark an epoch in the investigation of that still problematical phenomenon.

---

[1] Professor W. A. Norton, of Yale College, appears to have been the earliest formal advocate of the Expulsion Theory of the solar surroundings, in the second (1845) and later editions of his *Treatise on Astronomy.*

CHAPTER IV.

SPECTROSCOPIC WORK ON THE SUN.

THE new way struck out by Janssen and Lockyer was at once and eagerly followed. In every part of Europe, as well as in North America, observers devoted themselves to the daily study of the chromosphere and prominences. Foremost among these were Lockyer in England, Zöllner at Leipzig, Spörer at Anclam, Young at Hanover, New Hampshire, Secchi and Respighi at Rome. There were many others, but these names are conspicuous from the outset.

The first point to be cleared up was that of chemical composition. Leisurely measurements verified the presence above the sun's surface of hydrogen in prodigious masses, but showed that sodium had nothing to do with the orange-yellow ray identified with it in the haste of the eclipse. From its vicinity to the D pair (than which it is slightly more refrangible), the prominence-line was, however, designated D3, and the unknown substance emitting it was named by Frankland "helium." Young is inclined to associate with it two other faint but persistent lines in the spectrum of the chromosphere;[1] and Messrs. Liveing and Dewar pointed out, in 1879,[2] that the wave-lengths of all three are bound together with that of the coronal ray " 1474 " by numerical ratios virtually the same with those underlying the vibrations of hydrogen, and also conformed to by certain lines of lithium and magnesium. This obscure but interesting subject deserves further

[1] *Phil. Mag.*, vol. xlii. 1871, p. 380.
[2] *Proc. Roy. Soc.*, vol. xxviii. p. 475.

inquiry. It should be added that Mr. Lockyer attributes both the D3 and "1474" lines to a modification of hydrogen; but the actual relation would seem to be one of analogy rather than of identity.

Hydrogen and helium form the chief and unvarying materials of the solar sierra and its peaks; but a number of metallic elements make their appearance spasmodically under the influence of disturbances in the layers beneath. In September 1871, Young[1] drew up at Dartmouth College a list of 103 lines significant of injections into the chromosphere of iron, titanium, calcium, magnesium, and many other substances. During two months' observation in the pure air of Mount Sherman (8335 feet high) in the summer of 1872, these tell-tale lines mounted up to 273; and he believes their number might still be doubled by steady watching. Indeed, both Young and Lockyer have more than once seen the whole field of the spectroscope momentarily inundated with bright rays, as if the "reversing layer" seen at the beginning and end of eclipses had been suddenly thrust upwards into the chromosphere, and as quickly allowed to drop back again. It would thus appear that the two form one continuous region, of which the lower parts are habitually occupied by the heaviest vapours, but where orderly arrangement is continually overturned by violent eruptive disturbances.

The study of the *forms* of prominences practically began with Dr. Huggins's observation of one through an "open slit," February 13, 1869.[2] At first it had been thought possible to study them only in sections—that is, by admitting mere narrow strips or "lines" of their various kinds of light; while the actual shape of the objects emitting those lines had been arrived at by such imperfect devices as that of giving to the slit of the spectroscope a vibratory movement rapid enough to enable the eye to retain the impression of one part while others were successively presented to it. It was an immense gain to find their rays strong enough to bear so much of dilu-

[1] *Phil. Mag.*, vol. xlii. p. 377.　[2] *Proc. Roy. Soc.*, vol. xvii. p. 302.

tion with ordinary light as was involved in opening the spectroscopic shutter wide enough to exhibit the tree-like, or horn-
like, or flame-shaped bodies rising over the sun's rim in their
undivided proportions.   Three images of each prominence
are formed in the spectroscope—a crimson, a deep yellow, and
a bluish green.   The crimson, however (built up out of the C
line of hydrogen), is the most intense, and is commonly used
for purposes of observation and illustration.

Friedrich Zöllner was, by a few days, beforehand with
Huggins in describing the open-slit method, but was somewhat less prompt in applying it.   His first survey of a complete
prominence, pictured in, and not simply intersected by, the
slit of his spectroscope, was obtained July 1, 1869.[1]   Shortly
afterwards the plan was successfully adopted by the whole
band of investigators.

A difference in kind was very soon perceived to separate
these objects into two well-marked classes.   Its natural and
obvious character was shown by its having struck several
observers independently.   The distinction of "cloud-prominences" from "flame-prominences" was announced by Lockyer,
April 27, by Zöllner, June 2, and by Respighi, December 4,
1870.

The first description is tranquil and relatively permanent,
sometimes enduring without striking change for many days.
They mimic terrestrial cloud-scenery—now appearing like
fleecy cirrus transpenetrated with the red glow of sunset—
now like prodigious masses of cumulo-stratus hanging heavily
above the horizon.   These solar clouds, however, have the
peculiarity of possessing *stems*.   Slender columns can ordinarily
be seen to connect the surface of the chromosphere with its
outlying portions.   Hence the fantastic likeness to forest
scenery presented by the long ranges of fiery trunks and
foliage at times seeming to fringe the sun's limb.   But while
this formation suggests an actual outpouring of incandescent
material, certain facts require a different interpretation.   At

[1] *Astr. Nach.*, No. 1769.

a distance, and quite apart from the chromosphere, prominences have been perceived, both by Secchi and Young, to *form*, just as clouds form in a clear sky, condensation being replaced by ignition. Filaments were then thrown out downwards towards the chromosphere, and finally the usual appearance of a "stemmed prominence" was assumed. Still more remarkable was an observation made by Trouvelot at Harvard College Observatory, June 26, 1874.[1] A gigantic comma-shaped prominence, 82,000 miles high, vanished from before his eyes by a withdrawal of light as sudden as the passage of a flash of lightning. The same observer has frequently witnessed a gradual illumination or gradual extinction of such objects, testifying to changes in the thermal or electrical condition of matter already *in situ*.

The chemistry of "cloud-prominences" is very simple. Hydrogen and helium are their only constituents. "Flame-prominences," on the other hand, show, in addition, the characteristic rays of a number of metals, amongst which iron, titanium, barium, sodium, and magnesium are conspicuous. They are intensely brilliant; sharply defined in their varying forms of jets, spikes, fountains, waterspouts; of rapid formation and speedy dissolution, seldom attaining to the vast dimensions of the more tranquil kind. They are visibly of eruptive origin, and are closely connected with spots; the materials ejected as "flames" cooling and settling down, according to Father Secchi,[2] as dark, depressed patches of increased absorption. The two classes of phenomena, at any rate, stand in a most intimate relation; they obey the same law of periodicity, and are confined to the same portions of the sun's surface, while quiescent prominences may be found right up to the poles and close to the equator.

The general distribution of prominences, including both species, follows that of faculæ much more closely than that of spots. From Father Secchi and Professor Respighi's observations, 1869–71, were derived the first clear ideas on the subject,

---

[1] *Am. Jour. of Sc.*, vol. xv. p. 85.   [2] *Le Soleil*, t. ii. p. 294.

which have been supplemented and modified by the later researches of Professors Tacchini and Riccò at Rome and Palermo. The results are somewhat complicated, but may be stated broadly as follows. The district of greatest prominence-frequency covers and overlaps by several degrees that of greatest spot-frequency. That is to say, it extends to about 40° north and south of the equator.[1] There is a visible tendency to a second pair of maxima nearer the poles. The poles themselves, as well as the equator, are regions of minimum occurrence. Distribution in time is governed by the spot-cycle, but the maximum lasts longer for prominences than for spots.

The structure of the chromosphere was investigated in 1869 and subsequent years by Professor Respighi, director of the Capitoline Observatory, as well as by Spörer, and by Bredichin of the Moscow Observatory. They found this supposed solar envelope to be of the same eruptive nature as the vast protrusions from it, and to be made up of a congeries of minute flames[2] set close together like blades of grass. " The appearance," Professor Young writes,[3] "which probably indicates a fact, is as if countless jets of heated gas were issuing through vents and spiracles over the whole surface, thus clothing it with flame which heaves and tosses like the blaze of a conflagration."

The summits of these filaments of fire are commonly inclined, as if by a wind sweeping over them, when the sun's activity is near its height, but erect during his phase of tranquillity. Spörer, in 1871, inferred the influence of permanent polar currents,[4] but Tacchini showed in 1876 that the deflections upon which this inference was based, ceased to be visible as the spot-minimum drew near.[5]

Another peculiarity of the chromosphere, denoting the re-

[1] L'Astronomie, August 1884, p. 292 (Riccò).
[2] Averaging about 100 miles across and 300 high. Le Soleil, t. ii. p. 35.
[3] The Sun, p. 180.   [4] Astr. Nach., No. 1854.
[5] Mem. degli Spettroscopisti Italiani, t. v. p. 4. Restated by Secchi, Ibid., t. vi. p. 56.

moteness of its character from that of a true atmosphere,[1] is the irregularity of its distribution over the sun's surface. There are no signs of its bulging out at the equator, as the laws of fluid equilibrium in a rotating mass would require ; but there are some that the fluctuations in its depth are connected with the phases of solar agitation. At times of minimum it seems to accumulate and concentrate its activity at the poles ; while maxima probably bring a more equable general distribution, with local depressions at the base of great prominences and above spots.

The reality of the appearance of violent disturbance presented by the " flaming " kind of prominence can be tested in a very remarkable manner. Christian Doppler,[2] professor of mathematics at Prague, enounced in 1842 the theorem that the colour of a luminous body, like the pitch of a sonorous body, must be changed by movements of approach or recession. The reason is this. Both colour and pitch are physiological effects, depending, not upon absolute wave-length, but upon the number of waves entering the eye or ear in a given interval of time. And this number, it is easy to see, must be increased if the source of light or sound is diminishing its distance, and diminished if it is increasing it. In the one case, the vibrating body *pursues* and crowds together the waves emanating from it ; in the other, it *retreats* from them, and so lengthens out the space covered by an identical number. The principle may be thus illustrated. Suppose shots to be fired at a target at fixed intervals of time. If the marksman advances, say twenty paces between each discharge of his rifle, it is evident that the shots will fall faster on the target than if he stood still ; if, on the contrary, he retires by the same amount, they will strike at correspondingly longer intervals. The result will of course be the same whether the target or the marksman be in movement.

[1] Its non-atmospheric character was early defined by Proctor, *Month. Not.*, vol. xxxi. p. 196.

[2] *Abh. d. Kön. Böhm. Ges. d. Wiss.*, Bd. ii. 1841–42, p. 467.

So far Doppler was altogether right.   As regards sound, any
one can convince himself that the effect he predicted is a real
one, by listening to the alternate shrilling and sinking of the
steam-whistle when an express train rushes through a station.
But in applying this principle to the colours of stars he went
widely astray ; for he omitted from consideration the double
range of invisible vibrations which ¸ partake of, and to the eye
exactly compensate, changes of refrangibility in the visible rays.
There is, then, no possibility of finding a criterion of velocity
in the hue of bodies shining, like the sun and stars, with continuous light.   There is a slight shift of the entire spectrum
up or down in the scale of refrangibility ; certain rays normally
visible become exalted or degraded (as the case may be) into
invisibility, and certain other rays at the opposite end undergo
the converse process ; but the sum-total of impressions on the
retina continues the same.

We are not, however, without the means of measuring this
sub-sensible transportation of the light-gamut.   Once more
the wonderful Fraunhofer lines came to the rescue.   They
were called by the earlier physicists " fixed lines ; " but it is
just because they are *not* fixed that, in this instance, we find
them useful.   They share, and in sharing betray, the general
shift of the spectrum.   This aspect of Doppler's principle was
adverted to by Fizeau in 1848,[1] and the first tangible results
in the estimation of movements of approach and recession
between the earth and the stars, were communicated by Dr.
Huggins to the Royal Society, April 23, 1868.   Eighteen
months later, Zöllner devised his " reversion-spectroscope " [2] for
doubling the measurable effects of line-displacements ; aided by
which ingenious instrument, and following a suggestion of its
inventor, Professor H. C. Vogel succeeded at Bothkamp, June 9
1871,[3] in detecting effects of that nature due to the solar rota

---

[1] In a paper read before the Société Philomathique de Paris, December
23, 1848, and first published *in extenso* in *Ann. de Chim. et de Phys.*, t. xix.
p. 211 (1870).

[2] *Astr. Nach.*, No. 1772.   [3] *Ibid.*, No. 1864.

tion. This application constitutes at once the test and the triumph of the method.

The eastern edge of the sun is continually moving towards us with an equatorial speed of about a mile and a quarter per second, the western edge retreating at the same rate. The displacements—towards the violet on the east, towards the red on the west—corresponding to this velocity are very small; so small that it seems hardly credible that they should have been laid bare to perception. They amount to but $\frac{1}{150}$th part of the interval between the two constituents of the D line of sodium; and the D line of sodium itself can be separated into a pair only by a powerful spectroscope. Nevertheless, Professor Young[1] was able to show quite satisfactorily, in 1876, not only deviations in the solar lines from their proper places indicating a velocity of rotation (1.42 miles per second) slightly in excess of that given by observations of spots, but the exemption of terrestrial lines (those produced by absorption in the earth's atmosphere) from the general push upwards or downwards. Shortly afterwards, Professor Langley, director of the Allegheny Observatory, having devised a means of comparing with great accuracy light from different portions of the sun's disc, found that while the obscure rays in two juxtaposed spectra derived from the solar poles were absolutely continuous, no sooner was the instrument rotated through ninety degrees, so as to bring its luminous supplies from opposite extremities of the equator, than the same rays became perceptibly "notched." The telluric lines, meanwhile, remained unaffected, so as to be "virtually mapped" by the process.[2] This rapid and unfailing mode of distinction was used by Cornu with perfect ease during his investigation of atmospheric absorption near Loiret in August and September 1883.[3]

A beautiful experiment of the same kind was performed by M. Thollon, of M. Bischoffsheim's observatory at Nice, in the summer of 1880.[4] He confined his attention to one delicately

---

[1] *Am. Jour. of Sc.*, vol. xii. p. 321.   [2] *Ibid.*, vol. xiv. p. 140.
[3] *Bull. Astronom.*, Feb. 1884, p. 77.   [4] *Comptes Rendus*, t. xci. p. 368.

defined group of four lines in the orange, of which the inner
pair are solar (iron) and the outer terrestrial. At the centre
of the sun the intervals separating them were sensibly equal;
but when the light was taken alternately from the right and
left limbs, a relative shift in alternate directions of the solar,
towards and from the stationary telluric rays became apparent.
This amounts to a demonstration that results of this kind are
worthy of confidence; and since they are, in certain cases, such
as to startle it, it is important to make sure of their foundations.

Mr. Lockyer[1] was the first to perceive the applicability of
this subtle and surprising discovery to the study of prominences,
the discontinuous light of which affords precisely the same
means of detecting movement without seeming change of
place, as do lines of absorption in a continuous spectrum.
Indeed, his observations at the sun's edge almost compelled
him to have recourse to an explanation made available just
when the need of it began to be felt. He saw bright lines, not
merely pushed aside from their normal places by a barely
perceptible amount, but bent, torn, broken, as if by the stress
of some tremendous violence. These remarkable appearances
were quite simply interpreted as the effects of movements
varying in amount and direction in the different parts of the
extensive mass of incandescent vapours falling within a single
field of view. Very commonly they are of a cyclonic character.
The opposite distortions of the same coloured rays betray the
fury of "counter-gales" rushing along at the rate of 120 miles
a second; while their undisturbed sections prove the persistence
of a "heart of peace" in the midst of that unimaginable fiery
whirlwind. Velocities up to 250 *miles a second*, or 15,000
times that of an express train at the top of its speed, were
thus observed by Young during his trip to Mount Sherman,
August 3, 1872.

Motions ascertainable in this way near the limb are, of
course, horizontal as regards the sun's surface; the analogies

[1] *Proc. Roy. Soc.*, vols. xvii. p. 415; xviii. p. 120.

they present might, accordingly, be styled *meteorological* rather than *volcanic.* But vertical displacements on a scale no less stupendous can also be shown to exist. Observations of the spectra of spots centrally situated (where motions in the line of sight are vertical) disclose the progress of violent uprushes and downrushes of ignited gases, for the most part in the penumbral or outlying districts. They appear to be occasioned by fitful and irregular disturbances, and have none of the systematic quality which would be required for the elucidation of sun-spot theories. Indeed, they almost certainly take place at a great height above the actual opening in the photosphere.

As to vertical motions above the limb, on the other hand, we have direct visual evidence of a truly amazing kind. The projected glowing matter has, by the aid of the spectroscope, been watched in transit. On September 7, 1871, Young examined at noon a vast hydrogen-cloud, 100,000 miles long, as it showed to the eye, and 54,000 high. It floated tranquilly above the chromosphere at an elevation of some 15,000 miles, and was connected with it by three or four upright columns, presenting the not uncommon aspect compared by Lockyer to that of a grove of banyans. Called away for a few minutes at 12.30, on returning at 12.55 the observer found—

"That in the meantime the whole thing had been literally blown to shreds by some inconceivable uprush from beneath. In place of the quiet cloud I had left, the air, if I may use the expression, was filled with flying débris—a mass of detached, vertical, fusiform filaments, each from 10″ to 30″ long by 2″ or 3″ wide,[1] brighter and closer together where the pillars had formerly stood, and rapidly ascending. They rose, with a velocity estimated at 166 miles a second, to fully 200,000 miles above the sun's surface, then gradually faded away like a dissolving cloud, and at 1.15 only a few filmy wisps, with some brighter streamers low down near the photosphere, remained to mark the place."[2]

[1] At the sun's distance, one second of arc represents about 450 miles.
[2] *Am. Jour. of Sc.*, vol. ii. 1871, p. 468.

A velocity of projection of *at least* 500 miles per second has been calculated by Proctor [1] to be necessary in order to account for this extraordinary display. It was marked by the simultaneous record at Greenwich of a magnetic disturbance, and was succeeded, the same evening, by a fine aurora. It has proved by no means an isolated occurrence. Young saw its main features repeated, October 7, 1881,[2] on a still vaster scale; for the exploded prominence attained, this time, an altitude of 350,000 miles—the highest yet chronicled. Mr. Lockyer, moreover, has seen a prominence 40,000 miles high blown to pieces in ten minutes; while uprushes have been witnessed by Respighi, of which the initial velocities were judged by him to be 400 or 500 miles a second. When it is remembered that a body starting from the sun's surface at the rate of 379 miles a second would, if it encountered no resistance, escape for ever from his control, it is obvious that we have, in the enormous forces of eruption or repulsion manifested in the outbursts just described, the means of accounting for the vast diffusion of matter in the solar neighbourhood. Nor is it possible to explain them away, as Cornu,[3] Faye,[4] and others have sought to do, by substituting for the rush of matter in motion, progressive illumination through electric discharges, or even through the mere reheating of gases cooled by expansion.[5] All the appearances are against such evasions of the difficulty presented by velocities stigmatised as "fabulous" and "improbable," but which, there is the strongest reason to believe, really exist.

On the 12th of December 1878, Mr. Lockyer formally expounded before the Royal Society his now famous hypothesis of the compound nature of the "chemical elements."[6] He was led to it by several converging lines of research. In a

[1] *Month. Not.*, vol. xxxii. p. 51.   [2] *Nature*, vol. xxiii. p. 281.
[3] *Comptes Rendus*, t. lxxxvi. p. 532.   [4] *Ibid.*, t. xcvi. p. 359.
[5] Such prominences as have been seen to grow by the spread of incandescence are of the quiescent kind, and present no deceptive appearance of violent motion.
[6] *Proc. Roy. Soc.*, vol. xxviii. p. 157.

letter to M. Dumas, dated December 3, 1873, he had sketched out the successive stages of "celestial dissociation" which he conceived to be represented in the sun and stars. The absence from the solar spectrum of metalloidal absorption he explained by the separation, in the fierce solar furnace, of such substances as oxygen, nitrogen, sulphur, carbon, &c., into simpler constituents possessing unknown spectra; while metals were at that time still admitted to be capable of existing there in a state of integrity. Three years later he made a further step. He announced, as the result of a comparative study of the Fraunhofer and electric-arc spectra of calcium, that the "molecular grouping" of that metal, which at low temperatures gives a spectrum with its chief line in the blue, is nearly broken up in the sun into another or others with lines in the violet.[1] The further progress of his work showed him this discrepancy between solar and terrestrial spectra as no exception, but "a truly typical case."[2]

From 1875 onwards this unwearied student of nature was engaged in the construction of a map of the solar spectrum on a scale of magnitude such that, when completed down to the infra-red, it will be 315 feet, or about *half a furlong* in length. The attendant laborious investigation, by the aid of photography, of metallic spectra, afforded him the supposed discovery of "basic lines." These are lines occurring in the spectra of two or more metals after all possible "impurities" have been eliminated, and were held to attest the presence of a common substràtum of matter in a simpler state of aggregation than any with which we are ordinarily acquainted. Now it is a singular fact that these "basic lines" are precisely those which appear, with a persistence altogether out of proportion to their actual numbers, in the spectrum of the chromosphere

[1] *Proc. Roy. Soc.*, vol. xxiv. p. 353. The remarkable pair of lines in the violet (H and K) attributed to calcium stand urgently in need of being cleared up. Vogel discovered in 1879 a hydrogen line coincident with H (*Monatsb. Preuss. Ak.*, Feb. 1879, p. 115). Young attributes both H and K to that substance, on the ground of their anomalous behaviour in prominences (*Nature*, vol. xxiii. p. 281). [2] *Proc. Roy. Soc.*, vol. xxviii. p. 444.

when agitated by eruptive injections. The presence of iron, for example, instead of being signified by the flashing out of some of the strong representative lines which are the first to appear and the last to disappear in its laboratory-spectrum, makes itself known by the brightening of some inconspicuous ray, claimed, moreover, with an equal title, by (say) calcium or titanium. What more natural than to conclude, with Mr. Lockyer, that the erupted substance is not really iron at all, but some more elementary form of matter entering into the composition of iron as well as of calcium and titanium, the reduction having been brought about by the inconceivable heat of the sub-photospheric regions?

There is, nevertheless, a difficulty in accepting this plausible view. The foundation of fact upon which it rests is insecure. The lines called *basic* are probably not really identical, but only very closely coincident. They are formed of doublets or triplets merged together by insufficient dispersion. Out of Thalèn's original list of seventy rays common to several spectra,[1] only seven (besides about five which by their situation elude scrutiny) have so far resisted Thollon's and Young's powerful spectroscopes; and the process of resolution will almost certainly be carried farther. Thus the argument from community of lines to community of substance may be regarded as already half extinct. The circumstance, however, still requires explanation, that these twin-lines—these spots of rendezvous, it might be said, for different sets of vibrations— are specially selected for display in solar disturbances—are predominantly brightened in flames and thickened in spots.

But the really strong point of the "dissociation theory" has yet to be mentioned. It is that the contortions or displacements due to motion are frequently seen to affect a single line belonging to a particular substance, while the other lines of *that same substance* remain imperturbable. Now, how is this most singular fact, which seems at first sight to imply that a

---

[1] Many of these were shown by Mr. Lockyer, who was the first to sift the matter, to be due to very slight admixtures of the several metals concerned.

body may be at rest and in motion at one and the same instant, to be accounted for? It is accounted for, on Mr. Lockyer's hypothesis, easily enough, by supposing that the rays thus discrepant in their testimony, do *not* belong to one kind of matter, but to several, combined, at ordinary temperatures, to form a body in appearance "elementary." Of these different vapours, one or more may of course be rushing rapidly towards or from the observer, while the others remain still ; and since the line of sight across the average prominence region pene-trates, at the sun's edge, a depth of about 300,000 miles,[1] all the incandescent materials separately occurring along which line are projected into a single "flame" or "cloud," it will be perceived that there is ample room for diversities of behaviour.

The alternative mode of escape from the perplexity consists in assuming that the vapour in motion is rendered luminous under conditions which reduce its spectrum to one or two rays, the unaffected lines being derived from a totally distinct mass of the same substance shining with its ordinary emis-sions.[2] The supposition is by no means a violent one, since both hydrogen and nitrogen can readily be brought, in the laboratory, into the state of monochromatic radiation ; and even sodium has, by careful manipulation, been induced to give a spectrum from which the all but ubiquitous D line is missing.[3] The results to the eye would, on either supposi-tion, be the same.

Mr. Lockyer's view has the argument from continuity in its favour. It only asks us to believe that processes which we know to take place on the earth under certain conditions, are carried further in the sun, where the same conditions are, it may be presumed, vastly exalted. We find that the bodies we

---

[1] Thollon's estimate (*Comptes Rendus*, t. xcvii. p. 902) of 300,000 *kilo-metres* seems considerably too low. Limiting the "average prominence region" to a shell 54,000 miles deep (2' of arc as seen from the earth), the visual line will, at mid-height (27,000 miles from the sun's surface), travel through (in round numbers) 320,000 miles of that region.

[2] Liveing and Dewar, *Phil. Mag.*, vol. xvi. (5th ser.), p. 407.

[3] Lockyer, *Proc. Roy. Soc.*, vol. xxix. p. 140.

call "compound" split asunder at fixed degrees of heat *within* the range of our resources. Why should we hesitate to admit that the bodies we call "simple" do likewise at degrees of heat *without* the range of our resources? There is no intrinsic difference separating them. The term "element" simply expresses terrestrial incapability of reduction. That, in celestial laboratories, the means and their effect here absent should be present, would be an inference challenging, in itself, no expression of incredulity.

Yet there are grave objections to assent when the actual circumstances of the case are attentively considered. Of these objections we need at present advert to only one; but it is fundamental. Far from being a simplification, the hypothesis in question introduces an enormous complication into the workings of nature. We now recognise sixty-four "elements" provisionally so called; for no chemist supposes them to be essentially and *ab origine* distinct kinds of matter. But, if Mr. Lockyer's reasoning be admitted as valid, these sixty-four should be multiplied many times; for it asserts that each body known to us upon the earth is broken up in the sun into several constituents, and the evidence in favour of the "basic" nature of any of these constituents has, as we have seen, virtually collapsed. Thus hydrogen is "dissociated" into at least three separate substances entirely independent of any others, and the components of iron should be counted by the score. Nay, if the principle be admitted, which is the implied postulate underlying the arguments used, that a truly elementary body can radiate but one kind of light—gives, in other words, a spectrum of one bright line—it is difficult to stop short of the conclusion that each of the multitudinous coloured rays in the spectra of our sixty-four "elements" is the individual representative of a distinct species of matter.

There can be no doubt that the spectra of bodies are an index to changes in their molecular constitution of every kind and degree, from a complete disruption of the molecule into atoms, homogeneous or heterogeneous, to some unspeakably minute, yet orderly and harmonious rearrangement of parts in the complex little system of which the movements are the source of light. Mr. Lockyer's " working hypothesis " thus raises questions which science is not yet prepared to answer. It brings us face to face with the mysteries of the ultimate constitution of matter, and of its relations to the vibrating medium filling space. It makes our ignorance on the subject seem at once more dense and more definite. Nevertheless, this in itself (though the saying appear paradoxical) constitutes an advance and gives hope of progress. The mustering, drafting, and drilling of facts due to Mr. Lockyer's diligence, must in the end tell for truth, although their interpretation be for a time doubtful.

Professor A. J. Ångström of Upsala takes rank after Kirchhoff as a subordinate founder, so to speak, of solar spectroscopy. His anticipation of its fundamental principle (equivalence of emission and absorption) had, perhaps, scarcely the absolute character claimed for it ; but his work in the development of that principle was of extraordinary value. His great map of the " normal " solar spectrum [1] was published in 1868, two years before he died. Robert Thalèn was his coadjutor in its execution, and the immense labour which it cost was amply repaid by its eminent and lasting usefulness. It is still the universal standard of reference in all spectroscopic inquiries within the range of the *visible* emanations.

The discovery that hydrogen exists in the atmosphere of the sun was made by Ångström in 1862. His list of solar elements published in that year,[2] the result of an investigation separate from, though conducted on the same principle as Kirchhoff's, included the substance which we now know to be predominant amongst them. Dr. Plücker of Bonn had identified in 1859

[1] The normal spectrum is that depending exclusively upon wave-length—the fundamental constant given by nature as regards light. It is obtained by the interference of rays, in the manner first exemplified by Fraunhofer, and affords the only unvarying standard for measurement. In the refraction-spectrum (upon which Kirchhoff's map was founded), the relative positions of the lines vary with the material of the prisms.

[2] *Ann. d. Phys.*, Bd. cxvii. p. 296.

the Fraunhofer line F with the green ray of hydrogen, but drew no inference from his observation. The agreement was verified by Ångström; two further coincidences were established; and in 1866 a fourth hydrogen line in the extreme violet (named *h*) was detected in the solar spectrum. With Thalèn, he besides added manganese, aluminium, and titanium to the constituents of the sun enumerated by Kirchhoff, and raised the number of identical rays in the solar and terrestrial spectra of iron to no less than 460.[1]

Thus, when Mr. Lockyer entered on that branch of inquiry in 1872, fourteen substances were recognised as common to the earth and sun. Early in 1878 he was able, by applying the test of *length* in lieu of that of *strength* in the comparison of lines (looking, that is, rather to their persistence through a wide range of temperature, than to their brilliancy at any one temperature), to increase the list provisionally to thirty-three.[2] All these are metals; for there is strong reason to believe that hydrogen presents a solitary instance of an ordinarily gaseous metal, just as mercury does of an ordinarily liquid one. Up to 1877 the fourteen metalloids (non-metallic elements) were conspicuous by absence.

But in that year the late Dr. Henry Draper of New York announced a discovery of very wide significance. As the upshot of an investigation lasting several years, he found oxygen to be revealed in the sun, not, like the metals, by the *reversal* of its spectral rays, but by their *direct* presence. Each one of eighteen bright lines in its photographed spectrum was seen to be represented by a strictly corresponding brilliant band in the analysed light of the sun.[3] The reality of these coincidences having been doubted, Dr. Draper set to work afresh, and on the 13th of June 1879[4] laid before the Royal Astronomical Society photographs on a scale four times that of the original ones, in which the solar counterpart of the laboratory-spectrum

[1] *Comptes Rendus*, t. lxiii. p. 647.   [2] *Ibid.*, t. lxxxvi. p. 317.
[3] *Am. Jour. of Sc.*, vol. xiv. p. 89; *Nature*, vol. xvi. p. 364.
[4] *Month. Not.*, vol. xxxix. p. 440.

of oxygen was no less apparent than before. Mr. Ranyard remarked that, by this fourfold dispersion, the evidential value of the eighteen observed coincidences was increased 4^{18}, or (in round numbers) 68,719 million times ; but the rigid numerical test of probability does not in this case carry its full weight of conviction. The discrimination of *bright* lines from a very slightly less lucid background must, it is plain, be always a matter of much delicacy and some uncertainty, especially when the lines to be discriminated are not sharp, but more or less blurred and widened. Nevertheless the correspondences in Dr. Draper's photographs are far too striking to be overlooked, and afford strong ground for accepting his conclusion (recommended, besides, by our innate tendency to complete an analogy) that the most widely prevalent superficial constituent of the earth is not missing from the sun.

The peculiarity of its showing bright, instead of dark lines may be said to have given a new turn to the spectrum analysis of the heavenly bodies. It illustrates the endless variety in nature's modes of proceeding, and accentuates the danger of negative inferences. That a substance displays none of its distinctive beams in the spectrum of the sun or of a star, no longer affords even a presumption against its presence there. For it may be situated below the level where absorption occurs, or under a pressure such as to efface lines by continuous lustre ; it may be at a temperature so high that it gives out more light than it takes up, and yet its incandescence may be masked by the absorption of other bodies ; finally, it may just balance absorption by emission, with the result of complete spectral neutrality. An instructive example is that of helium, the enigmatical chromospheric element. Father Secchi remarked in 1868 [1] that there is no dark line in the solar spectrum matching its light ; and the faint traces of D3 absorption since detected would probably never have been observed, had not the substance producing them been otherwise known to exist.

Indications are not altogether wanting as to the cause of the

[1] *Comptes Rendus*, t. lxvii. p. 1123.

sun's oxygen attesting its presence as it does. The inner
organisation of the oxygen molecule is a considerably *plastic*
one. It is readily modified by heat, and these modifications
are reflected in its varying modes of radiating light. Dr.
Schuster enumerated in 1879 [1] four distinct oxygen spectra,
corresponding to various stages of temperature, or phases of
electrical excitement; and a fifth has been added by M.
Egoroff's discovery in 1883 [2] that certain well-known groups of
dark lines in the red end of the solar spectrum (Fraunhofer's
A and B) are due to absorption by the cool oxygen of our air.

Now, of these five different systems of luminous emission,
three are, in all probability, represented—one, as just stated,
through terrestrial, the others through solar action—in analysed
sunlight. The brilliant range of lines detected by Dr. Draper
belong to the maximum heat developed by high-tension electricity. The oxygen producing it certainly lies at a low level
in the sun, since its lines never appear in the spectrum of the
chromosphere; and we may conclude that it forms part of the
hottest layers of which we receive the radiations. The next, or
"compound-line spectrum," produced at a considerably lower
stage of thermal excitement, Dr. Schuster has found, with
evidence "little short of absolute certainty," to be dark in the
sun.[3] And here (as he pointed out) some prospect seems to
open of meeting with a definite criterion of the solar temperature. For evidently the degree of heat (whatever that may
be) at which spectrum No. 1 changes to spectrum No. 2
occurs somewhere between the stratum giving Draper's bright
lines and the stratum giving Schuster's dusky lines. This
brings us to the subject of the next chapter.

[1] *Phil. Trans.*, vol. clxx. p. 46.   [2] *Comptes Rendus*, t. xcvii. p. 555.
[3] *Nature*, vol. xvii. p. 148.

CHAPTER V.

TEMPERATURE OF THE SUN.

NEWTON was the first who attempted to measure the quantity
of heat received by the earth from the sun. His object in
making the experiment was to ascertain the temperature encountered by the comet of 1680 at its passage through perihelion. He found it, by multiplying the observed heating
effects of direct sunshine according to the familiar rule of the
"inverse squares of the distances," to be about 2000 times
that of red-hot iron.[1]

Determinations of the sun's thermal power made with some
scientific exactness, date, however, from 1837. A few days
previous to the beginning of that year, Herschel began observing at the Cape of Good Hope with an "actinometer,"
and obtained results agreeing quite satisfactorily with those
derived by Pouillet from experiments made in France some
months later with a "pyrheliometer."[2] Pouillet found that
the vertical rays of the sun falling on each square centimetre
of the earth's surface are competent (apart from atmospheric
absorption) to raise the temperature of 1.7633 grammes of
water one degree centigrade per minute. This number (1.7633)
he called the "solar constant;" and the unit of heat chosen is
known as the "calorie." Hence it was computed that the
total amount of solar heat received during a year would suffice
to melt a layer of ice covering the entire earth to a depth of
30.89 metres, or 100 feet; while the heat emitted would melt,

[1] *Principia*, p. 498 (1st ed.)   [2] *Comptes Rendus*, t. vii. p. 24.

R

at the sun's surface, a stratum 11.80 metres thick each minute.
A careful series of observations showed that nearly half the
heat incident upon our atmosphere is stopped in its passage
through it.

Herschel got somewhat larger figures, though he assigned
only a third as the spoil of the air.   Taking a mean between
his own and Pouillet's, he calculated that the ordinary expenditure of the sun per minute would have power to ʿmelt a
cylinder of ice 184 feet in diameter, reaching from his surface
to that of α Centauri ; or, putting it otherwise, that an ice-rod
45.3 miles across, continually darted into the sun with the
velocity of light, would scarcely consume, in dissolving, the
thermal supplies now poured abroad into space.[1]   It is nearly
certain that this estimate should be increased by about two-
thirds in order to bring it up to the truth.

Nothing would, at first sight, appear simpler than to pass
from a knowledge of solar emission—a strictly measurable
quantity—to a knowledge of the solar temperature ; this being
defined as the temperature to which a surface thickly coated
with lamp-black (that is, of standard radiating power) should
be raised to enable it to send us, from the sun's distance, the
amount of heat actually received from the sun.   Sir John
Herschel showed that heat-rays at the sun's surface must be
192,000 times as dense as when they reach the earth ; but it
by no means follows that either the surface emitting, or a body
absorbing those heat-rays must be 192,000 times hotter than a
body exposed here to the full power of the sun.   The reason
is, that the rate of emission—consequently the rate of absorption, which is its correlative—increases very much faster than
the temperature.   In other words, a body radiates or cools at
a continually accelerated pace as it becomes more and more
intensely heated above its surroundings.

Newton, however, took it for granted that radiation and
temperature advance *pari passu*—that you have only to ascertain
the quantity of heat received from, and the distance of a remote

[1] *Results of Astr. Observations,* p. 446.

body in order to know how hot it is.[1] And this principle, which is known as "Newton's Law" of cooling, has still a limited number of adherents. Its validity was never questioned until De la Roche pointed out, in 1812,[2] that it was approximately true only over a low range of temperature; and five years later, Dulong and Petit generalised experimental results into the rule, that while temperature grows by arithmetical, radiation increases by geometrical progression.[3] Adopting this formula, Pouillet derived from his observations on solar heat a solar temperature of somewhere between 1461° and 1761° Cent. Now, the higher of these points—which is nearly that of melting platinum—is undoubtedly surpassed at the focus of certain burning-glasses which have been constructed of such power as virtually to bring objects placed there within a quarter of a million of miles of the photosphere. In the rays thus concentrated, platinum and diamond become rapidly vaporised, notwithstanding the great loss of heat by absorption, first in passing through the air, and again in traversing the lens. Pouillet's maximum is then manifestly too low, since it involves the absurdity of supposing a radiating mass capable of heating a distant body more than it is itself heated.

Less demonstrably, but scarcely less surely, Mr. J. J. Waterston, who attacked the problem in 1860, erred in the opposite direction. Working up, on Newton's principle, data collected by himself in India and at Edinburgh, he got for the "potential temperature" of the sun 12,880,000° Fahr.,[4] equivalent to 7,156,093° Cent. The phrase *potential temperature* (for which Violle substituted, in 1876, *effective temperature*) was designed to express the accumulation in a single surface, postulated for the sake of simplicity, of the radiations not improbably received from a multitude of separate

---

[1] "Est enim calor solis ut radiorum densitas, hoc est, reciproce ut quadratum distantiæ locorum a sole."—*Principia*, p. 508 (3d ed. 1726).

[2] *Jour. de Physique*, t. lxxv. p. 215.

[3] *Ann. de Chimie*, t. vii. 1817, p. 365.

[4] *Phil. Mag.*, vol. xxiii. (4th ser.), p. 505.

solar layers reinforcing each other; and might thus (it was explained) be considerably higher than the *actual* temperature of any one stratum.

At Rome, in 1861, Father Secchi repeated Waterston's experiments, and reaffirmed his conclusion;[1] while Soret's observations, made on the summit of Mont Blanc in 1867,[2] furnished him with materials for a fresh and even higher estimate of ten million degrees centigrade.[3] Yet from the very same data, substituting Dulong and Petit's for Newton's law, Vicaire deduced in 1872 a *provisional* solar temperature of 1398°.[4] This is below that at which iron melts, and we know that iron-vapour exists high up in the sun's atmosphere. The matter was taken into consideration on the other side of the Atlantic by Ericsson in 1871. He attempted to re-establish the shaken credit of Newton's principle, and arrived, by its means, at a temperature of four million degrees of Fahrenheit.[5] More recently, what he considers an "underrated computation," based upon observation of the quantity of heat received by his "sun motor," has given him three million degrees. This, he rightly thinks, *must* be accepted, if it be granted that the temperature produced by radiant heat is proportional to its density, or inversely as its diffusion.[6] Could this be granted, the question would be much simplified; but there is little doubt that the case is far otherwise when heat becomes intensified.

In 1876 the sun's temperature was proposed as the subject of a prize by the Paris Academy of Sciences; but although the essay of M. Jules Violle was crowned, the problem was declared to remain unsolved. Violle (who adhered to Dulong and Petit's formula) arrived at an *effective* temperature of 1500° C., but considered that it might *actually* reach 2500° C., owing

---

[1] *Nuovo Cimento*, t. xvi. p. 294.   [2] *Comptes Rendus*, t. lxv. p. 526.

[3] The direct result of 5½ million degrees was doubled in allowance for absorption in the sun's own atmosphere. *Comptes Rendus*, t. lxxiv. p. 26.

[4] *Ibid.*, p. 31.   [5] *Nature*, vols. iv. p. 204; v. p. 505.

[6] *Nature*, vol. xxx. p. 467.

to a probable inferiority in emissive power of the photospheric clouds to the lamp-black standard.[1] Experiments made in April and May 1881 giving a somewhat higher result, he raised this figure to 3000° C.[2]

Appraisements so outrageously discordant as those of Waterston, Secchi, and Ericsson on the one hand, and those of the French *savans* on the other, served only to show that all were based upon a vicious principle. Professor F. Rosetti,[3] accordingly, of the Paduan University, at last perceived the necessity for getting out of the groove of "laws" plainly in contradiction with facts. The temperature, for instance, of the oxyhydrogen flame was fixed by Bunsen at 2800° C.—an estimate certainly not very far from the truth. But if the two systems of measurement applied to the sun be used to determine the heat of a solid body rendered incandescent in this flame, it comes out, by Newton's mode of calculation, 45,000° C.; by Dulong and Petit's, 870° C.[4] Both, then, are justly discarded, the first as convicted of exaggeration, the second of undervaluation. The formula substituted by Rosetti was tested successfully up to 2000° C.; but since it is, like its predecessors, a purely empirical rule, is guaranted by no principle, and can, in consequence, not be trusted out of sight, it may, like them, break down at still higher elevations. All that can be said is that it gives the most plausible results. Radiation, so far as it obeys this new prescription, increases as the *square* of the *absolute* temperature—that is, of the number of degrees counted from the "absolute zero" of —273° C. Its employment gives for the sun's radiating surface an effective temperature of 20,380° C. (including a supposed loss of one-half in the solar atmosphere); and when a probable deficiency in emission (as compared with lamp-black) is set against a probable mutual reinforcement of superposed strata, Professor

[1] *Ann. de Chim.*, t. x. (5th ser.), p. 361.
[2] *Comptes Rendus*, t. xcvi. p. 254.
[3] *Phil. Mag.*, vol. viii. 1879, p. 324.
[4] *Ibid.*, p. 325.

Rosetti thinks that "effective" may be taken as nearly equivalent to "actual" temperature.

A new line of inquiry was struck out by Zöllner in 1870. Instead of tracking the solar radiations backwards with the dubious guide of empirical formulæ, he investigated their intensity at their source. He showed[1] that, considering prominences as simple effects of the escape of powerfully compressed gases, it was possible, from the known mechanical laws of heat and gaseous constitution, to deduce minimum values for the temperatures prevailing in the area of their development. These came out 27,700° C. for the strata lying immediately above, and 68,400° C. for the strata lying immediately below the photosphere, the former being regarded as the region *into* which, and the latter as the region *from* which the eruptions took place. In this calculation, no prominences exceeding 50,000 miles (1.5′) in height were included. But in 1884, G. A. Hirn of Colmar, taking into account the enormous velocities of projection observed in the interim, fixed two million degrees centigrade as the lowest *internal* temperature by which they could be accounted for; although of opinion that the condensations, presumed to give origin to the photospheric clouds, were incompatible with a higher *external* temperature than 50,000° to 100,000° C.[2]

This method of going straight to the sun itself, observing what goes on there, and inferring conditions, has much to recommend it; but its profitable use demands knowledge we are still very far from possessing. We are quite ignorant, for instance, of the actual circumstances attending the birth of the solar flames. The assumption that they are nothing but phenomena of elasticity is a purely gratuitous one. Spectroscopic indications, again, give hope of eventually affording a fixed point of comparison with terrestrial heat-sources; but their interpretation is still beset with uncertainties; nor can, indeed, the expression of transcendental temperatures in degrees of impossible thermometers be, at the best, other than

[1] *Astr. Nach.*, Nos. 1815–16.   [2] *L'Astronomie*, Sept. 1884, p. 334.

a futile attempt to convey notions respecting a state of things altogether outside the range of our experience.

A more tangible, as well as a less disputable proof of solar radiative intensity than any mere estimates of temperature, was provided in some experiments made by Professor Langley in 1878.[1] Using means of unquestioned validity, he found the sun's disc to radiate 87 times as much heat, and 5300 times as much light as an equal area of metal in a Bessemer converter after the air-blast had continued about twenty minutes. The brilliancy of the incandescent steel, nevertheless, was so blinding, that melted iron, flowing in a dazzling white-hot stream into the crucible, showed "deep brown by comparison, presenting a contrast like that of dark coffee poured into a white cup." Its temperature was estimated (not quite securely, as Young has pointed out)[2] at 1800° to 2000° C.; and no allowances were made, in computing relative intensities, for atmospheric ravages on sunlight, for the extra impediments to its passage presented by the smoke-laden air of Pittsburg, or for the obliquity of its incidence. Thus a very large balance of advantage lay on the side of the metal.

A further element of uncertainty in estimating the intrinsic strength of the sun's rays has still to be considered. From the time that his disc first began to be studied with the telescope, it was perceived to be less brilliant near the edges. Lucas Valerius of the Lyncean Academy seems to have been the first to note this fact, which, strangely enough, was denied by Galileo in a letter to Prince Cesi of January 25, 1613.[3] Father Scheiner, however, fully admitted it, and devoted some columns of his bulky tome to the attempt to find an appropriate explanation.[4] In 1729, Bouguer measured, with much accuracy, the amount of this darkening; and from his data, Laplace, adopting a principle of emission now known to be erroneous, concluded that the sun loses eleven-twelfths of his light through

---

[1] *Jour. of Science*, vol. i. (3d ser.), p. 653.
[2] *The Sun*, p. 269.   [3] *Op.*, t. vi. p. 198.
[4] *Rosa Ursina*, lib. iv. p. 618.

absorption in his own atmosphere.[1] The real existence of this atmosphere, which is totally distinct from the beds of ignited vapours producing the Fraunhofer lines, is not open to doubt, although its nature is still a matter of conjecture. The separate effects of its action on luminous, thermal, and chemical rays were carefully studied by Father Secchi, who in 1870,[2] inferred the total absorption to be $\frac{88}{100}$ of all radiations taken together, and added the important observation that the light from the limb is no longer white, but reddish-brown. *Selective* absorption was thus seen to be at work ; and this could evidently be studied to advantage only by taking the various rays of the spectrum separately, and finding out how much each had suffered in transmission.

This was done by H. C. Vogel in 1877.[3] Using a polarising photometer, he found that only 13 per cent. of the violet rays escape at the edge of the solar disc, 16 of the blue and green, 25 of the yellow, and 30 per cent. of the red. Midway between centre and limb, 88.7 of violet light and 96.7 of red penetrate the absorbing envelope, the removal of which would leave the sun's visible spectrum of just three times its present intensity in the most, and once and a half times in the least refrangible parts. The nucleus of a small spot was ascertained to be of the same luminous intensity as a portion of the unbroken surface about two and a half minutes from the limb. These experiments having been made during a spot-minimum, when there is reason to think that absorption is below its average strength, Vogel suggested their repetition at a time of greater activity.

Professor Langley went farther in the same direction. Reliable determinations of the "energy" of the individual spectral rays were, for the first time, rendered possible by his invention of the "bolometer" in 1880.[4] This exquisitely sensitive instrument affords the means of measuring heat,

---

[1] *Méc. Cél.*, liv. x. p. 323.   [2] *Le Soleil* (1st ed.), p. 136.
[3] *Monatsber.*, Berlin, 1877, p. 104.
[4] *Am. Jour. of Sc.*, vol. xxi. p. 187.

not directly, like the thermopile, but in its effects upon the conduction of electricity. It represents, in the phrase of the inventor, the finger laid upon the throttle-valve of a steam-engine. A minute force becomes the modulator of a much greater force, and thus from imperceptible becomes conspicuous. By locally raising the temperature of an inconceivably fine strip of platinum serving as the conducting-wire in a circuit, the flow of electricity is impeded at that point, and the included galvanometer records a disturbance of the electrical flow. Amounts of heat have, in this way, been detected in less than ten seconds, which, expended during a thousand years on the melting of a kilogramme of ice, would leave a part of the work still undone.

The heat contained in the diffraction spectrum is, with equal dispersions, barely one-tenth of that in the prismatic spectrum. It had, accordingly, never previously been found possible to measure it in detail—that is, ray by ray. But it is only from the diffraction, or normal spectrum that any true idea can be gained as to the real distribution of energy amongst the various constituents, visible and invisible, of a sunbeam. The effect of passage through a prism is to crowd together the red rays very much more than the blue. To this prismatic distortion was owing the establishment of a pseudo-maximum of heat in the infra-red, which disappeared when the natural arrangement by wave-length was allowed free play. Professor Langley's bolometer has shown that the hottest part of the normal spectrum virtually coincides with its most luminous part, both lying in the orange, close to the D line.[1] Thus the last shred of evidence in favour of the threefold division of solar radiations vanished, and it became obvious that the varying effects —thermal, luminous, or chemical—produced by them are due, not to any distinction of quality in themselves, but to the different properties of the substances they impinge upon. They are simply bearers of *vis viva*, conveyed in shorter or longer

---

[1] For J. W. Draper's partial anticipation of this result, see *Am. Jour. of Sc.*, vol. iv. 1872, p. 174.

vibrations; upon the capacity of the material particles meeting them for taking up those shorter or longer vibrations, and turning them variously to account in their inner economy, depends the result in each separate case.

A long series of experiments at Allegheny was completed in the summer of 1881 on the crest of Mount Whitney in the Sierra Nevada. Here, at an elevation of 14,887 feet, in the driest and purest air, perhaps, in the world, atmospheric absorptive inroads become less sensible, and the indications of the bolometer, consequently, surer and stronger. An enormous expansion was at once given to the invisible region in the solar spectrum below the red. Captain Abney had got chemical effects from undulations twelve ten-thousandths of a millimetre in length. These were the longest recognised as, or indeed believed, on theoretical grounds, to be capable of existing. Professor Langley now got heating effects from rays of above twice that wave-length, his delicate thread of platinum groping its way down to thirty ten-thousandths of a millimetre, or three "microms." The known extent of the solar spectrum was thus at once more than doubled. Its visible portion covers a range of about one octave; bolometric indications comprise between three and four. The great importance of the newly explored region appears from the fact that three-fifths of the entire energy of sunlight reside in the infra-red, while scarcely more than one-hundredth part of that amount is found in the better known ultra-violet space.[1]

Atmospheric absorption had never before been studied with such precision as it was by Professor Langley on Mount Whitney. Aided by simultaneous observations from Lone Pine, at the foot of the Sierra, he was able to calculate the intensity belonging to each ray before entering the earth's gaseous envelope—in other words, to construct an extra-atmospheric curve of energy in the spectrum. The result showed that the blue end suffered far more than the red, absorption varying

---

[1] *Phil. Mag.*, vol. xiv. p. 179 (March 1883).

inversely as wave-length. This property of stopping predominantly the quicker vibrations is shared, as both Vogel and Langley[1] have conclusively shown, by the solar atmosphere. The effect of this double absorption is as if two plates of reddish glass were interposed betwen us and the sun, the withdrawal of which would leave his orb, not only three or four times more brilliant, but in colour of a distinct greenish-blue, not very different from the tint of the second (F) line of hydrogen.[2]

The fact of the unveiled sun being *blue* has an important bearing upon the question of his temperature, to afford a somewhat more secure answer to which was the ultimate object of Professor Langley's persevering researches ; for it is well known that, as bodies grow hotter, the proportionate representation in their spectra of the more refrangible rays becomes greater. The lowest stage of incandescence is the familiar one of *red* heat. As it gains intensity, the quicker vibrations come in, and an optical balance of sensation is established at *white* heat. The final term of *blue* heat, as we now know, is attained by the photosphere. On this ground alone, then, of the large original preponderance of blue light, we must raise our estimate of solar heat ; and actual measurements show the same upward tendency. Until quite lately, Pouillet's figure of 1.7 calories per minute per square centimetre of terrestrial surface, was the received value for the "solar constant." Forbes had, it is true, got 2.85 from observations on the Faulhorn in 1842 ;[3] but they failed to obtain the confidence they merited. Pouillet's result was not definitively superseded until Violle, from actinometrical measures at the summit and base of Mont Blanc in 1875, computed the intensity of solar radiation at 2.54,[4] and Crova, about the same time, at Montpellier, showed it to be above two calories.[5] Langley went higher still. His pre-

[1] *Comptes Rendus*, t. xcii. p. 701.   [2] *Nature*, vol. xxvi. p. 589.
[3] *Phil. Trans.*, vol. cxxxii. p.1273.   [4] *Ann. de Chim.*, t. x. p. 321.
[5] *Ibid.*, t. xi. p. 505.

liminary estimate, December 30, 1882, agreed with Forbes's ; and his definitive one, when the results of the Mount Whitney expedition are fully worked out, is likely to fall scarcely short of three calories, as the amount of heat reaching the outskirts of our atmosphere.[1] Thus, modern inquiries, though they give no signs of agreement, within any tolerable limits of error, as to the probable temperature of the sun, tend, with growing certainty, to render more and more evident the vastness of the thermal stores contained in the great central reservoir of our system.

[1] *Phil. Mag.*, vol. xiv. p. 181.

CHAPTER VI.

THE SUN'S DISTANCE.

THE question of the sun's distance arises naturally from the consideration of his temperature, since the intensity of the radiations emitted as compared with those received and measured, depends upon it. But the knowledge of that distance has a value quite apart from its connection with solar physics. The semi-diameter of the earth's orbit is our standard measure for the universe. It is the great fundamental datum of astronomy—the unit of space, any error in the estimation of which is multiplied and repeated in a thousand different ways, both in the planetary and sidereal systems. Hence its determination has been called by Airy "the noblest problem in astronomy." It is also one of the most difficult. The quantities dealt with are so minute that their sure grasp tasks all the resources of modern science. An observational inaccuracy which would set the moon nearer to, or farther from us than she really is by one hundred miles, would vitiate an estimate of the sun's distance to the extent of sixteen million ![1] What is needed in order to attain knowledge of the desired exactness is no less than this : to measure an angle about equal to that subtended by a halfpenny 2000 feet from the eye, within a little more than a thousandth part of its value.

The angle thus represented is what is called the "horizontal parallax" of the sun. By this amount—the breadth of a halfpenny at 2000 feet—he is, to a spectator on the rotating earth, removed at rising and setting from his meridian place in the

[1] Airy, *Month. Not.*, vol. xvii. p. 210.

heavens. Such, in other terms, would be the magnitude of the terrestrial radius as viewed from the sun. If we knew this magnitude with certainty and precision, we should also know with certainty and precision—the dimensions of the earth being, as they are, well ascertained—the distance of the sun. In fact, the one quantity commonly stands for the other in works treating professedly of astronomy. But this angle of parallax or apparent displacement cannot be directly measured —cannot even be perceived with the finest instruments. Not from its smallness. The parallactic shift of the nearest of the stars as seen from opposite sides of the earth's orbit, is many times smaller. But at the sun's limb, and close to the horizon, where the visual angle in question opens out to its full extent, atmospheric troubles become overwhelming, and altogether swamp the far more minute effects of parallax.

There remain indirect methods. Astronomers are well acquainted with the proportions which the various planetary orbits bear to each other. They are so connected, in the manner expressed by Kepler's Third Law, that the periods being known, it only needs to find the interval between any two of them in order to infer at once the distances separating them all from one another and from the sun. The plan is given; what we want to discover is the scale upon which it is drawn; so that, if we can get a reliable measure of the distance of a single planet from the earth, our problem is solved.

Now some of our fellow-travellers in our unending journey round the sun, come at times well within the scope of celestial trigonometry. The orbit of Mars lies at one point not more than thirty-five million miles outside that of the earth, and when the two bodies happen to arrive together in or near the favourable spot—a conjuncture which recurs every fifteen years —the desired opportunity is granted. Mars is then "in opposition," or on the *opposite* side of us from the sun, crossing the meridian consequently at midnight.[1] It was from an opposition of Mars, observed in 1672 by Richer at Cayenne in

[1] Mars comes into opposition once in about 780 days; but owing to the

concert with Cassini in Paris, that the first scientific estimate of the sun's distance was derived. It appeared to be nearly eighty-seven millions of miles (parallax 9.5″); while Flamsteed deduced 81,700,000 (parallax 10″) from his independent observations of the same occurrence—a difference quite insignificant at that stage of the inquiry. But Picard's result was just half Flamsteed's (parallax 20″; distance forty-one million miles); and Lahire considered that we must be separated from the hearth of our system by an interval of *at least* 136 million miles.[1] So that uncertainty continued to be on a gigantic scale.

Venus, on the other hand, comes closest to the earth when she passes between it and the sun. At such times of "inferior conjunction" she is, however, still twenty-six million miles, or (in round numbers) 109 times as distant as the moon. Moreover, she is so immersed in the sun's rays that it is only when her path lies across his disc that the requisite facilities for measurement are afforded. These "partial eclipses of the sun by Venus" (as Encke terms them) are coupled together in pairs,[2] of which the components are separated by eight years, recurring at intervals alternately of 105½ and 121½ years. Thus, the first calculated transit took place in December 1631, and its companion (observed by Horrocks) in the same month (N.S.) 1639. Then, after the lapse of 121½ years, came the

eccentricity of both orbits, his distance from the earth at those epochs varies from thirty-five to sixty-two million miles.

[1] J. D. Cassini, *Hist. Abrégée de la Parallaxe du Soleil,* p. 122, 1772.

[2] The present period of coupled eccentric transits will, in the course of ages, be succeeded by a period of single, nearly central transits. The alignments by which transits are produced, of the earth, Venus, and the sun, close to the place of intersection of the two planetary orbits, now occur, the first a little in front of, the second, after eight years less two and a half days, a little behind the node. But when the first of these two meetings takes place very near the node, giving a nearly central transit, the second falls too far from it, and the planet escapes projection on the sun. The reason of the liability to an eight-yearly recurrence is that eight revolutions of the earth are accomplished in only a very little more time than thirteen revolutions of Venus.

June couple of 1761 and 1769; and again, after $105\frac{1}{2}$, the two recently observed December 8, 1874, and December 6, 1882. Throughout the twentieth century there will be no transit of Venus; but the astronomers of the twenty-first will only have to wait four years for the first of a June pair. The rarity of these events is due to the fact that the orbits of the earth and Venus do not lie in the same plane. If they did, there would be a transit each time that our twin-planet overtakes us in her more rapid circling—that is, on an average, every 584 days. As things are actually arranged, she passes above or below the sun, except when she happens to be very near the line of intersection of the two tracks.

Such an occurrence as a transit of Venus seems, at first sight, full of promise for solving the problem of the sun's distance. For nothing would appear easier than to determine exactly either the duration of the passage of a small, dark orb across a large brilliant disc, or the instant of its entry upon or exit from it. And the differences in these times (which, owing to the comparative nearness of Venus, are quite considerable), as observed from remote parts of the earth, can be translated into differences of space—that is, into apparent or parallactic displacements, whereby the distance of Venus becomes known, and thence, by a simple sum in proportion, the distance of the sun. But in that word "exactly" what snares and pitfalls lie hid! It is so easy to think and to say; so indefinitely hard to realise. The astronomers of the eighteenth century were full of hope and zeal. They confidently expected to attain, through the double opportunity offered them, to something like a permanent settlement of the statistics of our system. They were grievously disappointed. The uncertainty as to the sun's distance, which they had counted upon reducing to a few hundred thousand miles, remained at many millions.

In 1822, however, Encke, then director of the Seeberg Observatory near Gotha, undertook to bring order out of the confusion of discordant, and discordantly interpreted observations. His combined result for both transits (1761

and 1769) was published in 1824,[1] and met universal acquies-
cence. The parallax of the sun thereby established was
8.5776″, corresponding to a mean distance [2] of 95¼ million
miles. Yet this abolition of doubt was far from being so
satisfactory as it seemed. Serenity on the point lasted exactly
thirty years. It was disturbed in 1854 by Hansen's announcement [3] that the observed motions of the moon could be drawn
into accord with theory only on the terms of bringing the sun
considerably nearer to us than he was supposed to be.

Dr. Matthew Stewart, professor of mathematics in the University of Edinburgh, had made a futile attempt in 1763 to
deduce the sun's distance from his disturbing power over our
satellite.[4] Tobias Mayer of Göttingen, however, whose short
career was so fruitful of suggestions, struck out the right way
to the same end; and Laplace, in the seventh book of the
Mécanique Céleste,[5] gave a solar parallax derived from the lunar
"parallactic inequality" substantially identical with that issuing
from Encke's subsequent discussion of the eighteenth-century
transits. Thus two wholly independent methods—the trigonometrical, or method by survey, and the gravitational, or
method by perturbation—seemed to corroborate each the
upshot of the use of the other until the nineteenth century was
well past its meridian. It is singular how often errors conspire to lead conviction astray.

Hansen's note of alarm in 1854 was echoed by Leverrier in
1858.[6] He found that an apparent monthly oscillation of the

---

[1] *Die Entfernung der Sonne: Fortsetzung*, p. 108. Encke slightly corrected his result of 1824 in *Berlin Abh.*, 1835, p. 295.

[2] Owing to the ellipticity of its orbit, the earth is nearer to the sun in
January than in June by 3,100,000 miles. The quantity to be determined,
or "mean distance," is that lying midway between these extremes—is, in
other words, half the major axis of the ellipse in which the earth travels.

[3] *Month. Not.*, vol. xv. p. 9.

[4] *The Distance of the Sun from the Earth determined by the Theory of
Gravity*, Edinburgh, 1763.   [5] *Opera*, t. iii. p. 326.

[6] *Comptes Rendus*, t. xlvi. p. 882. The parallax 8.95″ derived by
Leverrier from the "parallactic inequality" in the earth's motion, was
corrected by Stone to 8.91″. *Month. Not.*, vol. xxviii. p. 25.

sun which reflects a real monthly movement of the earth round its common centre of gravity with the moon, and which depends for its amount solely on the mass of the moon and the distance of the sun, required a diminution in the admitted value of that distance by fully four million miles. Three years later he pointed out that certain perplexing discrepancies between the observed and computed places both of Venus and Mars, would vanish on the adoption of a similar measure.[1] Moreover, a favourable opposition of Mars gave the opportunity in 1862 for fresh observations, which, separately worked out by Stone and Winnecke, agreed with all the newer investigations in fixing the great unit at slightly over 91 million miles. In Newcomb's hands they gave $92\frac{1}{2}$ million.[2] The accumulating evidence in favour of a large reduction in the sun's distance was just then reinforced by an auxiliary result of a totally different and unexpected kind.

The discovery that light does not travel instantaneously from point to point, but takes some short time in transmission, was made by Olaus Römer in 1675, through observing that the eclipses of Jupiter's satellites invariably occurred later, by a considerable interval, when the earth was on the far side, than when it was on the near side of its orbit. Half this interval, or the time spent by a luminous vibration in crossing the "mean radius" of the earth's orbit, is called the "light-equation;" and the determination of its precise value has claimed the minute care distinctive of modern astronomy. Delambre in 1792 made it 493.2 seconds. Glasenapp, a Russian astronomer, raised the estimate in 1874 to 500.84 seconds; and this, from the extreme care employed, can hardly, at the outside, be more than a couple of seconds astray. Hence, if we had any independent means of ascertaining how fast light travels, we could tell at once how far off the sun is.

There is yet another way by which knowledge of the swiftness of light would lead us straight to the goal. The heavenly

---

[1] *Month. Not.*, vol. xxxv. p. 156.
[2] *Wash. Obs.*, 1865, App. ii. p. 28.

bodies are perceived, when carefully watched and measured, to be pushed forwards out of their true places, in the direction of the earth's motion, by a very minute quantity. This effect (already adverted to) has been known since Bradley's time as "aberration." It arises from a combination of the two movements of the earth round the sun and of the light-waves through the ether. If the earth stood still, or if light spent no time on the road from the stars, such an effect would not exist. Its amount represents the proportion between the velocities with which the earth and the light rays pursue their respective journeys. This proportion is, roughly, one to ten thousand. So that here again, if we knew the rate per second of luminous transmission, we should also know the rate per second of the earth's movement, consequently the size of its orbit and the distance of the sun.

But, until lately, instead of finding the distance of the sun from the velocity of light, there has been no means of ascertaining the velocity of light except through the imperfect knowledge possessed as to the distance of the sun. The first successful terrestrial experiments on the point date from 1849; and it is certainly no slight triumph of human ingenuity to have taken rigorous account of the delay of a sunbeam in flashing from one mirror to another. Fizeau led the way,[1] and he was succeeded, after a few months, by Léon Foucault,[2] who, in 1862, had so far perfected Wheatstone's method of revolving mirrors, as to be able to announce with authority that light travelled slower, and that the sun was in consequence nearer, than had been supposed.[3] Thus a third line of separate research was found to converge to the same point with the two others.

[1] *Comptes Rendus*, t. xxix. p. 90.
[2] *Ibid.*, t. xxx. p. 551.
[3] *Ibid.*, t. lv. p. 501. The previously admitted velocity was 308 million metres per second; Foucault reduced it to 298 million. Combined with Struve's "constant of aberration" this gave 8.86″ for the solar parallax, which exactly agreed with Cornu's result from a repetition of Fizeau's experiments in 1872. *Comptes Rendus*, t. lxxvi. p. 338.

Such a conspiracy of proof was not to be resisted, and at the anniversary meeting of the Royal Astronomical Society in February 1864, the correction of the solar distance took the foremost place in the annals of the year. Lest, however, a sudden bound of four million miles nearer to the centre of our system should shake public faith in astronomical accuracy, it was explained that the change in the solar parallax corresponding to that huge leap, amounted to no more than the breadth of a human hair 125 feet from the eye![1] From 1866 the improved value of 8.90" was adopted in the Nautical Almanac, while Newcomb's result of 8.85" has appeared since 1869 in the Berlin Ephemeris. In astronomical literature the change was initiated by Sir Edmund Beckett in the first edition (1865) of his *Astronomy without Mathematics*.

If any doubt remained as to the misleading character of Encke's deduction, so long implicitly trusted in, it was removed by Powalky's and Stone's rediscussions, in 1864 and 1868 respectively, of the transit observations of 1769. Using improved determinations of the longitude of the various stations, and a selective judgment in dealing with their materials, which, however indispensable, did not escape adverse criticism, they brought out results confirmatory of the no longer disputed necessity for largely increasing the solar parallax, and proportionately diminishing the solar distance.

Conclusions on the subject, however, were still regarded as purely provisional. A transit of Venus was fast approaching, and to its arbitrament, as to that of a court of final appeal, the pending question was to be referred. It is true that the verdict in the same case by the same tribunal a century earlier had proved of so indistinct a character as to form only a starting-point for fresh litigation; but that century had not passed in vain, and it was confidently anticipated that observational difficulties, then equally unexpected and insuperable, would yield to the elaborate care and skill of forewarned modern preparation.

[1] *Month. Not.*, vol. xxiv. p. 103.

The conditions of the transit of December 8, 1874, were sketched out by the then Astronomer-Royal (Sir George Airy) in 1857,[1] and formed the subject of eager discussions in this and other countries down to the very eve of the occurrence. In these Mr. Proctor took a leading part, supplying official omissions, and working out, with geometrical accuracy, the details of the relations between the different parts of the earth and Venus's shadow-cone; and it was due to his urgent representations that provision was made for the employment of the method identified with the name of Halley,[2] which had been too hastily assumed inapplicable to the first of each transit-pair. It depends upon the difference in the length of time taken by the planet to cross the sun's disc, as seen from various points of the terrestrial surface, and requires, accordingly, the visibility of both entrance and exit at the same station. Since these were, in 1874, about three and a half hours, and are often much farther apart, the choice of posts for the successful use of the "method of durations" is a matter of some difficulty.

The system described by Delisle in 1760, on the other hand, involves merely noting the instant of ingress or egress (according to situation) from opposite extremities of a terrestrial diameter; the disparity in time giving a measure of the planet's apparent displacement, hence of its actual rate of travel in miles per minute, from which its distances severally from earth and sun are immediately deducible. Its chief attendant difficulty is the necessity for accurately fixing the longitudes of the points of observation. This, however, was much more sensibly felt a century ago than it is now, and the improved facility and certainty of modern determinations have tended to give the Delislean plan a decided superiority over its rival.

These two traditional methods were supplemented in 1874 by the camera and the heliometer. From photography, above all, much was expected. Observations made by its means

[1] *Month. Not.*, vol. xvii. p. 208.
[2] Because closely similar to that proposed by him in *Phil. Trans.* for 1716.

would have the advantages of impartiality, multitude, and permanence. Peculiarities of vision and bias of judgment would be eliminated; the slow progress of the phenomenon would permit an indefinite number of pictures to be taken, their epochs fixed to a fraction of a second; while subsequent leisurely comparison and measurement could hardly fail, it was thought, to educe approximate truth from the mass of accumulated evidence. The use of the heliometer (much relied on by German observers) was so far similar to that of the camera that the object aimed at by both was the determination of the relative positions of the *centres* of the sun and Venus viewed, at the same absolute instant, from opposite sides of the globe. So that the two older methods seek to ascertain the exact times of meeting between the solar and planetary limbs; while the two modern methods work by measurement of the position of the dark body already thrown into complete relief by its shining background. The former are "methods by contact," the latter "methods by projection."

Every country which had a reputation to keep or to gain for scientific zeal was forward to co-operate in the great cosmopolitan enterprise of the transit. France and Germany each sent out six expeditions; twenty-six stations were in Russian, twelve in English, eight in American, three in Italian, one (equipped with especial care) in Dutch occupation. In all, at a cost of nearly a quarter of a million, some fourscore distinct posts of observation were provided; amongst them such inhospitable, and all but inaccessible rocks in the bleak Southern Ocean, as St. Paul's and Campbell Islands, swept by hurricanes, and fitted only for the habitation of seabirds, where the daring votaries of science, in the wise prevision of a long leaguer by the elements, were supplied with stores for many months, or even a whole year. Siberia and the Sandwich Islands were thickly beset with observers; parties of three nationalities encamped within the mists of Kerguelen Island, expressively termed the "Land of Desolation," in the sanguine, though not wholly frustrated hope of a glimpse of the sun at the right moment. M. Janssen narrowly escaped destruction from a typhoon in the China seas on his way to Nagasaki; Lord Lindsay (now Earl of Crawford and Balcarres) equipped, at his private expense, an expedition to the Mauritius, which was in itself an epitome of modern resource and ingenuity.

During several years the practical methods best suited to ensure success for the impending enterprise, formed a subject of "European debate." Official commissions were appointed to receive and decide upon evidence; and experiments were in progress for the purpose of defining the actual circumstances of the contacts, the accurate determination of which constituted the only tried, though by no means an assuredly safe road to the end in view. In England, America, France, and Germany, artificial transits were mounted, and the members of the various expeditions were carefully trained to unanimity in estimating the phases of junction and separation between a moving dark circular body and a broad illuminated disc. In the last century, a formidable and prevalent phenomenon had swamped all pretensions to rigid accuracy. This was an effect analogous to "Baily's Beads," which acquired notoriety as the "Black Drop" or "Black Ligament." It may be described as substituting adhesion for contact, the limbs of the sun and planet, instead of meeting and parting with the desirable clean definiteness, *clinging* together as if made of some glutinous material, and prolonging their connection by means of a dark band or dark threads stretched between them. Some astronomers ascribe this baffling appearance entirely to instrumental imperfections; others to atmospheric agitation; others again to the optical encroachment of light upon darkness known as "irradiation." It is probable that all these causes conspire, in various measure, to produce it; and it is certain that by suitable precautions, combined with skill in the observer and a reasonably tranquil air, its *conspicuous* appearance may, in most cases, be obviated.

The organisation of the British forces reflected the utmost credit on the energy and ability of Lieutenant-Colonel Tupman, of the Royal Marine Artillery, who was responsible for the whole. No useful measure was neglected. Each observer went out ticketed with his "personal equation," his senses drilled into a species of martial discipline, his powers absorbed, so far as possible, in the action of a cosmopolitan observing machine. Instrumental uniformity and uniformity of method were attainable, and were attained; but diversity of judgment unhappily survived the best-directed efforts for its extirpation.

The eventful day had no sooner passed than telegrams began to pour in, announcing an outcome of considerable, though not unqualified success. The weather had proved generally favourable; all the manifold arrangements had (save for some casual mishaps) worked well; contacts had been plentifully observed; photographs in lavish abundance had been secured; a store of materials, in short, had been laid up, of which it would take years to work out the full results by calculation. Gradually, however, it came to be known that the hope of a definitive issue must be abandoned. Unanimity was found to be as remote as ever. The dreaded "black ligament" gave, indeed, less trouble than was expected; but another appearance supervened which took most observers by surprise. This was the illumination due to the atmosphere of Venus. Astronomers, it is true, were not ignorant that the planet had, on previous occasions, been seen girdled with a lucid ring; but its power to mar observations by the distorting effect of refraction had scarcely been reckoned with. It proved, however, to be very great. Such was the difficulty of determining the critical instant of internal contact, that (in Colonel Tupman's words) "observers side by side, with adequate optical means, differed as much as twenty or thirty seconds in the times they recorded for phenomena which they have described in almost identical language." [1]

Such uncertainties in the data admitted of a corresponding variety in the results. From the British observations of ingress and egress Sir George Airy [2] derived, in 1877, a solar parallax

[1] *Month. Not.*, vol. xxxviii. p. 447.   [2] *Ibid.*, p. 11.

of 8.76″ (corrected to 8.754″), indicating a mean distance of
93,375,000 miles.   Mr. Stone obtained a value of ninety-two
millions (parallax 8.88″), and held any parallax less than 8.84″
or more than 8.93″ to be "absolutely negatived" by the docu-
ments available.[1]   Yet, from the same, Colonel Tupman deduced 8.81″,[2] implying a distance 700,000 miles greater than
Stone had obtained.   The French observations of contacts
gave (the best being selected) a parallax of about 8.88″; French
micrometric measures the obviously exaggerated one of 9.05″.[3]

Photography, as practised by most of the European parties,
was a total failure.   Utterly discrepant values of the microscopic displacements designed to serve as sounding-lines for
the solar system, issued from attempts to measure even the
most promising pictures.   "You might as well try to measure
the zodiacal light," it was remarked to Sir George Airy.
Those taken on the American plan (adopted by Lord Lindsay),
of using telescopes of so great focal length as to afford, without
further enlargement, an image of the requisite size, gave notably
better results.   From an elaborate comparison of these (some
dating from Vladivostock, Nagasaki, and Pekin, others from
Kerguelen and Chatham Islands), Mr. D. P. Todd, of the
American Nautical Almanac, deduced a solar distance of about
ninety-two million miles (parallax 8.883″ ± 0.034″),[4] a value, as
Mr. Stone has pointed out, favoured by a considerable accumulation of independent evidence.

On the whole, estimates of the great spatial unit cannot be
said to have gained any security from the combined effort of
1874.   A few months before the transit, Mr. Proctor considered
that the uncertainty then amounted to 1,448,000 miles;[5] five
years after the transit, Professor Harkness judged it to be still
1,575,950 miles;[6] yet it had been hoped that it would have
been brought down to 100,000.   As regards the end for which

[1] *Month. Not.*, vol. xxxviii. p. 294.   [2] *Ibid.*, p. 334.
[3] *Comptes Rendus*, t. xcii. p. 812.   [4] *Observatory*, No. 51, p. 205.
[5] *Transits of Venus*, p. 89 (1st ed.)
[6] *Am. Jour. of Sc.*, vol. xx. p. 393.

it had been undertaken, the grand campaign had come to nothing. Nevertheless, no sign of discouragement was apparent. There was a change of view, but no relaxation of purpose. The problem, it was seen, could be solved by no single heroic effort, but by the patient approximation of gradual improvements. Astronomers, accordingly, looked round for fresh means or more refined expedients for applying those already known. A new phase of exertion was entered upon.

On September 5, 1877, Mars came into opposition near the part of his orbit which lies nearest to that of the earth, and Dr. Gill (now Her Majesty's Astronomer at the Cape of Good Hope) took advantage of the circumstance to appeal once more to him for a decision on the *quæstio vexata* of the sun's distance. He chose, as the scene of his labours, the Island of Ascension, and for their plan a method recommended by Airy in 1857,[1] but never before fairly tried. This is known as the "diurnal method of parallaxes." Its principle consists in substituting successive morning and evening observations from the same spot, for simultaneous observations from remote spots, the rotation of the earth supplying the necessary difference in the points of view. Its great advantage is that of unity in performance. A single mind, looking through the same pair of eyes, reinforced with the same optical appliances, is employed throughout, and the errors inseparable from the combination of data collected under different conditions are avoided. There are many cases in which one man can do the work of two better than two men can do the work of one. The result of Dr. Gill's skilful determinations (made with Lord Lindsay's heliometer) was a solar parallax of 8.78″, corresponding to a distance of 93,080,000 miles.[2] The bestowal of the Royal Astronomical Society's gold medal stamped the merit of this distinguished service.

But there are other subjects for this kind of inquiry besides Mars and Venus. Professor Galle of Breslau suggested in

[1] *Month. Not.*, vol. xvii. p. 219.
[2] *Mem. Roy. Astr. Soc.*, vol. xlvi. p. 163.

1872 [1] that some of the minor planets might be got to repay astronomers for much disinterested toil spent in unravelling their motions, by lending aid to their efforts towards a correct celestial survey. Ten or twelve come near enough, and are bright enough for the purpose; and, in fact, the absence of sensible magnitude is one of their chief recommendations, since a point of light offers far greater facilities for exact measurement than a disc. The first attempt to work this new vein was made at the opposition of Phocæa in 1872; and from observations of Flora in the following year at twelve observatories in the northern and southern hemispheres, Galle deduced a solar parallax of 8.87″.[2] At the Mauritius in 1874, Lord Lindsay and Dr. Gill applied the "diurnal method" to Juno, then conveniently situated for the purpose; and the continued use of similar occasions affords, in the opinion of the latter, the best available means for improving knowledge of the sun's distance. They frequently recur; they need no elaborate preparation; a single astronomer armed with a heliometer can do all the requisite work. The recommendation of Dr. Gill was accordingly acted upon in 1882, when favourable oppositions of both Victoria and Sappho took place; and it is probable that each future event of the kind will be made to serve as a step towards the desired level of accuracy.

The second of the nineteenth-century pair of Venus-transits was looked forward to with much abated enthusiasm. Russia refused her active co-operation in observing it, on the ground that oppositions of the minor planets were trigonometrically more useful, and financially far less costly; and her example was followed by Austria, while Italian astronomers limited their sphere of action to their own peninsula. Nevertheless, it was generally held that a phenomenon which the world could not again witness until it was four generations older should, at the price of any effort, not be allowed to pass in neglect.

---

[1] *Astr. Nach.*, No. 1897.
[2] Hilfiker, *Bern Mittheilungen*, 1878, p. 109.

An International Conference, accordingly, met at Paris in 1881 with a view to concerting a plan of operations. America, however, preferring independent action, sent no representative ; and the European break-down of photography in recording transit-phases was admitted by its official abandonment. It was decided to give Delisle's method another trial; and the ambiguities attending and marring its use were sought to be obviated by careful regulations for ensuring agreement in the estimation of the critical moments of ingress and egress.[1] But, in fact (as M. Puiseux had shown [2]), contacts between the limbs of the sun and planet, so far from possessing the geo-metrical simplicity long attributed to them, are really made up of a prolonged succession of various and varying phases, im-possible either to predict or identify with anything like rigid exactitude. Dr. Ball compares the task of determining the pre-cise instant of their meeting or parting, to that of telling the hour with accuracy on a watch without a minute-hand ; and the comparison is admittedly inadequate. For not only is the apparent movement of Venus across the sun extremely slow, being but the excess of her real motion over that of the earth ; but three distinct atmospheres—the solar, terrestrial, and cytherean—combine to deform outlines and mask the geometrical relations which it is desired to connect with a strict count of time.

The result was very much what had been expected. The arrangements were excellent, and were only in a few cases disconcerted by bad weather. The British parties, under the experienced guidance of Mr. Stone, the Radcliffe observer, took up positions scattered (not at random) over the globe from Queensland to Bermuda, and accumulated an ample supply of skilful observations ; the Americans gathered in a whole library of photographs, amongst them a fine series taken at the new Lick Observatory on Mount Hamilton ; the Germans and Belgians trusted to the heliometer; the French used the

---

[1] *Comptes Rendus,* t. xciii. p. 569.   [2] *Ibid.,* t. xcii. p. 481.

camera as an adjunct to the method by contacts. Yet little or no approach was made to solving the problem. The range of doubt as to the sun's distance remained as wide as before. The value published by M. Houzeau, late director of the Brussels Observatory, in 1884,[1] forcibly illustrates this unwelcome conclusion. From 606 measures of Venus on the sun, taken with a new kind of heliometer at St. Jago in Chili, he derives a solar parallax of 8.911″, and a distance of 91,727,000 miles. But the "probable error" of this determination amounts to 0.084″ either way; that is, it is subject to a "more or less" of 900,000 miles, or to a total uncertainty of 1,800,000.

The state and progress of knowledge on this important subject have thus not materially altered since they were summed up by Faye and Harkness in 1881.[2] The methods employed in its investigation fall (as we have seen) into three separate classes—the trigonometrical, the gravitational, and the "phototachymetrical"—an ungainly adjective used to describe the method by the velocity of light. Each has its special difficulties and sources of error; each has counter-balancing advantages. The last of the three is that in which M. Faye places most confidence. As its mean result he finds a parallax of 8.813″, implying a distance of 92,750,000 miles—exact, in his opinion, to 104,000. And this agrees admirably with Todd's mean result, from the same method, of 92,800,000 miles (parallax 8.803″).[3] On the other hand, considerably divergent values have high authorities in their favour. Cornu (as already stated) obtained 8.86″; and Harkness, from a combination of Glasenapp's "light-equation" with Michelson's light-velocity, deduces 8.758″.

By a beautiful series of experiments on Foucault's principle, Master A. A. Michelson, of the United States Navy, fixed in 1879 the rate of luminous transmission at 299,930 kilometres a

[1] In *Annales de l' Obs.*, t. v. 1884. See *Observatory*, vol. vii. p. 212.

[2] *Comptes Rendus*, t. xcii. p. 375; *Am. Jour. of Sc.*, vol. xxii. p. 375.

[3] *Am. Jour. of Sc.*, vol. xix. 1880, p. 64.

second.[1]   This determination claims, and doubtless possesses, a high degree of accuracy.   Todd believes it to be entitled to four times as much confidence as any previous one ; and its credit extends to the values of parallax arrived at by its means. Nevertheless there are still difficulties.   Experiments on the velocity of light are necessarily made in air at the ordinary pressure; the results are then "corrected for a vacuum;" but we cannot be sure that even thus they give the precise rate at which it flies from planet to planet.   Further, an uncertainty of several hundredths of a second still prevails as to the precise amount of the aberration of light.[2]   Nor are the eclipses of Jupiter's satellites, upon which the value of the light-equation depends, by any means instantaneous phenomena ; so that the apparent times of their occurrence may easily be erroneous by one or two seconds.   All these sources of uncertainty are on an extremely minute scale ; they become, however, enormously magnified in the resulting distance of the sun.

On the whole, the most promising plan of investigation *at present* is the "diurnal method" applied to minor planets in opposition, as exemplified by Lord Lindsay and Dr. Gill in 1874.   But the method by lunar and planetary disturbances is unlike all the others in having time on its side.   It is this which Leverrier declared with emphasis must inevitably prevail, because its accuracy is continually growing.[3]   The scarcely perceptible errors which still impede its application are of such a nature as to accumulate year by year; eventually, then, they will challenge, and must receive, a more and more perfect correction.

The best authorities now concur in placing the sun somewhere between ninety-two and ninety-three millions of miles from us.   Mr. Stone abides by 92,000,000 ; Professor Harkness prefers 92,365,000 ; M. Faye, 92,750,000 ; Professor Young,

[1] *Am. Jour. of Sc.*, vol. xviii. p. 393.

[2] Struve's "constant of aberration," 20.445″, has lately been increased to 20.517″ by M. Magnus Nyrèn of St. Petersburg.

[3] *Month. Not.*, vol. xxxv. p. 401.

92,885,000 ; [1] Dr. Ball, 93,000,000.[2] If the accord is not all
that could be desired, it is encouraging to remember that
throughout the first half of the last century doubt claimed a
margin of fully twenty million miles ; now *possible* error
amounts to little more than one and a half millions, and *probable* error is of even less extent.

[1] *The Sun*, p. 278.   See, however, his lecture on "Pending Problems in
Astronomy," *Nature*, September 18, 1884, in which he admits his previous
confidence to be somewhat shaken by the results of the last transit of
Venus.

[2] In his discourse "On the Sun's Distance" at the Southport Meeting
of the British Association, September 21, 1883.

CHAPTER VII.

PLANETS AND SATELLITES.

JOHANN HIERONYMUS SCHRÖTER was the Herschel of Germany. He did not, it is true, possess the more brilliant gifts of his rival. Herschel's piercing discernment, comprehensive intelligence, and inventive splendour were wanting to him. He was, nevertheless, the founder of descriptive astronomy in Germany, as Herschel was in England.

Born at Erfurt in 1745, he prosecuted legal studies at Göttingen, and there imbibed from Kästner a life-long devotion to science. From the law, however, he got the means of living, and, what was to the full as precious to him, the means of observing. Entering the sphere of Hanoverian officialism in 1788, he settled a few years later at Lilienthal near Bremen, as "Oberamtmann," or chief magistrate. Here he built a small observatory, enriched in 1785 with a seven-foot reflector by Herschel, then one of the most powerful instruments to be found anywhere out of England. It was soon surpassed, through his exertions, by the first-fruits of native industry in that branch. Schrader of Kiel transferred his workshops to Lilienthal in 1792, and constructed there, under the superintendence and at the cost of the astronomical Oberamtmann, a thirteen-foot reflector, declared by Lalande to be the finest telescope in existence, and one twenty-seven feet in focal length, probably as inferior to its predecessor in real efficiency as it was superior in size.

Thus, with instruments of gradually increasing power, Schröter studied during thirty-four years the topography of the moon and planets. The field was then almost untrodden; he had but few and casual predecessors, and has since had no equal in the sustained and concentrated patience of his hourly watchings. Both their prolixity and their enthusiasm are faithfully reflected in his various treatises. Yet the one may be pardoned for the sake of the other, especially when it is remembered that he struck out a substantially new line, and that one of the main lines of future advance. Moreover, his infectious zeal communicated itself; he set the example of observing when there was scarcely an observer in Germany; and under his roof Harding and Bessel received their training as practical astronomers.

But he was reserved to see evil days. Early in 1813 the French under Vandamme occupied Bremen. On the night of April 20, the Vale of Lilies was, by their wanton destructiveness, laid waste with fire; the Government offices were destroyed, and with them the chief part of Schröter's property, including the whole stock of his books and writings. There was worse behind. A few days later, his observatory, which had escaped the conflagration, was broken into, pillaged, and ruined. His life was wrecked with it. He survived the catastrophe three years without the means to repair, or the power to forget it, and gradually sank from disappointment into decay, terminated by death, August 29, 1816. He had, indeed, done all the work he was capable of; and though not of the first quality, it was far from contemptible. He laid the foundation of the *comparative* study of the moon's surface, and the descriptive particulars of the planets laboriously collected by him constituted a store of more or less reliable information hardly added to during the ensuing half century. They rested, it is true, under some shadow of doubt; but the most recent observations have tended on several points to rehabilitate the discredited authority of the Lilienthal astronomer. We may now briefly resume, and pursue in its further progress the course of his studies, taking the planets in the order of their distances from the sun.

T

In April 1792 Schröter first saw reason to conclude, from the gradual degradation of light on its partially illuminated disc, that Mercury possesses a tolerably dense atmosphere.[1] During the transit of May 7, 1799, he was, moreover, struck with the appearance of a ring of softened luminosity encircling the planet to an apparent height of three seconds, or about a quarter of its own diameter.[2] Although a "mere thought" in texture, yet it remained persistently visible both with the seven-foot and the thirteen-foot reflectors, armed with powers up to 288. It had a well-marked greyish boundary, and reminded him, though indefinitely fainter, of the penumbra of a sun-spot. A similar appendage, but more distinctly bright, had been noticed by De Plantade at Montpellier, November 11, 1736, and again in 1786 and 1789 by Prosperin and Flaugergues. Mercury projected on the sun, November 9, 1802, appeared to Ljunberg at Copenhagen surrounded with a dark zone; but Herschel, on the same day, saw its "preceding limb cut the luminous solar clouds with the most perfect sharpness."[3] The presence, however, of a "halo," appearing to some observers a little darker, to others a little brighter than the solar surface, was unmistakable in 1832. Professor Moll of Utrecht described it as "a nebulous ring of a darker tinge, approaching to the violet colour."[4] To Huggins and Stone, November 5, 1868, it showed as lucid and most distinct. No change in the colour of the glasses used, or the powers applied, could get rid of it, and it lasted throughout the transit.[5] It was again well seen by Christie and Dunkin at Greenwich, May 6, 1878,[6] and with much precision of detail by Trouvelot at Cambridge (U.S.)[7] No observations of much interest were made during the transit of November 8, 1881. Dr. Little, at Shanghai, perceived an unvarying "darkish halo," of which,

---

[1] *Neueste Beyträge zur Erweiterung der Sternkunde*, Bd. iii. p. 14 (1800).
[2] *Ibid.*, p. 24.   [3] *Phil. Trans.*, vol. xciii. p. 215.
[4] *Mem. Roy. Astr. Soc.*, vol. vi. p. 116.
[5] *Month. Not.*, vol. xxix. pp. 11, 25.   [6] *Ibid.*, vol. xxxviii. p. 398.
[7] *Am. Jour. of Sc.*, vol. xvi. p. 124.

however, neither Mr. Ellery at Melbourne, nor Mr. Tebbutt at Windsor, New South Wales, saw any trace.[1] They, on the other hand, took note of a certain whitish spot on the planet's disc, which, ever since 1697, when it was detected by Wurzelbauer at Erfurt, has been one of the most frequent attendant phenomena of a transit of Mercury. It is not always centrally situated, and is sometimes seen in duplicate, so that Powell's explanation by diffraction is obviously insufficient. Nevertheless there can scarcely be a doubt that it is an optical effect of some kind.

As to the "halo," it is less easy to decide. That Mercury possesses a considerably refractive atmosphere is certified by the observation of De Plantade in 1736,[2] and the still more definite observation of Simms in 1832,[3] of a luminous edge to the part of the disc *outside* the sun at ingress or egress. The natural complement to this appearance would be a dusky annulus round the planet *on* the sun—precisely such as was seen by Moll and Little—due to the imperfect transparency of its gaseous envelope. But the brilliant ring vouched for by others is not so readily explicable. Airy has shown that it cannot possibly be caused by refraction, and must accordingly be set down as "strictly an ocular nervous phenomenon."[4] It is the less easy to escape from this conclusion that we find the virtually airless moon capable of exhibiting a like appendage. Professor Stephen Alexander of the United States Survey, with two other observers, perceived, during the eclipse of the sun of July 18, 1860, the advancing lunar limb to be bordered with a bright band;[5] and photographic effects of the same kind appear in pictures of transits of Venus and partial solar eclipses.

In the case of Mercury, a real effect is perhaps complicated with an illusory one. Different eyes are very differently sensitive to degrees of light and shade. Absorption by a Mercurian atmosphere is doubtless in some degree present, and it may be

[1] *Month. Not.*, vol. xlii. pp. 101–104.
[2] *Mém. de l'Ac.*, 1736, p. 440.   [3] *Month. Not.*, vol. ii. p. 103.
[4] *Ibid.*, vol. xxiv. p. 18.   [5] *Ibid.*, vol. xxiii. p. 234 (Challis).

that the faintly shadowed ring produced by it impresses some observers, by contrast with the ink-black disc of the planet, as bright. The further investigation of this curious subject must wait for the next transit of Mercury, May 9, 1891.

As to the constitution of this planet, the spectroscope has little to tell. Its light is of course that of the sun reflected, and its spectrum is consequently a faint echo of the Fraunhofer spectrum. Dr. H. C. Vogel, who first examined it in April 1871, *suspected* traces of the action of an atmosphere like ours,[1] but, it would seem, on slight grounds. It is, however, certainly very poor in blue rays.

On March 26, 1800, Schröter, observing with his 13-foot reflector in a peculiarly clear sky, perceived the southern horn of Mercury's crescent to be quite distinctly blunted.[2] Interception of sunlight by a Mercurian mountain rather more than eleven English miles high, explained the effect to his satisfaction. By carefully timing its recurrence, he concluded rotation on an axis in a period of 24 hours 4 minutes. This was the first determination of the kind, and was the reward of twenty years' unceasing vigilance. It was confirmed by watching the successive appearances of a dusky streak and blotch in May and June 1801.[3] These, however, were inferred to be no permanent markings on the body of the planet, but atmospheric formations, the streak at times drifting forwards (it was thought) under the fluctuating influence of Mercurian breezes. From a rediscussion of these observations Bessel inferred that Mercury rotates on an axis inclined 70° to the plane of its orbit in 24 hours 53 seconds. A close analogy would thus exist between the alternations of its seasons and those of the earth, save that their effects must (except within the polar circles) be well-nigh swallowed up in the larger vicissitudes produced by the considerable eccentricity of its path, causing its distance from the sun to vary from 29 to 43 million miles, and the

[1] *Untersuchungen über die Spectra der Planeten*, p. 9.
[2] *Neueste Beyträge*, Bd. iii. p. 50.
[3] *Astr. Jahrbuch*, 1804, pp. 97–102.

light and heat received, from four to ten times the amount reaching our planet.

The rounded appearance of the southern horn seen by Schröter was more or less doubtfully caught by Noble (1864), Burton, and Franks (1877);[1] but was obvious to Mr. W. F. Denning at Bristol on the morning of November 5, 1882.[2] He also discerned brilliant and dusky spaces, the displacements of which, during four days, indicated rotation in about twenty-five hours. The general aspect of the planet reminded him of that of Mars;[3] but the difficulties in the way of its observation are enormously enhanced by its constant close attendance on the sun.

The theory of Mercury's movements has always given trouble. In Lalande's,[4] as in Mästlin's time, the planet seemed to exist for no other purpose than to throw discredit on astronomers; and even to Leverrier's powerful analysis it long proved recalcitrant. On the 12th of September 1859, however, he was able to announce before the Academy of Sciences[5] the terms of a compromise between observation and calculation. They involved the addition of a new member to the solar system. The hitherto unrecognised presence of a body about the size of Mercury itself, revolving at somewhat less than half its mean distance from the sun (or, if farther, then of less mass, and *vice versa*), would, it was pointed out, produce exactly the effect required, of displacing the perihelion of the former planet 38 seconds a century more than could otherwise be accounted for. The planes of the two orbits, however, should not lie far apart, as otherwise a nodal disturbance would arise not perceived to exist. It was added that a ring of asteroids similarly placed would answer the purpose equally well, and was more likely to have escaped notice.

Upon the heels of this forecast followed promptly a seeming verification. Dr. Lescarbault, a physician residing at Orgères,

[1] Webb, *Celestial Objects*, p. 46 (4th ed.)
[2] *L'Astronomie*, t. ii. p. 141.   [3] *Observatory*, No. 82, p. 40.
[4] *Hist. de l'Astr.*, p. 682.   [5] *Comptes Rendus*, t. xlix. p. 379.

whose slender opportunities had not blunted his hopes of
achievement, had, ever since 1845, when he witnessed a transit
of Mercury, cherished the idea that an unknown planet might
be caught thus projected on the solar background.   Unable to
observe continuously until 1858, he, on March 26, 1859, saw
what he had expected—a small, perfectly round object slowly
traversing the sun's disc.   The fruitless expectation of re-
observing the phenomenon, however, kept him silent, and it
was not until December 22, after the news of Leverrier's prediction had reached him, that he wrote to acquaint him with
his supposed discovery.[1]   The Imperial Astronomer thereupon
hurried down to Orgères, and by personal inspection of the
simple apparatus used, by searching cross-examination and
local inquiry, convinced himself of the genuine character and
substantial accuracy of the reported observation.   He named
the new planet "Vulcan," and computed elements giving it a
period of revolution slightly under twenty days.[2]   But it has
never since been seen.   M. Liais, director of the Brazilian
Coast Survey, thought himself justified in asserting that it
never had been seen.   Observing the sun for twelve minutes
after the supposed ingress recorded at Orgères, he noted those
particular regions of its surface as "très uniformes d'in-
tensité."[3]   He subsequently, however, admitted Lescarbault's
good faith, at first rashly questioned.   The planet-seeking doctor
was, in truth, only one among many victims of similar illusions.

Waning interest in the subject was revived by a fresh announcement of a transit witnessed, it was asserted, by Weber
at Peckaloh, April 4, 1876.[4]   The pseudo-planet, indeed, was
detected shortly afterwards on the Greenwich photographs,
and was found to have been seen by M. Ventosa at Madrid
in its true character of a sun-spot without penumbra ; but
Leverrier had meantime undertaken the investigation of a list
or twenty similar dubious appearances, collected by Haase, and

[1] *Comptes Rendus*, t. l. p. 40.   [2] *Ibid.*, p. 46.
[3] *Astr. Nach.*, Nos. 1248 and 1281.
[4] *Comptes Rendus*, t. lxxxiii. pp. 510, 561.

republished by Wolf in 1872.[1] From these five were picked
out as referring in all likelihood to the same body, the reality
of whose existence was now confidently asserted, and of which
more or less probable transits were fixed for March 22, 1877,
and October 15, 1882.[2] But, widespread watchfulness notwithstanding, no suspicious object came into view at either
epoch.

The next announcement of the discovery of "Vulcan" was
on the occasion of the total solar eclipse of July 29, 1878.[3]
This time it was stated to have been seen at some distance
south-west of the obscured sun, as a ruddy star with a minute
planetary disc ; and its simultaneous detection by two observers
—the late Professor James C. Watson, stationed at Rawlins
(Wyoming Territory), and Professor Lewis Swift at Denver
(Colorado)—was at first readily admitted. But their separate
observations could, on a closer examination, by no possibility
be brought into harmony, and, if valid, certainly referred to
two distinct objects, if not to four ; each astronomer eventually
claiming a pair of planets. Nor could any one of the four be
identified with Lescarbault's and Leverrier's Vulcan, which,
if a substantial body revolving round the sun, must then (as
Oppolzer showed)[4] have been found on the *east* side of that
luminary. The most feasible explanation of the puzzle seems
to be that Watson and Swift merely saw each the same two
stars in Cancer : haste and excitement doing the rest.[5] Nevertheless they strenuously maintained their opposite conviction.[6]

Intra-Mercurial planets have since been diligently searched
for when the opportunity of a total eclipse offered, especially
during the long obscuration at Caroline Island. Not only did
Professor Holden "sweep" in the solar vicinity, but Palisa

[1] *Handbuch der Mathematik*, Bd. ii. p. 327.
[2] *Comptes Rendus*, t. lxxxiii. p. 721.
[3] *Nature*, vol. xviii. pp. 461, 495, 539.   [4] *Astr. Nach.*, No. 2239.
[5] *Astr. Nach.*, Nos. 2253-2254 (C. H. F. Peters).
[6] *Ibid.*, Nos. 2263 and 2277. See also Tisserand in *Ann. Bur. des Long.*,
1882, p. 729.

and Trouvelot agreed to divide the field of exploration, and thus make sure of whatever planetary prey there might be within reach; yet with only negative results. Belief in the presence of any considerable body or bodies within the orbit of Mercury is, accordingly, now at a low ebb. Yet the existence of the anomaly in the Mercurian movements indicated by Leverrier has been made only surer by further research.[1] Its elucidation constitutes one of the "pending problems" of astronomy. It need only be remarked that, owing to the absence of extra-disturbance of the nodes, neither a condensation inwards of the matter showing to us as the zodiacal light, nor any accumulation of meteors revolving far from the plane of Mercury's orbit, will meet the requirements of the situation.

From the óbservation at Bologna in 1666–67 of some very faint spots, Domenico Cassini concluded a rotation or libration of Venus—he was not sure which—in about twenty-three hours.[2] By Bianchini in 1726 the period was augmented to twenty-four *days* eight hours. J. J. Cassini, however, in 1740, showed that the data collected by both observers were consistent with rotation in twenty-three hours twenty minutes.[3] So the matter rested until Schröter's time. After watching nine years in vain, he at last, February 28, 1788, perceived the ordinarily uniform brightness of the planet's disc to be marbled with a filmy streak, which returned periodically to the same position in about twenty-three hours twenty-eight minutes. This approximate estimate was corrected by the application of a more definite criterion. On December 28, 1789, the southern horn of the crescent Venus was seen truncated, an outlying lucid point interrupting the darkness beyond. Precisely the same appearance recurred two years later, giving for the planet's

[1] See J. Bauschinger's *Untersuchungen* (1884), summarised in *Bull. Astr.*, t. i. p. 506. Newcomb finds the anomalous motion of the perihelion to be even larger (43″ instead of 38″) than Leverrier made it. *Month. Not.*, Feb. 1884, p. 187.

[2] *Jour. des Sçavans*, Dec. 1667, p. 122.

[3] *Élémens d'Astr.*, p. 525.

rotation a period of twenty three hours twenty-one minutes.[1] To this only twenty-two seconds were added by De Vico, as the result of over 10,000 observations made with the Cauchoix refractor of the Collegio Romano, 1839–41. The axis of rotation was found to be much more bowed towards the orbital plane than that of the earth, the equator making with it an angle of 53° 11′ 26″. Of fundamental importance as regards our views of the planet's constitution, is the fact that De Vico plainly identified the individual markings drawn by Bianchini 113 years earlier.[2] They cannot, then (if this conclusion be accurate), possess the evanescent atmospheric character attributed to them by Schröter, but must be inherent peculiarities of surface.

Of the frequently mountainous nature of that surface there appears to be no reasonable doubt. Francesco Fontana at Naples in 1643 noticed irregularities along the inner edge of the crescent.[3] De la Hire in 1700 considered them—regard being had to difference of distance—to be much more strongly marked than those visible in the moon.[4] Schröter's assertions to the same effect, though scouted with some unnecessary vehemence by Herschel,[5] have since been repeatedly confirmed ; amongst others by Mädler, De Vico, Langdon, who in 1873 saw the broken line of the " terminator " (the boundary between light and darkness) with peculiar distinctness through a veil of auroral cloud ;[6] by Denning,[7] March 30, 1881, despite preliminary impressions to the contrary, as well as by C. V. Zenger at Prague, January 8, 1883. The great mountain mass, presumed to occasion the periodical blunting of the southern horn, was precariously estimated by the Lilienthal observer to rise to the prodigious height of nearly twenty-seven miles, or just five times the elevation of Mount Everest ! Yet the phenomenon persists, whatever may be thought of the explanation. More-

---

[1] *Beobachtungen über die sehr beträchtlichen Gebirge und Rotation der Venus*, 1793, p. 45. Schröter's final result in 1811 was 23h. 21m. 7.977s. *Monat. Corr.*, Bd. xxv. p. 367.

[2] *Astr. Nach.*, No. 404.   [3] *Novæ Observationes*, p. 92.

[4] *Mém. de l'Ac.*, 1700, p. 296.   [5] *Phil. Trans.*, vol. lxxxiii. p. 201.

[6] Webb, *Cel. Objects*, p. 58.   [7] *Month. Not.*, vol. xlii. p. 111.

298   *HISTORY OF ASTRONOMY.*

over, the speck of light beyond, interpreted as the visible sign
of a detached peak rising high enough above the encircling
shadow to catch the first and last rays of the sun, was frequently
discerned by Baron van Ertborn in 1876;[1] while an object
near the northern horn of the crescent, strongly resembling a
lunar ring-mountain, was delineated both by De Vico in 1841
and by Denning forty years later. Another curious circumstance, first observed by Schröter in August 1793, and since
abundantly verified, is that the phases of the crescent Venus
are continually retarded, and of the waning Venus accelerated
by several days. In both cases the disc is illuminated over a
much more restricted area than it ought to be from its position.
The same applies to Mercury. Schröter's explanation by the
arrest of nearly level sunlight through the intervention of lofty
ranges is far from satisfactory; but no other has been offered.

We are almost equally sure that Venus, as that the earth
is encompassed with an atmosphere. Yet, notwithstanding
luminous appearances plainly due to refraction during the
transits both of 1761 and 1769, Schröter, in 1792, took the initiative in coming to a definite conclusion on the subject.[2] It was
founded, first, on the rapid diminution of brilliancy towards
the terminator, attributed to atmospheric absorption; next, on
the extension beyond a semicircle of the horns of the crescent;
lastly, on the presence of a bluish gleam illuminating the early
hours of the Cytherean night with what was taken to be
genuine twilight. Even Herschel admitted that sunlight, by
the same effect through which the heavenly bodies show *visibly
above* our horizons while still *geometrically below* them, appeared
to be bent round the shoulder of the globe of Venus. Ample
confirmation of the fact has since been afforded. At Dorpat
in May 1849, the planet being within 3° 26' of inferior conjunction, Mädler found the arms of waning light upon the disc
to embrace no less than 240° of its extent;[3] and in December

---

[1] *Bull. Ac. de Bruxelles*, t. xliii. p. 22.
[2] *Phil. Trans.*, vol. lxxxii. p. 309; *Aphroditographische Fragmente*, p.
85 (1796).   [3] *Astr. Nach.*, No. 679.

1842, Mr. Guthrie, of Bervie, N.B., actually observed, under similar conditions, *the whole circumference* to be lit up with a faint nebulous glow.[1] Here the solar rays evidently pierced the planet's atmosphere *from behind,* pursuing a curved path, as if through a lens. The same curious phenomenon was intermittently seen by Mr. Leeson Prince at Uckfield in September 1863;[2] but with more satisfactory distinctness by Mr. C. S. Lyman of Yale College,[3] before and after the conjunction of December 11, 1866, and during nearly five hours previous to the transit of 1874, when the yellowish ring of refracted light showed at one point an approach to interruption, it might be presumed through the intervention of a bank of clouds. These effects can be accounted for, as Mr. Neison pointed out,[4] only by supposing the atmosphere of Venus to be nearly twice as dense at the surface of its globe, and to possess nearly twice as much refractive power as that of the earth.

Similar appearances are conspicuous during transits. But while the Mercurian halo is characteristically seen *on* the sun, the "silver thread" round the limb of Venus commonly shows on the part *off* the sun. There are, however, instances of each description in both cases. Mr. Grant, in collecting the records of physical phenomena accompanying the transits of 1761 and 1769, remarks that no one person saw both kinds of annulus, and argues thence a dissimilarity in their respective modes of production.[5] Such a dissimilarity probably exists, in the sense that the inner section of the ring is due to absorption, the outer to refraction by the same planetary atmosphere; but the distinction of separate visibility has not been borne out by recent experience. Several of the Australian observers during the transit of 1874 witnessed the complete phenomenon. Mr. J. Macdonnell, at Eden, saw a "shadowy nebulous ring" surround the whole disc when ingress was two-thirds accomplished;

[1] *Month. Not.*, vol. xiv. p. 169.   [2] *Ibid.*, vol. xxiv. p. 25.
[3] *Am. Jour. of Sc.*, vol. xliii. p. 129 (2d ser.); vol. ix. p. 47 (3d ser.)
[4] *Month. Not.*, vol. xxxvi. p. 347.   [5] *Hist. Phys. Astr.*, p. 431.

Mr. Tornaghi, at Goulburn, perceived a halo, entire and un-mistakable, at half egress.[1]   Similar observations were made at Sydney,[2] and were renewed in 1882 by Lescarbault at Orgères, by Metzger in Java, and by Barnard at Vanderbilt University.[3]

Spectroscopic indications of aqueous vapour as present in the atmosphere of Venus, were obtained in 1874 and 1882, by Tacchini and Riccò in Italy, and by Young in New Jersey.[4] Janssen, however, who made a special study of the point subsequently to the transit of 1882, found them much less certain than his earlier expectations led him to expect;[5] and Vogel, by repeated examinations, 1871–73, could detect only the very slightest variations from the pattern of the solar spectrum.   Some additions there indeed seem to be in the thickening of certain water-lines, and also of a group (B) since shown by Egoroff to be developed through the absorptive action of cool oxygen; but so nearly evanescent as to induce the persuasion that the light we receive from Venus is reflected from a heavy cloud-stratum, and has traversed, consequently, only the rarer upper portion of its atmosphere.[6]   This would also account for the extreme brilliancy of the planet.   On the 26th and 27th of September 1878, a close conjunction gave Mr. James Nasmyth the rare opportunity of watching Venus and Mercury for several hours side by side in the field of his reflector; when the former appeared to him like clean silver, the latter as dull as lead or zinc.[7]   Yet the light *incident* upon Mercury is, on an average, three and a half times as strong as the light reaching Venus.   Thus, the reflective power of Venus must be singularly strong.   And we find accordingly, from

[1] *Mem. Roy. Astr. Soc.*, vol. xlvii. pp. 77, 84.

[2] *Astr. Reg.*, vol. xiii. p. 132.

[3] *L'Astronomie*, t. ii. p. 27; *Astr. Nach.*, No. 2021; *Am. Jour. of Sc.*, vol. xxv. p. 430.

[4] *Mem. Spettr. Ital.*, Dicembre 1882; *Am. Jour. of Sc.*, vol. xxv. p. 328.

[5] *Comptes Rendus*, t. xcvi. p. 288.

[6] Vogel, *Unters. über die Spectra der Planeten*, p. 15.

[7] *Nature*, vol. xix. p. 23.

a combination of Zöllner's with Pickering's results, that its
"albedo" is but little inferior to that of new-fallen snow; in
other words, it gives back $72\frac{1}{2}$ per cent. of the luminous rays
impinging upon it.

This view, that we see only the cloud-canopy of Venus, is
manifestly inconsistent with the supposed permanency of its
spots, or with the perception of shadow effects on a rugged
crust. It is, however, with some reservation, shared by Mr. E.
L. Trouvelot, who since 1875 has pursued a diligent telescopic
study of the planet at Cambridge (U.S.) Not the least surprising fact about this sister-globe is that the axis on which it
rotates is *hooded* at each end with some shining substance.
These polar appendages were discovered in 1813 by Gruithuisen,[1] who set them down as polar snow-caps like those
of Mars. Nor is it altogether certain that he was wrong.
Trouvelot, indeed, in January 1878, perceived (or thought that
he perceived) the southern one to be composed of isolated
peaks thrown into relief against the sky, and hence concluded
both to represent lofty groups of mountains penetrating the
vapour-stratum supposed to form the greater part of the visible
disc. He pointed out, moreover, that the place of the southern
spot might be called identical with that of a projection above
the limb detected by MM. Bouquet de la Grye and Arago in
measuring photographs of Venus in transit taken at Puebla
and Port-au-Prince in 1882.[2] This projection corresponded
to a real elevation of about sixty-five miles. But it was more
probably due to "photographic irradiation" from a local
excess of brilliancy, the result—according to the French investigators' conjecture—of accumulations of ice and snow, or
the continuous formation of vast cloud-masses.

The same photographs show that in figure Venus very
closely resembles our earth, the equatorial bulging produced
by rotation being $\frac{1}{303}$ of its mean radius.

The "secondary," or "ashen light" of Venus was first

---

[1] *Nova Acta Acad. Naturæ Curiosorum,* Bd. x. p. 239.
[2] *Observatory,* vols. iii. p. 416, vii. p. 239.

noticed by Riccioli in 1643; it was seen by Derham about 1715, by Kirch in 1721, by Schröter and Harding in 1806;[1] and the reality of the appearance has since been authenticated by numerous and trustworthy observations. It is precisely similar to that of the "old moon in the new moon's arms;" and Zenger, who witnessed it with unusual distinctness, January 8, 1883,[2] supposes it due to the same cause—namely, to the faint gleam of reflected earth-light from the night-side of the planet. When we remember, however, that "full earth-light" on Venus, at its nearest, has little more than $\frac{1}{12,000}$ its intensity on the moon, we see at once that the explanation is inadequate. Nor can Professor Schafarik's,[3] by phosphorescence of the warm and teeming oceans with which Zöllner[4] regarded the globe of Venus as mainly covered, be seriously entertained. Vogel's suggestion is more plausible. He and Lohse, at Bothkamp, November 3-11, 1871, saw the dark hemisphere *partially* illuminated by secondary light, extending 30° from the terminator, and thought the effect might be produced by a very extensive twilight.[5] An atmospheric diffusion of sunlight seems, in fact, the best answer to the riddle. It involves difficulties, but probably none that are insuperable.

The third planet encountered in travelling outwards from the sun is the abode of man. He has in consequence opportunities of studying its physical habitudes altogether different from the baffling glimpses afforded to him of the other members of the solar family. Regarding the earth, then, a mass of knowledge so varied and comprehensive has been accumulated as to form a science—or rather several sciences—apart. But underneath all lie astronomical relations, the recognition and investi-

---

[1] *Astr. Jahrbuch*, 1809, p. 164.   [2] *Month. Not.*, vol. xliii. p. 331.
[3] *Report Brit. Ass.*, 1873, p. 407. The paper contains a valuable record of observations of the phenomenon.
[4] *Photom. Untersuchungen*, p. 301.
[5] *Beobachtungen zu Bothkamp*, Heft ii. p. 126.

gation of which constitute one of the most significant intellectual events of the present century.

It is indeed far from easy to draw a line of logical distinction between items of knowledge which have their proper place here, and those which should be left to the historian of geology. There are some, however, of which the cosmical connections are so close that it is impossible to overlook them. Amongst these is the ascertainment of the solidity of the globe. At first sight it seems difficult to conceive what the apparent positions of the stars can have to do with subterranean conditions; yet it was from star measurements alone that Hopkins, in 1839, concluded the earth to be solid to a depth of at least 800 or 1000 miles.[1] His argument was, that if it were a mere shell filled with liquid, precession and nutation would be much larger than they are observed to be. For the shell alone would follow the pull of the sun and moon on its equatorial girdle, leaving the liquid behind; and being thus so much the lighter, would move the more readily. There is, it is true, grave reason to doubt whether this reasoning corresponds with the actual facts of the case;[2] but the conclusion to which it led has been otherwise affirmed and extended.

Indications to an identical effect have been derived from another kind of external disturbance, affecting our globe through the same agencies. Sir William Thomson pointed out in 1862[3] that tidal influences are brought to bear on land as well as on water, although obedience to them is perceptible only in the mobile element. Some bodily distortion of the earth's figure *must* however take place, unless we suppose it of absolute or "preternatural" rigidity, and the amount of such distortion

---

[1] *Phil. Trans.*, 1839, 1841, 1842.

[2] Delaunay objected (*Comptes Rendus*, t. lxvii. p. 65) that the viscosity of the contained liquid (of which Hopkins took no account) would, where the movements were so excessively slow as those of the earth's axis, almost certainly cause it to behave like a solid. Sir W. Thomson, however (*Report Brit. Ass.*, 1876, ii. p. 1), considers Hopkins's argument valid as regards the comparatively quick solar semi-annual and lunar fortnightly nutations,

[3] *Phil. Trans.*, vol. cliii. p. 573.

can be determined from its effect in diminishing oceanic tides below their calculated value. For if the earth were perfectly plastic to the stresses of solar and lunar gravity, tides—in the ordinary sense—would not exist. Continents and oceans would swell and subside together. It is to the *difference* in the behaviour of solid and liquid terrestrial constituents that the ebb and flow of the waters are due.

Six years later, the distinguished Glasgow professor suggested that this criterion might, by the aid of a prolonged series of exact tidal observations, be practically applied to test the interior condition of our planet.[1] In 1882, accordingly, suitable data extending over thirty-three years having at length become available, Mr. G. H. Darwin performed the laborious task of their analysis, with the general result that the "effective rigidity" of the earth's mass must be *at least* as great as that of steel.[2]

In a paper read before the Geological Society, December 15, 1830,[3] Sir John Herschel threw out the idea that the perplexing changes of climate revealed by the geological record might be explained through certain slow fluctuations in the eccentricity of the earth's orbit, produced by the disturbing action of the other planets. Shortly afterwards, however, he abandoned the position as untenable;[4] and it was left to Mr. James Croll, in 1864 and subsequent years, to reoccupy and convert it into a strong, if not an impregnable one. Within restricted limits (as Lagrange, and, more certainly and definitely, Leverrier proved), the path pursued by our planet round the sun alternately contracts, in the course of ages, into a moderate ellipse, and expands almost to a circle, the major axis, and consequently the mean distance, remaining invariable. Even at present, when the eccentricity approaches a minimum, the sun is nearer to us in January than in July by above three million miles,

[1] *Report Brit. Ass.*, 1868, p. 494.   [2] *Ibid.*, 1882, p. 474.
[3] *Trans. Geol. Soc.*, vol. iii. (2d ser.), p. 293.
[4] See his *Treatise on Astronomy*, p. 199 (1833).
[5] *Phil. Mag.*, vol. xxviii. (4th ser.), p. 121.

and some 850,000 years ago this difference was more than four times as great. Mr. Croll has brought together [1] a mass of evidence to support the view that, at epochs of considerable eccentricity, the hemisphere of which the winter, occurring at aphelion, was both intensified and prolonged, must have undergone extensive glaciation ; while the opposite hemisphere, with a short, mild winter, and long, cool summer, enjoyed an approach to perennial spring. These conditions were exactly reversed at the end of 10,500 years, through the shifting of the perihelion combined with the precession of the equinoxes, the frozen hemisphere blooming into a luxuriant garden as its seasons came round to occur at the opposite sides of the terrestrial orbit, and the vernal hemisphere subsiding simultaneously into ice-bound rigour. Thus a plausible explanation was offered of the anomalous alternations of glacial and semi-tropical periods, attested, on incontrovertible geological evidence, as having succeeded each other in times past over what are now temperate regions. The most recent glacial epoch is placed by Mr. Croll about 200,000 years ago, when the eccentricity of the earth's orbit was 3.4 times as great as it now is. At present, a faint representation of such a state of things is afforded by the southern hemisphere. One condition of glaciation in the coincidence of winter with the maximum of remoteness from the sun, is present ; the other— a high eccentricity—is deficient. Yet the ring of ice-bound territory hemming in the southern pole is well known to be far more extensive than the corresponding region in the north.

This ingenious hypothesis has certainly made good its footing among the better-warranted speculations of science. The precise nature of the connection between geological and astronomical events indicated by it may be questioned, but there can no longer be any doubt that, in some form, such a relation exists. Its ascertainment marks one further step in that process of unification between things celestial and things ter-

[1] *Climate and Time*, 1875.

restrial which forms, it might be said, the vast presiding idea of astronomical history during the last three centuries.

The first attempt at an experimental estimate of the " mean density" of the earth was Maskelyne's observation in 1774 of the deflection of a plumb-line through the attraction of Schehallien. The conclusion thence derived, that our globe weighs $4\frac{1}{2}$ times as much as an equal bulk of water,[1] was not very exact. It was considerably improved upon by Cavendish, who, in 1798, brought into use the "torsion-balance" constructed for the same purpose by John Michell. The resulting estimate of 5.48 was raised to 5.66 by Francis Baily's elaborate repetition of the process in 1838–42. From the latest experiments on the subject—those made in 1872–73 by Cornu and Baille—the slightly inferior value of 5.56 was derived; and it was further shown that the data collected by Baily, when corrected for a systematic error, gave a practically identical result (5.55).[2] Newton's guess at the average weight of the earth as five or six times that of water has thus been curiously verified.

Operations for determining the figure of the earth have been carried out during the present century on an unprecedented scale. The Russo-Scandinavian arc, of which the measurement was completed under the direction of the elder Struve in 1855, reached from Hammerfest to Ismailia on the Danube, a length of 25° 20′. But little inferior to it was the Indian arc, begun by Lambton in the first years of the century, continued by Everest, revised and extended by Walker. The general upshot is to show that the polar compression of the earth is somewhat greater than had been supposed. The admitted fraction until lately was $\frac{1}{300}$; that is to say, the thickness of the protuberant equatorial ring was taken to be $\frac{1}{300}$ of the mean radius. But Sabine's pendulum experiments, discussed by Airy in 1826, gave $\frac{1}{289}$;[3] and arc measurements tend more and more towards agreement with this figure. A

---

[1] *Phil. Trans.*, vol. lxviii. p. 783.   [2] *Comptes Rendus*, t. lxxvi. p. 954.
[3] *Phil. Trans.*, vol. cxvi. p. 548.

fresh investigation led the late J. B. Listing in 1878 [1] to state
the dimensions of the terrestrial spheroid as follows : equatorial
radius = 6,377,377 metres ; polar radius = 6,355,270 metres ;
ellipticity = $\frac{1}{288.48}$.

It is, however, far from certain that the figure of the earth is
one of strict geometrical regularity. Nay, it is by no means
clear that even its main outlines are best represented by what
is called an " ellipsoid of revolution "—in other words, by a
globe flattened at top and bottom, but symmetrical on every
side. From a survey of geodetical results all over the world,
Colonel Clarke concludes that different meridians possess
different amounts of curvature ; [2] so that the equator, instead
of being a circle, as it should be—apart from perturbing causes
—in a rotating body, must, on this view, be itself an ellipse,
and our planet be correctly described as in shape "an ellipsoid of three unequal axes." But the point is still *sub judice*.
Operations towards its decision are in active progress both in
Europe and India.

The moon possesses for us an unique interest. She in all probability shared the origin of the earth ; she perhaps prefigures
its decay. She is at present its minister and companion. Her
existence, so far as we can see, serves no other purpose than to
illuminate the darkness of terrestrial nights, and to measure,
by swiftly-recurring and conspicuous changes of aspect, the
long span of terrestrial time. Inquiries stimulated by visible
dependence, and aided by relatively close vicinity, have resulted in a wonderfully minute acquaintance with the features
of the single lunar hemisphere open to our inspection.

Selenography, in the modern sense, is not yet a hundred
years old. It originated with the publication in 1791 of
Schröter's *Selenotopographische Fragmente*. [3] Not but that the
lunar surface had already been diligently studied, chiefly by

[1] *Astr. Nach.*, No. 2228.
[2] *Phil. Mag.*, vol. vi. (5th ser.), p. 92.
[3] The second volume was published at Göttingen in 1802.

Hevelius, Cassini, and Tobias Mayer; the idea, however, of investigating the moon's physical condition, and detecting symptoms of the activity there of natural forces through minute topographical inquiry, first obtained effect at Lilienthal. Schröter's delineations, accordingly, imperfect though they were, afforded a starting-point for a *comparative* study of the superficial features of our satellite.

The first of the curious objects which he named "rills" was noted by him in 1787. Before 1801 he had found eleven; Lohrmann added 75; Mädler 55; Schmidt published in 1866 a catalogue of 425, of which 278 had been detected by himself;[1] and he eventually brought the number up to nearly 1000. They are, then, a very persistent lunar feature, though wholly without terrestrial analogue. There is no difference of opinion as to their nature. They are quite obviously clefts in a rocky surface, 100 to 500 yards deep (the depression of the great rill near Aristarchus was estimated by Schmidt at 554 yards), usually a couple of miles across, and pursuing straight, curved, or branching tracks up to 150 miles in length. As regards their origin, the most probable view is that they are fissures produced in cooling; but Neison inclines to consider them rather as dried watercourses.[2]

On February 24, 1792, Schröter perceived what he took to be distinct traces of a lunar twilight, and continued to observe them during nine ensuing years.[3] They indicated, he thought, the presence of a shallow atmosphere (not reaching a height of more than 8400 feet), about $\frac{1}{29}$th as dense as our own. Bessel, on the other hand, considered that the only way of "saving" a lunar atmosphere was to deny it any refractive power, the sharpness and suddenness of star-occultations negativing the possibility of gaseous surroundings exceeding in density (as he computed on an extreme supposition) $\frac{1}{500}$th that of terrestrial air.[4] Newcomb places the maximum at $\frac{1}{400}$. Sir

[1] *Ueber Rillen auf dem Monde*, p. 13.   [2] *The Moon*, p. 73.
[3] *Selen. Fragm.*, Th. ii. p. 399.
[4] *Astr. Nach.*, No. 263 (1834); *Pop. Vorl.*, pp. 615–620 (1838).

John Herschel concluded "the non-existence of any atmosphere at the moon's edge having one-1980th part of the density of the earth's atmosphere."[1]

This decision was fully borne out by Dr. Huggins's spectroscopic observation of the disappearance behind the moon's limb of the small star ε Piscium, January 4, 1865.[2] Not the slightest sign of selective absorption or unequal refraction was discernible. The entire spectrum went out at once, as if a slide had suddenly dropped over it from above. The spectroscope has uniformly told the same tale; for M. Thollon's observation during the total solar eclipse at Sohag of a supposed thickening at the moon's rim, of certain dark lines in the solar spectrum, is now all but admitted to have been illusory. Moonlight, analysed with the prism, is found to be pure reflected sunlight, diminished in *quantity*, owing to the low reflective capability of the lunar surface, to about one-sixth its incident intensity, but wholly unmodified in *quality*.

Yet there is little or no doubt that the diameter of the moon, as determined from occultations, is 4″ smaller than it appears by direct measurement. This fact, which emerged from Sir George Airy's discussion, in 1865,[3] of an extensive series of Greenwich and Cambridge observations, would naturally result from lunar atmospheric refraction. He showed, however, that even if the entire effect were thus produced (a certain share is claimed by irradiation) the atmosphere involved would be 2000 times thinner than our own air at the sea-level. A gaseous stratum of such extreme tenuity could scarcely produce any spectroscopic effect. It is certain (as Mr. Neison has pointed out[4]) that a lunar atmosphere of very great extent and of no inconsiderable mass would possess, owing to the low power of lunar gravity, a very small surface density, and might thus escape direct observation while playing a very important part in the economy of our satellite. Some renewed evidence

[1] *Outlines of Astr.*, par. 431. [2] *Month. Not.*, vol. xxv. p. 61.
[3] *Month. Not.*, vol. xxv. p. 264. [4] *The Moon*, p. 25.

of actual crepuscular gleams on the moon has, besides, been lately furnished to MM. Paul and Prosper Henry of the Paris Observatory by their skilful use of a powerful telescope.[1]

The first to emulate Schröter's selenographical zeal was Wilhelm Gotthelf Lohrmann, a land-survevor of Dresden, who, in 1824, published four out of twenty-five sections of the first scientifically executed lunar chart, on a scale of 37½ inches to a lunar diameter. His sight, however, began to fail three years later, and he died in 1840, leaving materials from which the work was completed and published in 1878 by Dr. Julius Schmidt, late director of the Athens Observatory. Much had been done in the interim. Beer and Mädler began at Berlin in 1830 their great trigonometrical survey of the lunar surface, as yet neither revised nor superseded. A map, issued in four parts, 1834–36, on nearly the same scale as Lohrmann's, but more detailed and authoritative, embodied the results. It was succeeded, in 1837, by a descriptive volume bearing the imposing title, *Der Mond; oder allgemeine vergleichende Selenographie.* This summation of knowledge in that branch, though in truth leaving many questions open, had an air of finality which tended to discourage further inquiry.[2] It gave form to a reaction against the sanguine views entertained by Hevelius, Schröter, Herschel, and Gruithuisen as to the possibilities of agreeable residence on the moon, and relegated the "Selenites," one of whose cities Schröter thought he had discovered, and of whose festal processions Gruithuisen had not despaired of becoming a spectator, to the shadowy land of the Ivory Gate. All examples of change in lunar formations were, moreover, dismissed as illusory. The light contained in the work was, in short, a "dry light," not stimulating to the imagination. "A mixture of a lie," Bacon shrewdly remarks, "doth ever add pleasure." For many years, accordingly, Schmidt had the field of selenography almost to himself.

Reviving interest in the subject was at once excited and displayed by the appointment, in 1864, of a Lunar Committee of

---

[1] Webb, *Cel. Objects,* p. 79.   [2] Neison, *The Moon,* p. 104.

the British Association. The indirect were of greater value than the direct fruits of its labours. An English school of seleno-graphy rose into importance. Popularity was gained for the subject by the diffusion of works conspicuous for ingenuity and research. Messrs. Nasmyth's and Carpenter's beautifully illus-trated volume (1874) was succeeded, after two years, by a still more weighty contribution to lunar science. Mr. Neison's book was accompanied by a map, based on the survey of Beer and Mädler, but adding some 500 measures of position, besides the representation of several thousand new objects. With Schmidt's *Charte der Gebirge des Mondes*, Germany once more took the lead. This splendid delineation—the result of thirty-two years' labour—was built upon Lohrmann's foundation, but embraces the detail contained in upwards of 3000 original drawings. No less than 32,856 craters are represented in it. The scale is seventy-five inches to a diameter. An additional help to lunar inquiries was provided at the same time in this country by the establishment, through the initiative of the late Mr. W. R. Birt, of the Selenographical Society.

But the strongest incentive to diligence in studying the rugged features of our celestial helpmate has been the idea of probable or actual variation in them. A change always seems to the inquisitive intellect of man like a breach in the defences of Nature's secrets, through which it may hope to make its way to the citadel. What is desirable easily becomes credible; and thus statements and rumours of lunar convulsions have suc-cessively, during the last hundred years, obtained credence, and successively, on closer investigation, been rejected. The subject is one as to which illusion is peculiarly easy. Our view of the moon's surface is a bird's-eye view. Its conforma-tion reveals itself indirectly through irregularities in the dis-tribution of light and darkness. The forms of its elevations and depressions can be inferred only from the shapes of the black, unmitigated shadows cast by them. But these shapes are in a state of perpetual and bewildering fluctuation, partly through changes in the angle of illumination, partly through changes in our point of view, caused by what are called the moon's "librations." [1] The result is, that no single observation can be *exactly* repeated by the same observer, since identical conditions recur only after the lapse of a great number of years.

Local peculiarities of surface, besides, are liable to produce perplexing effects. The reflection of earth-light at a particular angle from certain bright summits completely, though temporarily deceived Herschel into the belief that he had witnessed, in 1783 and 1787, volcanic outbursts on the dark side of the moon. The persistent recurrence, indeed, of similar appearances under circumstances less amenable to explanation, inclined Webb to the view that effusions of native light actually occur. [2] More cogent proofs, must, however, be adduced before a fact so intrinsically improbable can be admitted as true.

But from the publication of Beer and Mädler's work until 1866, the received opinion was that no genuine sign of activity had ever been seen, or was likely to be seen, on our satellite ; that her face was a stereotyped page, a fixed and irrevisable record of the past. A profound sensation, accordingly, was produced by Schmidt's announcement, in October 1866, that the well-known crater "Linné" had disappeared, [3] effaced, as it was supposed, by an igneous outflow. The case seemed undeniable, and is still dubious. Linné had been known to Lohrmann and Mädler, 1822–32, as a deep crater, five or six miles in diameter, the third largest in the dusky plain known as the "Mare Serenitatis;" and Schmidt had observed and

---

[1] The combination of a uniform rotational, with an unequal orbital movement causes a slight swaying of the moon's globe, now east, now west, by which we are enabled to see round the edges of the averted hemisphere. There is also a "parallactic" libration, depending on the earth's rotation ; and a species of nodding movement—the "libration in latitude"—is produced by the inclination of the moon's axis to her orbit, and by her changes of position with regard to the terrestrial equator. Altogether, about $\frac{2}{11}$ of the *invisible* side come into view.   [2] *Cel. Objects,* p. 58 (4th ed.)

[3] *Astr. Nach.,* No. 1631.

drawn it, 1840–43, under a practically identical aspect. Now it appears under high light as a whitish spot, in the centre of which, as the rays begin to fall obliquely, a pit, probably under two miles across, emerges into view. The crateral character of this comparatively minute depression was detected by Father Secchi, February 11, 1867.

This, however, is not all. Schröter's description of Linné, as seen by him November 5, 1788, tallies quite closely with modern observation;[1] while its inconspicuousness in 1797 is shown by its omission from Russell's lunar globe and maps.[2] We are thus driven to adopt one of two suppositions: either Lohrmann, Mädler, and Schmidt were entirely mistaken in the size and importance of Linné, or a real change in its outward semblance supervened during the first half of this century, and has since passed away, perhaps again to recur. The latter hypothesis seems the more probable; and its probability is strengthened by much evidence of actual obscuration or variation of tint in other parts of the lunar surface, more especially on the floor of the great "walled plain" named "Plato."[3]

An instance of an opposite kind of change was alleged by Dr. Hermann J. Klein of Cologne in March 1878.[4] In Linné, the obliteration of an old crater had been assumed; in "Hyginus N.," the formation of a new crater was asserted. Yet, quite possibly, the same cause may have produced the effects thought to be apparent in both. It is, however, far from certain that any real change has affected the neighbourhood of Hyginus. The novelty of Klein's observation of May 19, 1877, may have consisted simply in the detection of a hitherto unrecognised feature. The region is one of complex formation, consequently of more than ordinary liability to deceptive variations in aspect under rapid and entangled

---

[1] Respighi, *Les Mondes*, t. xiv. p. 294; Huggins, *Month. Not.*, vol. xxvii. p. 298.   [2] Birt, *ibid.*, p. 95.

[3] *Report Brit. Ass.*, 1872, p. 245.

[4] *Astr. Reg.*, vol. xvi. p. 265; *Astr. Nach.*, No. 2275.

fluctuations of light and shade.[1]   Moreover, it seems to be
certain, from Messrs. Pratt and Capron's attentive study, that
"Hyginus N." is no true crater, but a shallow, saucer-like
depression, difficult of clear discernment.[2]   Under suitable
illumination, nevertheless, it contains, and is marked by, an
ample shadow.[3]

In both these controverted instances of change, lunar photography was invoked as a witness; but, notwithstanding the
great advances made in the art by Mr. De la Rue in this
country, by Dr. Henry Draper, and above all by Mr. Lewis
M. Rutherfurd, in America, without decisive results.   Autographic records, it may be expected, will gain increasing
authority on such points in the future.

Melloni was the first to get undeniable heating effects from
moonlight.   His experiments were made at Naples early in
1846,[4] and were repeated with like result by Zantedeschi at
Venice four years later.   A rough measure of the intensity of
those effects was arrived at by Piazzi Smyth at Guajara, on the
Peak of Teneriffe, in 1856.   At a distance of fifteen feet from
the thermomultiplier, a Price's candle was found to radiate
just twice as much heat as the full moon.[5]   But by far the
most exact and extensive series of observations on the subject
were those made by the present Earl of Rosse, 1869–72.   The
lunar radiations, from the first to the last quarter, displayed,
when concentrated with the Parsonstown three-foot mirror, appreciable thermal energy, increasing with the phase, and largely
due to "dark heat," distinguished from the quicker-vibrating
sort by inability to traverse a plate of glass.   This was supposed to indicate an actual heating of the surface, during the
long lunar day of 300 hours, to about 500° F.,[6] the moon thus

---

[1] See Lord Lindsay and Dr. Copeland in *Month. Not.*, vol. xxxix. p.
195.

[2] *Observatory*, vols. ii. p. 296 ; iv. p. 373.   Mr. N. G. Green (*Astr. Reg.*,
vol. xvii. p. 144) concludes the object a mere "spot of colour," dark under
oblique light.   [3] Webb, *Cel. Objects*, p. 101.

[4] *Comptes Rendus*, t. xxii. p. 541.   [5] *Phil. Trans.*, vol. cxlviii. p. 502.

[6] *Proc. Roy. Soc.*, vol. xvii. p. 443.

acting as a direct radiator no less than as a reflector of heat. These results, though not fully borne out by further and more careful trials executed at Parsonstown by Dr. Copeland,[1] have lately received some countenance from Professor Langley's experiments with the bolometer, showing that moonlight undeniably contains a proportion of obscure thermal rays.

This implies some kind of atmospheric *clothing*. For, entirely denuded of such, Professor Langley has shown [2] that even under the fiercest sunshine the lunar surface must abide frostbound at somewhere below 50° Fahr. ; that is to say, mercury, and *à fortiori* water, could never liquefy on an airless moon. That it is capable of sending us any perceptible heat *on its own account*—that is, apart from its office as a reflector of solar radiations—proves conclusively that it is preserved from immediate contact with the cold of space by the survival of some thin remnant of aerial covering.

Although that fundamental part of astronomy known as "celestial mechanics" lies outside the scope of this work, and we must therefore pass over in silence the immense labours of Plana, Damoiseau, Hansen, Delaunay, and Airy in reconciling the observed and calculated motions of the moon, there is one slight, but significant discrepancy which is of such importance to the physical history of the solar system, that some brief mention must be made of it.

Halley discovered in 1693, by examining the records of ancient eclipses, that the moon was going faster then than 2000 years previously—so much faster, as to have got ahead of the place in the sky she would otherwise have occupied, by about two of her own diameters. It was one of Laplace's highest triumphs to have found an explanation of this puzzling fact. He showed, in 1787, that it was due to a very slow change in the ovalness of the earth's orbit, tending, during the present age of the world, to render it more nearly circular. The pull of the sun upon the moon is thereby lessened; the

---

[1] *Phil. Trans.*, vol. clxiii. p. 625.
[2] *Nature*, vol. xxvi. p. 316.

counter-pull of the earth gets the upper hand ; and our satellite, drawn nearer to us by something less than an inch each year,[1] proportionally quickens her pace. Many thousands of years hence the process will be reversed ; the terrestrial orbit will close in at the sides, the lunar orbit will open out under the growing stress of solar gravity, and our celestial chronometer will lose instead of gaining time.

This is all quite true as Laplace put it ; but it is not enough. Adams, the virtual discoverer of Neptune, found with surprise in 1853 that the received account of the matter was "essentially incomplete," and explained, when the requisite correction was introduced, only half the observed acceleration.[2] What was to be done with the remaining half? Here Delaunay, the eminent French mathematical astronomer, unhappily drowned at Cherbourg in 1872 by the capsizing of a pleasure-boat, came to the rescue.[3]

It is obvious to any one who considers the subject a little attentively, that the tides must act to some extent as a friction-brake upon the rotating earth. In other words, they must bring about an almost infinitely slow lengthening of the day. For the two masses of water piled up by lunar influence on the hither and farther sides of our globe, strive, as it were, to detach themselves from the unity of the terrestrial spheroid, and to follow the movements of the moon. The moon, accordingly, holds them *against* the whirling earth, which revolves like a shaft in a fixed collar, wasting its momentum as heat dissipated through space. This must go on (so far as we can see) until the periods of the earth's rotation and of the moon's revolution coincide. Nay, the process will be continued— should our oceans survive so long—by the feebler tide-raising power of the sun, ceasing only when day and night cease to alternate, when one side of our planet is plunged in perpetual darkness and the other seared by unchanging light.

---

[1] Airy, *Observatory*, No. 37, p. 420.
[2] *Phil. Trans.*, vol. cxliii. p. 397 ; *Proc. Roy. Soc.*, vol. vi. p. 321.
[3] *Comptes Rendus*, t. lxi. p. 1023.

Here, then, we have the secret of the moon's turning always the same face towards the earth. It is that in primeval times, when the moon was liquid or plastic, an earth-raised tidal wave rapidly and forcibly reduced her rotation to its present exact agreement with her period of revolution. This was divined by Kant[1] nearly a century before the necessity for such a mode of action presented itself to any other thinker. In a weekly paper published at Königsberg in 1754, the modern doctrine of "tidal friction" was clearly outlined by him, both as regards its effects actually in progress on the rotation of the earth, and as regards its effects already consummated on the rotation of the moon—the whole forming a preliminary attempt at what he called a "natural history" of the heavens. His sagacious suggestion, however, remained entirely unnoticed until revived—it would seem independently—by Julius Robert Mayer in 1848 ;[2] while similar, and probably original conclusions were reached by William Ferrel of Allensville, Kentucky, in 1853.[3]

Delaunay was not then the inventor or discoverer of tidal friction ; he merely displayed it as an effective cause of change. He showed reason for believing that its action in checking the earth's rotation, far from being, as Ferrel had supposed, completely neutralised by its contraction through cooling, was a fact to be reckoned with in computing the movements, as well as in speculating on the history of the heavenly bodies. The outstanding acceleration of the moon was thus at once explained. It was explained as apparent only—the reflection of a real lengthening, by one second in 100,000 years, of the day. But on this point the last word has not yet been spoken.

Professor Newcomb undertook in 1870 the formidable task of a complete rediscussion of the lunar theory. The results, published in 1878,[4] have proved somewhat perplexing. They

[1] *Sämmtl. Werke* (ed. 1839), Th. vi. pp. 5-12. See also Mr. C. J. Monro's useful indications in *Nature*, vol. vii. p. 241.

[2] *Dynamik des Himmels*, p. 40.

[3] *Gould's Astr. Jour.*, vol. iii. p. 138.

[4] *Wash. Obs.* for 1875, vol. xxii. App. ii.

tend, in general, to reduce the amount of acceleration left unaccounted for by Laplace's gravitational theory, and proportionately to diminish the importance of the part played by tidal friction.   But, in order to bring about this diminution, and at the same time conciliate Alexandrian and Arabian observations, it is necessary to reject *as total* the ancient solar eclipses known as those of Thales and Larissa.   This may be a necessary, but it must be admitted to be a hazardous expedient.

It was further shown that small residual irregularities are still found in the movements of our satellite, inexplicable either by any known gravitational influence, or by any *uniform* value that could be assigned to secular acceleration.[1]   If set down to the account of imperfections in the "time-keeping" of the earth, it could only be on the arbitrary supposition of fluctuations in its rate of going themselves needing explanation. This, it is true, might be found, as Sir W. Thomson pointed out in 1876,[2] in very slight changes of figure, not altogether unlikely to occur.   But into this cloudy and speculative region astronomers for the present decline to penetrate.   They prefer, if possible, to deal only with calculable causes, and thus to preserve for their "most perfect of sciences" its special prerogative of assured prediction.

[1] Newcomb, *Pop. Astr.* (4th ed.), p. 101.
[2] *Report Brit. Ass.*, 1876, p. 12.

CHAPTER VIII.

PLANETS AND SATELLITES (continued).

"THE analogy between Mars and the earth is perhaps by far the greatest in the whole solar system." So Herschel wrote in 1783,[1] and so it may safely be repeated to-day, after an additional hundred years of scrutiny. This circumstance lends a particular interest to inquiries into the physical habitudes of our exterior planetary neighbour.

Fontana was the first to catch glimpses, at Naples in 1636 and 1638,[2] of dusky stains on the ruddy disc of Mars. They were next seen by Hooke and Cassini in 1666, and this time with sufficient distinctness to serve as indexes to the planet's rotation, determined by the latter as taking place in a period of twenty-four hours forty minutes.[3] Increased confidence was given to this result through Maraldi's precise verification of it in 1719.[4] Amongst the spots observed by him, he distinguished two as stable in position, though variable in size. They were of a peculiar character, showing as bright patches round the poles, and had already been noticed during sixty years back. A current conjecture of their snowy nature obtained validity when Herschel connected their fluctuations in extent with the progress of the Martian seasons. It was hard to resist the inference of frozen precipitations when once it was clearly perceived that the shining polar zones did actually diminish alternately and grow with the alternations of summer and winter in the corresponding hemisphere.

---

[1] *Phil. Trans.*, vol. lxxiv. p. 260. [2] *Novæ Observationes*, p. 105.
[3] *Phil. Trans.*, vol. i. p. 243. [4] *Mém. de l'Ac.*, 1720, p. 146.

This, it may be said, was the opening of our acquaintance with the state of things prevailing on the surface of Mars. It was accompanied by a steady assertion, on Herschel's part, of permanence in the dark markings, notwithstanding partial obscurations by clouds and vapours floating in a "considerable but moderate atmosphere." Hence the presumed inhabitants of the planet "probably enjoy a situation in many respects similar to ours." [1]

Schröter, on the other hand, went altogether wide of the truth as regards Mars. He held that the surface visible to us is a mere shell of drifting cloud, deriving a certain amount of apparent stability from the influence on evaporation and condensation of subjacent but unseen areographical features; [2] and his opinion prevailed with his contemporaries. It was, however, rejected by Kunowsky in 1822, and finally overthrown by Beer and Mädler's careful studies during five consecutive oppositions, 1830–39. They identified at each the same dark spots, frequently blurred with mists, especially when the local winter prevailed, but fundamentally unchanged. [3] In 1862 Mr. Lockyer established a "marvellous agreement" with Beer and Mädler's results of 1830, leaving no doubt as to the complete fixity of the main features, amid "daily, nay, hourly," variations of detail through transits of clouds. [4] On seventeen nights of the same opposition, F. Kaiser of Leyden obtained drawings in which nearly all the markings noted in 1830 at Berlin reappeared, besides spots frequently seen respectively by Arago in 1813, by Herschel in 1783, and one sketched by Huygens in 1672 with a writing-pen in his diary. [5] From these data the Leyden observer arrived at a period of rotation of 24h. 37m. 22.62s., being just one second shorter than that deduced, exclusively from their own observations, by Beer

---

[1] *Phil. Trans.*, vol. lxxiv. p. 273.

[2] A large work, entitled *Areographische Fragmente*, in which Schröter embodied the results of his labours on Mars, 1785–1803, narrowly escaped the conflagration of 1813, and was published at Leyden in 1881.

[3] *Beiträge*, p. 124.   [4] *Mém. R. A. Soc.*, vol. xxxii. p. 183.

[5] *Astr. Nach.*, No. 1468.

and Mädler. But the exactness of even this result has been surpassed. Taking a drawing by Hooke of March 12, 1666 (N.S.), as a starting-point, and delineations by Browning in 1867 and 1869 as termini, Mr. Proctor was enabled to measure the rotation of Mars by means of an interval of about 203 years.[1] Provided that the right count be kept in the number of entire rotations performed (which is easily secured by comparison with intermediate observations), extraordinary accuracy can in this way be obtained; for an almost infinitesimal error becomes multiplied by frequent repetition into something so considerable as to compel correction. Mr. Proctor, for instance, showed that an estimate astray by so much as the tenth of a second would, when carried back to Hooke's time, throw the planet out of its true position by 2 hours 20 seconds. The period then adopted of 24h. 37m. 22.735s. is possibly one or two hundredths of a second too long, but is undoubtedly of a precision unapproached in the case of any other heavenly body save the earth itself.

Two facts bearing on the state of things at the surface of Mars were, then, fully acquired to science in or before the year 1862. The first was that of the seasonal fluctuations of the polar spots; the second, that of the permanence of certain dark grey or greenish patches, perceived with the telescope as standing out from the deep yellow ground of the disc. The opinion has steadily gained consistency during the last half-century that these varieties of tint correspond to the real diversities of a terraqueous globe, the " ripe cornfield"[2] sections representing land, the dusky spots and streaks, oceans and straits. Sir J. Herschel in 1830 led the way in ascribing the redness of the planet's light to an inherent peculiarity of soil.[3] Previously it had been assimilated to our sunset glows rather than to our red sandstone formations—set down, that is, to an atmospheric stoppage of blue rays. But the extensive Martian

---

[1] *Month. Not.*, vols. xxviii. p. 37 ; xxix. p. 232 ; xxxiii. p. 552.
[2] Flammarion, *L'Astronomie*, t. i. p. 266,
[3] Smyth, *Cel. Cycle*, vol. i. p. 148 (1st ed.)

atmosphere, implicitly believed in on the strength of some erroneous observations by Cassini and Römer in the seventeenth century, vanished before the sharp occultation of a small star in Leo, witnessed by Sir James South in 1822;[1] and Dawes's observation in 1865,[2] that the ruddy tinge is deepest near the central parts of the disc, certified its non-atmospheric origin. The absolute whiteness of the polar snow-caps was alleged in support of the same inference by Dr. Huggins in 1867.[3]

All recent observations tend to show that the atmosphere of Mars is much thinner than our own. This was to have been expected *à priori*, since the same proportionate mass of air would, owing to the small size and inferior specific gravity of Mars, as compared with the earth, form a very much sparser covering over each square mile of his surface.[4] Besides, gravity there possesses much less than half its force here, so that this sparser covering would weigh less, and be less condensed than if it enveloped the earth. Atmospheric pressure would accordingly be of about two and a quarter, instead of fifteen terrestrial pounds per square inch. This corresponds with what the telescope shows us. It is extremely doubtful whether any features of the earth's actual surface could be distinguished by a planetary spectator, however well provided with optical assistance. Professor Langley's inquiries[5] have led him to conclude that fully twice as much light is absorbed by our air as had previously been supposed—say forty per cent. of vertical rays in a clear sky. Of the sixty reaching the earth, less than a quarter would be reflected even from white sandstone; and this quarter would again pay its toll of forty per cent. in escaping back to space. Thus not more than eight or nine out of the original hundred sent by the sun would, under the most favourable circumstances, and

---

[1] *Phil. Trans.*, vol. cxxi. p. 417.   [2] *Month. Not.*, vol. xxv. p. 227.
[3] *Phil. Mag.*, vol. xxxiv. p. 75.
[4] Proctor, *Quart. Jour. of Science*, vol. x. p. 185 ; Maunder, *Sunday Mag.*, Jan., Feb., March, 1882.   [5] *Am. Jour. of Sc.*, vol. xxviii. p. 163.

from the very centre of the earth's disc, reach the eye of a Martian or lunar observer. The light by which he views our world is, there is little doubt, light reflected from the various strata of our atmosphere, cloud- or mist-laden or serene, as the case may be, with an occasional snow-mountain figuring as a permanent white spot.

This consideration at once shows us how much more tenuous the Martian air must be, since it admits of topographical delineations of the Martian globe. The clouds, too, that form in it seem to be rather of the nature of ground-mists than of heavy cumulus.[1] There is, indeed, plenty of aqueous vapour present. A characteristic group of dark rays, due to its absorptive action, was detected by Dr. Huggins in the analysed light of the planet in 1867,[2] and serves to raise the conjecture of "snowy poles" to a verisimilitude scarcely to be distinguished from certainty.

The climate of Mars seems to be unexpectedly mild. The polar snows are both less extensive and less permanent than those on the earth. The southern white hood, always eccentrically situated, was noticed by Schiaparelli in 1877 to have survived the summer only as a small lateral patch, the pole itself being quite free from snow. But we might expect to see the whole wintry hemisphere, at any rate, frostbound, since the sun radiates less than half as much heat on Mars as on the earth. Water seems, nevertheless, to remain, as a rule, uncongealed everywhere outside the polar regions. We are at a loss to imagine by what beneficent arrangement the rigorous conditions naturally to be looked for, can be modified into a climate which might be found tolerable by creatures constituted like ourselves.

Martian topography may be said to form now-a-days a separate sub-department of descriptive astronomy. The amount of detail become legible by close scrutiny on a little disc which, once in fifteen years, attains a maximum of about

---

[1] Burton, *Trans. Roy. Dublin Soc.*, vol. i. 1880, p. 169.
[2] *Month. Not.*, vol. xxvii. p. 179.

$\frac{1}{5000}$ the area of the full moon, must excite surprise, and might provoke incredulity. Spurious discoveries, however, have little chance of holding their own where there are so many competitors quite as ready to dispute as to confirm.

The first really good map of Mars was constructed in 1869 by Mr. Proctor from drawings by Dawes. Kaiser of Leyden followed in 1872 with a representation founded upon data of his own providing in 1862–64; and M. Terby, in his valuable *Aréographie*, presented to the Brussels Academy in 1874[1] a careful discussion of all important observations from the time of Fontana downwards, thus virtually adding to knowledge by summarising and digesting it. The memorable opposition of September 5, 1877, marked a fresh epoch in the study of Mars. While executing a trigonometrical survey (the first attempted) of the disc, then of the unusual size of 25″ across, Signor G. V. Schiaparelli, director of the Milan Observatory, detected a novel and curious feature. What had been taken for Martian continents were found to be, in point of fact, agglomerations of islands, separated from each other by a network of so-called "canals." These are obviously extensions of the "seas," originating and terminating in them, and sharing their grey-green hue, but running sometimes to a length of three or four thousand miles in a straight line, and preserving throughout a nearly uniform breadth of about sixty miles. Further inquiries have fully substantiated the discovery made at the Brera Observatory. "The "canals" of Mars are an actually existent and permanent phenomenon. An examination of the drawings in his possession showed M. Terby that they had been seen, though not distinctively recognised, by Dawes, Secchi, and Holden; several were independently traced out by Burton at the opposition of 1879; and all were recovered by Schiaparelli himself in 1879 and 1881–82.

When the planet culminated at midnight, and was therefore in opposition, December 26, 1881, its distance was greater, and its apparent diameter less than in 1877, in the proportion

[1] *Mémoires Couronnés*, t. xxxix.

of sixteen to twenty-five. Its atmosphere was, however, more transparent, and ours of less impediment to northern observers, the object of scrutiny standing considerably higher in northern skies. Never before, at any rate, had the true aspect of Mars come out so clearly as at Milan with the 8¾-inch Merz refractor of the observatory, between December 1881 and February 1882. The canals were all again there, but this time they were —in as many as twenty cases—*seen in duplicate.* That is to say, a twin-canal ran parallel to the original one at an interval of 200 to 400 miles.[1]

We are here brought face to face with an apparently insoluble enigma. Schiaparelli regards the "gemination" of his canals as a periodical phenomenon depending on the Martian seasons; but it is as yet premature to form an opinion. Fresh evidence will, it is to be hoped, become available during the next favourable opposition in 1892.

Meanwhile, the closeness of the terrestrial analogy remains somewhat impaired. The distribution of land and water on Mars, at any rate, appears to be of a completely original type. The interlacing everywhere of continents with arms of the sea (if that be the correct interpretation of the visual effects) implies that their levels scarcely differ;[2] and it is held by Schiaparelli and others that their outlines are not absolutely constant, encroachments of dusky upon bright tints suggesting the possibility of extensive inundations. Mr. N. E. Green's noteworthy observations at Madeira in 1877 seem to indicate, on the other hand, a rugged south polar region. The contour of the snow-cap not only appeared indented, as if by valleys and promontories, but brilliant points were discerned outside the white area, attributed to isolated snow-peaks.[3] Still more elevated, if similarly explained, must be the "ice island" first seen in a comparatively low latitude by Dawes in January 1865.

Mars was gratuitously supplied with a pair of satellites long

---

[1] *Mem. Spettr. Italiani*, t. xi. p. 28.

[2] Flammarion, *L'Astronomie*, t. i. p. 206.

[3] *Month. Not.*, vol. xxxviii. p. 41.

before he was found actually to possess them.   Kepler interpreted Galileo's anagram of the "triple" Saturn in this sense ;
they were perceived by Micromégas on his long voyage through
space ; and the Laputan astronomers had even arrived at a
knowledge, curiously accurate under the circumstances, of
their distances and periods.   But terrestrial observers could
see nothing of them until the night of August 11, 1877.   The
planet was then within one month of its second nearest approach to the earth during this century ; and in 1845 the
Washington 26-inch refractor was not in existence.[1]   Professor
Asaph Hall, accordingly, determined to turn the conjuncture
to account for an exhaustive inquiry into the surroundings of
Mars.   Keeping his glaring disc just outside the field of view,
a minute attendant speck of light was "glimpsed" August 11.
Bad weather however intervened, and it was not until the 16th
that it was ascertained to be what it appeared—a satellite.   On
the following evening a second, still nearer to the primary,
was discovered, which, by the bewildering rapidity of its passages hither and thither, produced at first the effect of quite a
crowd of little moons.[2]

Both these delicate objects have since been repeatedly
observed, both in Europe and America, even with comparatively small instruments.   But at each opposition since that of
1877 the distance of the planet has been increasing, and in
1884 was too great to permit of their detection elsewhere than
at Washington.   It is unlikely that they will be again seen
before 1888 or 1890.

The names chosen for them were taken from the Iliad, where
"Deimos" and "Phobos" (Fear and Panic) are represented
as the companions in battle of Ares.   In several respects, they
are interesting and remarkable bodies.   As to size, they may
be said to stand midway between meteorites and satellites.
From careful photometric measures executed at Harvard in
1877 and 1879, Professor Pickering concluded their diame-

---

[1] See Mr. Wentworth Erck's remarks in *Trans. Roy. Dublin Soc.*, vol.
i. p. 29.   [2] *Month. Not.*, vol. xxxviii. p. 206.

ters to be respectively six and seven miles.[1] This is on the assumption that they reflect the same proportion of the light incident upon them that their primary does. But it may very well be that they are less reflective, in which case they would be more extensive. The albedo of Mars, according to Zöllner, is 0.2762; his surface, in other words, returns 27.62 per cent. of the rays striking it. If we put the albedo of his satellites equal to that of our moon, 0.1736, their diameters will be increased from six and seven to $9\frac{1}{2}$ and $11\frac{1}{8}$ miles, Phobos, the inner one, being the larger. Their actual dimensions do not, in all probability, exceed this estimate. It is interesting to note that Deimos, according to Professor Pickering's very distinct perception, does not share the reddish tint of Mars.

Both satellites move quickly in small orbits. Deimos completes a revolution in thirty hours eighteen minutes, at a distance from the surface of its ruling body of 12,500 miles; Phobos in seven hours thirty-nine minutes twenty-two seconds, at a distance of only 3760 miles. This is the only known instance of a satellite circulating faster than its primary rotates, and is a circumstance of some importance as regards theories of planetary development. To a Martian spectator the curious effect would ensue of a celestial object, seemingly exempt from the general motion of the sphere, rising in the west, setting in the east, and culminating three, or even four times a day.

The detection of new members of the solar system has come to be one of the most ordinary of astronomical events. Since 1846 no single year has passed without bringing its tribute of asteroidal discovery. In the last of the seventies alone, a full score of miniature planets were distinguished from the thronging stars amid which they seem to move; 1875 brought seventeen such recognitions; their number touched a minimum of one in 1881; it rose in 1882 to eleven, dropped to four in 1883, and remounted as far as nine in 1884. At the present date (September 1885), 250 asteroids are known to revolve between the orbits of Mars and Jupiter. Of these, no less than forty-

---

[1] *Annals Harvard Coll. Obs.*, vol. xi. pt. ii. p. 317.

eight are claimed by a single observer—Professor J. Palisa of Vienna; Dr. C. H. F. Peters of Clinton, N.Y., comes in a good second with forty-three; Watson, Borrelly, Luther, Hind, Goldschmidt, Tempel, and many others, have each contributed numerously to swell the sum-total.   The construction by Chacornac and his successors at Paris, and more recently by Peters at Clinton, of ecliptical charts showing all stars down to the thirteenth and fourteenth magnitudes respectively, renders the picking out of moving objects above that brightness a mere question of time and diligence.   Far more onerous is the task of keeping them in view once discovered—of tracking out their paths, fixing their places, and calculating the disturbing effects upon them of the mighty Jovian mass.   These complex operations have come to be centralised at Berlin under the superintendence of Professor Tietjen, and their results are given to the public through the medium of the *Berliner Astronomisches Jahrbuch.*

The crowd of orbits thus disclosed invites attentive study. D'Arrest remarked in 1851,[1] when only thirteen minor planets were known, that supposing their paths to be represented by solid hoops, not one of the thirteen could be lifted from its place without bringing the others with it.   The complexity of interwoven tracks thus illustrated has grown almost in the numerical proportion of discovery.   Yet no two actually intersect, because no two lie exactly in the same plane, so that the chances of collision are at present *nil.*   There is only one case, indeed, in which it seems to be eventually possible. M. Lespiault has pointed out that the curves traversed by "Fidés" and "Maïa" approach so closely that a time may arrive when the bodies in question will either coalesce or unite to form a binary system.[2]

The maze threaded by the 250 asteroids contrasts singularly with the harmoniously ordered and rhythmically separated orbits of the larger planets.   Yet the seeming confusion is not

[1] *Astr. Nach.*, No. 752.
[2] L. Niesten, *Annuaire*, Bruxelles, 1881, p. 269.

without a plan. The established rules of our system are far from being totally disregarded by its minor members. The orbit of Vesta, with its inclination of 34° 42′, touches the limit of departure from the ecliptic-level; the average plane of the asteroidal paths differs by only about one degree from that of the sun's equator; their mean eccentricity is below that of the curve traced out by Mercury, and all without exception are pursued in the planetary direction—from west to east.

The zone in which these small bodies travel is about three times as wide as the interval separating the earth from the sun. It extends perilously near to Jupiter, and actually encroaches upon the sphere of Mars. In one of his lectures at Gresham College in 1879,[1] Mr. Ledger remarked that the minor planet Aethra, when in perihelion, *gets inside* Mars in aphelion by as much as five millions of miles, though at so different a level in space that there is no close approach.

The distribution of the asteroids over the zone frequented by them is very unequal. They are most densely congregated about the place where a single planet ought, by Bode's Law, to revolve; it may indeed be said that only stragglers from the main body are found more than fifty million miles within or without a mean distance from the sun 2.8 times that of the earth. Significant gaps, too, occur where some force prohibitive of their presence would seem to be at work. What the nature of that force may be, Professor Daniel Kirkwood of the Indiana University indicated, first in 1866 when the number of known asteroids was only eighty-eight, and again with more confidence in 1876 from the study of a list then run up to 172.[2] It appears that these bare spaces are found just where a revolving body would have a period connected by a simple relation with that of Jupiter. It would perform two or three circuits to his one, five to his two, nine to his five, and so on. Kirkwood's inference is that the gaps in question were cleared of asteroids by the attractive influence of Jupiter. For disturbances recurring time after

---

[1] *Sun and Planets,* p. 267.   [2] *Smiths. Report,* 1876, p. 358.

time—owing to commensurability of periods—nearly at the same part of the orbit, would have accumulated until the shape of that orbit was notably changed. The body thus displaced would have come in contact with other cosmical particles of the same family with itself—then, it may be assumed, more evenly distributed than now—would have coalesced with them, and permanently left its original track. In this way the regions of maximum perturbation would gradually have become denuded of their occupants.

We can scarcely doubt that this law of commensurability has largely influenced the present distribution of the asteroids. The correspondence of the facts with the hypothesis is in general striking. At the same time it is not perfect. The minor planet Menippe, for example, revolves almost exactly five times while Jupiter revolves once ; and (as Professor Newcomb has pointed out [1]) several of its companions have periods nearly three-eighths that of the disturbing planet. The clue offered by Professor Kirkwood is not therefore to be rejected ; but further inquiry, here as elsewhere, is needed.

Leverrier fixed, in 1853,[2] one-fourth of the earth's mass as the outside limit for the combined masses of all the bodies circulating between Mars and Jupiter; but it is far from probable that this maximum is at all nearly approached. M. Niesten estimated that the whole of the 216 asteroids discovered up to August 1880 amounted in *volume* to only $\frac{1}{4000}$ of our globe,[3] and we may safely add—since they are tolerably certain to be lighter, bulk for bulk, than the earth—that their proportionate *mass* is smaller still. Professor Pickering, from determinations of light-intensity, assigns to Vesta a diameter of 319 miles, to Pallas 167, to Juno 94, down to twelve and fourteen for the smaller members of the group.[4] An albedo equal to that of Mars is assumed as the basis of the calculation. Professor M. W. Harrington, director of the Ann Arbor

[1] *Pop. Astr.*, p. 338 (2d ed.)   [2] *Comptes Rendus*, t. xxxvii. p. 797.
[3] *Annuaire*, Bruxelles, 1881, p. 243.
[4] *Harvard Annals*, vol. xi. part ii. p. 294.

Observatory, on the other hand, concludes Vesta, from the size of her visible disc, to be as much as 520 miles across.[1] But if this be so, her surface is singularly absorptive of light, returning only ten per cent. of the rays striking it. The same observer holds Vesta and Flora to be together nearly equal in bulk to the whole of their remaining companions.[2] He has also ascertained, with much probability, the variability of Vesta to the extent of one stellar magnitude, and attributes the changes to a rapid axial rotation combined with an unequally reflective surface.

There is no good reason to suppose that any of the minor planets possess atmospheres. The aureolæ seen by Schröter to surround Ceres and Pallas have been dissipated by optical improvements. Vogel in 1872 thought he had detected an air-line in the spectrum of Vesta;[3] but admitted that its presence required confirmation, which has not been forthcoming.

Crossing the zone of asteroids on our journey outward from the sun, we meet with a group of bodies widely different from the "inferior" or terrestrial planets. Their gigantic size, low specific gravity, and rapid rotation, obviously from the first threw the "superior" planets into a class apart; and modern research has added qualities still more significant of a dissimilar physical constitution. Jupiter, a huge globe 86,000 miles in diameter, stands pre-eminent amongst them. He is, however, only *primus inter pares;* all the wider inferences regarding his condition may be extended, with little risk of error, to his fellows; and inferences in his case rest on surer grounds than in the case of the others, from the advantages offered for telescopic scrutiny by his comparative nearness.

Now the characteristic modern discovery concerning Jupiter is that he is a body midway between the solar and terrestrial stages of cosmical existence—a decaying sun or a developing earth, as we choose to put it—whose vast unexpended stores

[1] *Am. Jour. of Sc.*, vol. xxvi. (3d ser.), p. 464.
[2] *Observatory*, vol. vii. p. 339.   [3] *Spectra der Planeten*, p. 24.

of internal heat are mainly, if not solely, efficient in producing the interior agitations betrayed by the changing features of his visible disc. This view was anticipated in the last century. Buffon wrote in his *Époques de la Nature* (1778) :[1]— " La surface de Jupiter est, comme l'on sait, sujette a des changemens sensibles, qui semblent indiquer que cette grosse planète est encore dans un état d'inconstance et de bouillonnement."

Primitive incandescence, attendant, in his fantastic view, on planetary origin by cometary impacts with the sun, combined, he concluded, with vast bulk to bring about this result. Jupiter had not yet had time to cool. Kant thought similarly in 1785 ;[2] but the idea did not commend itself to the astronomers of the time, and dropt out of sight until Mr. Nasmyth arrived at it afresh in 1853.[3] Even still, however, terrestrial analogies held their ground. The dark belts running parallel to the equator, first seen at Naples in 1630, continued to be associated—as Herschel had associated them in 1781—with Jovian trade-winds, in raising which the deficient power of the sun was supposed to be compensated by added swiftness of rotation. But opinion was not permitted to halt here.

In 1860 G. P. Bond of Cambridge (U.S.) derived some remarkable indications from experiments on the light of Jupiter.[4] They showed that fourteen times more of the photographic rays striking it are reflected by the planet than by our moon, and that, unlike the moon, which sends its densest rays from the margin, Jupiter is brightest near the centre. But the most perplexing part of his results was that Jupiter actually seemed to give out more light than he received. The question of original luminosity thus definitely raised can assuredly not be answered with an unqualified negative. Bond, however, considered his data too uncertain for the support of so bold an assumption, and, even if the presence of native light were proved, thought that it might emanate from auroral clouds of

---

[1] Tom. i. p. 93.   [2] *Berlinische Monatsschrift,* 1785, p. 211.
[3] *Month. Not.,* vol. xiii. p. 40.   [4] *Mem. Am. Ac.,* vol. viii. p. 221.

the terrestrial kind. The conception of a sun-like planet was still a remote, and seemed an extravagant one.

Only since it was adopted and enforced by Zöllner in 1865,[1] can it be regarded as permanently acquired to science. The rapid changes in the cloud-belts both of Jupiter and Saturn, he remarked, attest a high internal temperature. For we know that all atmospheric movements on the earth are sun-heat transformed into motion. But sun-heat at the distance of Jupiter possesses but $\frac{1}{27}$, at that of Saturn $\frac{1}{100}$ of its force here. The large amount of energy, then, obviously exerted in those remote firmaments must have some other source, to be found nowhere else than in their own active and all-pervading fires, not yet banked in with a thick solid crust.

The same acute investigator dwelt, in 1871,[2] on the similarity between the modes of rotation of the great planets and of the sun, applying the same principles of explanation to each case. The fact of this similarity is undoubted. Cassini[3] and Schröter both noticed that markings on Jupiter travelled quicker the nearer they were to his equator; and Cassini even hinted at their possible assimilation to sun-spots.[4] It is now well ascertained that, as a rule (not altogether without exceptions), equatorial spots give a period some $5\frac{1}{2}$ minutes shorter than those in latitudes of about 30°. But, as Mr. Denning has pointed out,[5] no single period will satisfy the observations either of different markings at the same epoch, or of the same markings at different epochs. Accelerations and retardations, depending upon internal conditions, take place in very much the same kind of way as in solar maculæ, inevitably suggesting a similar eruptive origin.

Amongst popular writers, Mr. Proctor has been foremost in realising the highly primitive condition of these giant orbs, and in impressing the facts and their logical consequences upon the public mind. The inertia of ideas on the subject has been

[1] *Photom. Unters.*, p. 303.   [2] *Astr. Nach.*, No. 1851.
[3] *Mém. de l'Ac.*, t. x. p. 514.   [4] *Ibid.*, 1692. p. 7.
[5] *Month. Not.*, vol. xliv. p. 63.

overcome largely through the arguments reiterated in the
various and well-known works published by him since 1870.
It should be added that Mr. Mattieu Williams in his *Fuel of
the Sun* adopted, equally early, similar views.

The interesting query as to Jupiter's surface-incandescence
has been studied since Bond's time with the aid of all the
appliances furnished to physical inquirers by modern inventiveness, yet without bringing to it a categorical reply.
Zöllner in 1865 estimated his albedo at 0.62, that of fresh-
fallen snow being 0.78, and of white paper 0.70.[1]   But the
disc of Jupiter is by no means purely white.   The general
ground is tinged with ochre, the polar zones are leaden or
fawn-coloured, large spaces are at times stained or diffused
with chocolate-browns and rosy hues.   It is occasionally seen
ruled from pole to pole with dusky bars, and is never wholly
free from obscure markings.   The reflection then by it, as a
whole, of 62 per cent. of the rays impinging upon it, might
well suggest some original reinforcement.

Nevertheless, the spectroscope gives little countenance to
the supposition of any considerable permanent light-emission.
The spectrum of Jupiter, as examined by Huggins, 1862–64,
and by Vogel, 1871–73, shows the familiar Fraunhofer rays
belonging to reflected sunlight.   But it also shows lines of
native absorption.   Some of these are identical with those
produced by the action of our own atmosphere, especially one
or more groups due to aqueous vapour ; others are of unknown
origin, and it is remarkable that one amongst the latter—a
strong band in the red—agrees in position with a dark line in
the spectra of some ruddy stars.[2]   There is, besides, a general
absorption of blue rays, intensified—as Le Sueur observed at
Melbourne in 1869[3]—in the dusky markings, evidently through
an increase of depth in the atmospheric strata traversed by
the light proceeding from them.

All these observations, however (setting aside the stellar

[1] *Photom. Unters.*, pp. 165, 273.   [2] Vogel, *Sp. d. Planeten*, p. 33, *note.*
[3] *Proc. Roy. Soc.*, vol. xviii. p. 250.

line as of doubtful significance), point to a cool planetary atmosphere. There is, we believe, only one on record evincing unmistakably the presence of intrinsic light. On September 27, 1879, Dr. Henry Draper obtained a photograph of Jupiter's spectrum, in which a strengthening of the impression was visible in the parts corresponding to the planet's equatorial regions.[1] This is just the right sort of evidence, but it is altogether exceptional. We are driven then to conclude that native emissions from Jupiter's visible surface are local and fitful, not permanent and general. Indeed, the total disappearance of his satellites on entering his shadow-cone, sufficiently proves that they receive from him no sensible illumination. This conclusion, however, by no means invalidates that of his excessively high internal temperature.

The curious phenomena attending Jovian satellite-transits may be explained, partly as effects of contrast, partly as due to temporary obscurations of the small discs projected on the large disc of Jupiter. At their first entry upon its marginal parts, which are two or three times less luminous than those near the centre, they usually show as bright spots, then vanish, and re-emerge dusky against the more lustrous background met in their gradual advance. But they sometimes appear bright throughout; while, on the other hand, instances are not rare, more especially of the third and fourth satellites standing out in such inky darkness as to be mistaken for their own shadows. The earliest witness of a "black transit" was Cassini, September 2, 1665; Römer in 1677, and Maraldi in 1707 and 1713, made similar observations, which have been multiplied during the present century. In some cases, the process of darkening has been visibly attended by the formation, or emergence into view, of spots on the transiting body, as noted by the two Bonds at Harvard, March 18, 1848.[2] The third satellite was seen by Dawes, half dark, half bright, when

---

[1] *Month. Not.*, vol. xl. p. 433.

[2] Engelmann, *Ueber die Helligkeitsverhältnisse der Jupiterstrabanten*, p. 59.

crossing Jupiter's disc, August, 21, 1867 ;[1] one-third dark by Davidson of California, January 15, 1884, under the same circumstances ;[2] and unmistakably spotted, both on and off the planet, by Schröter, Secchi, Dawes, and Lassell.

The different effects produced by Jupiter's satellites in transit result then intelligibly from the marked variability of their light ;[3] and their variability seems, in some degree, to depend upon their orbital positions. This amounts to saying that, as Herschel concluded in 1797, they always, like our moon, turn the same face towards their primary, thus always presenting to us when in the same relative situations, the same obscure or brilliant sections of their globes. As regards the outer satellite, Engelmann's researches in 1871, and the late C. E. Burton's in 1873, make this almost certain ; and there is a strong probability that it also applies to the other three. The phenomena, however, are quite too irregular to be completely rationalised on so simple and obvious a principle. We are also driven to assume changes in the power of reflecting light of the satellites themselves, which Vogel's detection of lines in their spectra—or traces of such—indicative of gaseous envelopes similar to that of Jupiter, entitle us to regard as possibly of atmospheric production.

In the course of his observations on Jupiter at Brussels in 1878, M. Niesten was struck with a rosy cloud attached to a whitish zone beneath the dark southern equatorial band.[4] Its size was enormous. At the distance of Jupiter, its measured dimensions of 13″ by 3″ implied a real extension in longitude of 30,000, in latitude of something short of 7000 miles. The earliest record of its appearance seems to be by Professor Pritchett, director of the Morrison Observatory (U.S.), who figured and described it July 9, 1878.[5] It was again delineated

---

[1] *Month. Not.*, vol. xxviii. p. 11.   [2] *Observatory*, vol. vii. p. 175.

[3] There is a consensus among observers as to the marked variability of all Jupiter's satellites, though Pickering strangely finds no trace of it in his exact measures of their light. See *Harvard Annals*, vol. xi. pt. ii. p. 245.

[4] *Bull. Ac. R. Bruxelles*, t. xlviii. p. 607.   [5] *Astr. Nach.*, No. 2294.

August 9, by Tempel at Florence.[1] In the following year it attracted the wonder and attention of almost every possessor of a telescope. Its colour had by that time deepened into a full brick-red, and was set off by contrast with a white equatorial spot of unusual brilliancy. During three ensuing years these remarkable objects continued to offer a visible and striking illustration of the compound nature of the planet's rotation. The red spot completed a circuit in nine hours fifty-five minutes thirty-six seconds ; the white spot in about five and a half minutes less. Their *relative* motion was thus no less than 260 miles an hour, bringing them together in the same meridian at intervals of forty-four days ten hours forty-two minutes. Neither, however, preserved continuously the same uniform rate of travel. The period of each had lengthened by some seconds in 1883, while sudden displacements, associated with the recovery of lustre after recurrent semi-effacements, were observed in the position of the white spot,[2] recalling the leap forwards of a reviving sun-spot. The analogy was extended to the red spot by a shining aureola of " faculæ," described by Bredichin at Moscow, and by Lohse at Potsdam, as encircling it in September 1879.[3]

The conspicuous visibility of this astonishing object lasted three years, and may, it is thought, shortly recur. When the planet returned to opposition in 1882–83, it had faded so considerably that Riccò's uncertain glimpse of it at Palermo, May 31, 1883, was expected to be the last. It had, nevertheless, begun to recover in December, and was seen, "reduced to a mere skeleton" by internal wasting of substance and colour, by Mr. Denning, February 18, 1885.[4] In this emaciated condition it presented a striking likeness to an "elliptical ring" observed in the same latitude by Mr. Gledhill at Halifax in 1869–70. This, indeed, might be called the preliminary sketch for the

---

[1] *Astr. Nach.*, No. 2284.
[2] Denning, *Month. Not.*, vol. xliv. pp. 64, 66; *Nature*, vol. xxv. p. 226.
[3] *Astr. Nach.*, Nos. 2280, 2282.
[4] *Observatory*, vol. viii. p. 95 ; *Nature*, July 4, 1885.

famous object brought to perfection ten years later, but which Mr. H. C. Russell of Sydney saw and drew in June 1876,[1] in what might be called an unfinished condition, before it had separated from its matrix, the dusky south-tropical belt. In earlier times, too, a marking "at once fixed and transient" had been repeatedly perceived attached to the southernmost of the central belts. It gave Cassini in 1665 a rotation-period of nine hours fifty-six minutes,[2] reappeared and vanished eight times during the next forty-three years, and was last seen by Maraldi in 1713. It was, however, very much smaller than the recent object, and showed no unusual colour.

The assiduous observations made by Mr. Denning at Bristol and by Professor Hough at Chicago on the "Great Red Spot" of 1879–82 afforded grounds only for negative conclusions as to its nature. It certainly did *not* represent the outpourings of a Jovian volcano ; it was in no sense attached to the Jovian soil—if the phrase have any application to that planet ; it was *not* a mere disclosure of a glowing mass elsewhere seethed over by rolling vapours. To say that its origin was in some way eruptive is to say almost nothing ; yet this is about all that can safely be affirmed on the subject. It might be described, again, with some probability as an accidental excrescence on the general circulatory system of a strongly heated and cooling body. There is some reason to suppose that its surface was depressed below the average cloud level, and that the cavity was filled with vapours. But it was almost certainly not self-luminous, a satellite projected upon it in transit having been seen to show as bright as upon the dusky equatorial bands.

In 1870, Mr. Ranyard,[3] acting upon an earlier suggestion of Dr. Huggins, collected records of unusual appearances on the disc of Jupiter, with a view to investigate the question of their recurrence at regular intervals. He concluded that the development of the deeper tinges of colour, and of the equatorial "port-hole" markings girdling the globe in regular

[1] *Proc. Roy. Soc. N. S. Wales*, vol. xiv. p. 68.
[2] *Phil. Trans.*, vol. i. p. 143.   [3] *Month. Not.*, vol. xxxi. p. 34.

alternations of bright and dusky, agreed, so far as could be ascertained, with epochs of sun-spot maximum. The further inquiries of Dr. Lohse at Bothkamp in 1873 [1] went to strengthen the coincidence, which had been anticipated *à priori* by Zöllner in 1871.[2] Yet subsequent experience has rather added to than removed doubts as to the validity of that first conclusion. It may, indeed, be taken for granted that what Hahn terms the universal pulse of the solar system [3] affects the vicissitudes of Jupiter; but the law of those vicissitudes is far from being so obviously subordinate to the rhythmical flow of central disturbance as are certain terrestrial phenomena. The fundamental agreement which probably exists is confused in its display by secondary causes.

It is likely that Saturn is in a still earlier stage of planetary development than Jupiter. He is the lightest for his size of all the planets. In fact, he would float in water. And since his density is shown, by the amount of his equatorial bulging, to increase centrally,[4] it follows that his superficial materials must be of a specific gravity so low as to be inconsistent, on any probable supposition, with the solid or liquid states. Moreover, the chief arguments in favour of the high temperature of Jupiter apply, with increased force, to Saturn; so that it may be concluded, without much risk of error, that a large proportion of his bulky globe, 70,000 miles in diameter, is composed of heated vapours, kept in active and agitated circulation by the process of cooling.

His unique set of appendages has, since the middle of the century, formed the subject of searching and fruitful inquiries, both theoretical and telescopic. The mechanical problem of the stability of Saturn's rings was left by Laplace in a very unsatisfactory condition. Considering them as rotating solid bodies, he pointed out that they could not maintain their

[1] *Beobachtungen,* Heft ii. p. 99.
[2] *Ber. Sächs. Ges. der Wiss.,* 1871, p. 553.
[3] *Beziehungen der Sonnenfleckenperiode,* p. 175.
[4] A. Hall, *Astr. Nach.,* No. 2269.

position unless their weight were in some way unsymmetrically distributed; but made no attempt to determine the kind or amount of irregularity needed to secure this end. Some observations by Herschel gave astronomers an excuse for taking for granted the fulfilment of the condition thus vaguely postulated; and the question remained in abeyance until once more brought prominently forward by the discovery of the inner dusky ring in 1850.

The younger Bond led the way, among modern observers, in denying the solidity of the structure. The fluctuations in its aspect were, he asserted in 1851,[1] inconsistent with such an hypothesis. The fine dark lines of division, frequently detected in both bright rings, and as frequently relapsing into imperceptibility, were due, in his opinion, to the real mobility of their particles, and indicated a fluid formation. Professor Benjamin Peirce of Harvard University immediately followed with a demonstration, on abstract grounds, of their non-solidity.[2] Streams of some fluid denser than water were, he maintained, the physical reality giving rise to the anomalous appearance first disclosed by Galileo's telescope.

The mechanism of Saturn's rings, proposed as the subject of the Adams Prize, was dealt with by the late James Clerk Maxwell in 1857. His investigation forms the groundwork of all that is at present known in the matter. Its upshot was to show that neither solid nor fluid rings could continue to exist, and that the only possible composition of the system was by an aggregated multitude of unconnected particles, each revolving independently in a period corresponding to its distance from the planet.[3] This idea of a satellite-formation had been, remarkably enough, several times entertained and lost sight of. It was first put forward by Roberval in the seventeenth century, again by Jacques Cassini in 1715, and with perfect definiteness by Wright of Durham in 1750.[4] Little heed, however, was taken

[1] *Astr. Jour.* (Gould's), vol. ii. p. 17.   [2] *Ibid.*, p. 5.
[3] *On the Stability of the Motion of Saturn's Rings*, p. 67.
[4] *Mém. de l'Ac.*, 1715, p. 47; Montucla, *Hist. des Math.*, t. iv. p. 19; *An Original Theory of the Universe*, p. 115.

of these casual anticipations of a truth which reappeared, a virtual novelty, as the legitimate outcome of the most refined modern methods.

The details of telescopic observation accord, on the whole, admirably with this hypothesis. The displacements or disappearance of secondary dividing-lines—the singular striated appearance, first remarked by Short in the eighteenth century, last by Perrotin and Lockyer at Nice, March 18, 1884,[1]—show the effects of waves of disturbance traversing a moving mass of gravitating particles;[2] the broken and changing line of the planet's shadow on the ring gives evidence of variety in the planes of the orbits described by those particles. There is but one serious discrepancy.

On the satellite-theory, the obscure inner ring is formed of similar small bodies to those aggregated in the lucid members of the system, only much more thinly strewn, and reflecting, consequently, much less light. It is not, however, easy to see why these sparser flights should show as a dense dark shading on the body of Saturn. Yet this is invariably the case. The objection, long felt, has recently been urged by Professor Hastings of Baltimore. The brightest parts of these appendages, he remarks,[3] are more lustrous than the globe they encircle; but if the inner ring consist of identical materials, possessing presumably an equal reflective capacity, the mere fact of their scanty distribution would not cause them to show as *dark* against the same globe. The conclusion seems inevitable, that the bright and dark rings are *not* composed of identical materials.

A question of singular interest, and one which we cannot refrain from putting to ourselves, is—whether we see in the rings of Saturn a finished structure, destined to play a permanent part in the economy of the system; or whether they represent merely a stage in the process of development out of

---

[1] *Comptes Rendus*, t. xcviii. p. 718.

[2] Proctor, *Saturn and his System* (1865), p. 125.

[3] *Smiths. Report*, 1880 (Holden).

the chaotic state in which it is impossible to doubt that the
materials of all planets were originally merged. M. Otto
Struve has attempted to give a definite answer to this important query.

A study of early and later records of observations disclosed
to him, in 1851, an apparent progressive approach of the inner
edge of the bright ring to the planet. The rate of approach he
estimated at about fifty-seven English miles a year, or 11,000
miles during the 194 years elapsed since the time of Huygens.[1]
Were it to continue, a collapse of the system must be far
advanced within three centuries. But was the change real
or illusory—a plausible, but deceptive inference from insecure data? M. Struve resolved to put it to the test. A
set of minutely careful micrometrical measures of the dimensions of Saturn's rings, executed by himself at Pulkowa in the
autumn of 1851, was provided as a standard of future comparison ; and he was enabled to renew them, under closely
similar circumstances, in 1882.[2] But the expected diminution
of the space between Saturn's globe and his rings had not
taken place. There was, indeed, a slight extension in the
width of the system, both outwards and inwards ; but so slight
that it could hardly be considered to lie outside the limits
of probable error. Still it is worth notice that just such a
separation of the rings was indicated by Clerk Maxwell's theory,
so that there is an *à priori* likelihood of its being in progress.
Moreover, since 1657, when Huygens described the interval
between the ring and the planet as rather exceeding the width
of the ring, it is all but certain that a growth inwards has
actually occurred. For the two bright rings together, instead
of being narrower than the interval, are now more than one
and a half times as broad. Hence the expressions used by
Huygens, no less than most of the old drawings, are glaringly
inconsistent with the planet's present appearance.

---

[1] *Mém. de l'Ac. Imp.* (St. Petersb.), t. vii. 1853, p. 464.
[2] *Astr. Nach.*, No. 2498.

There seems reason to admit that Kirkwood's law of commensurability has had some effect in bringing about the present distribution of the matter composing these appendages. Here the disturbing bodies are Saturn's moons, while the divisions and boundaries of the rings represent the spaces where their disturbing action conspires to eliminate revolving particles. Kirkwood, in fact, showed, in 1867,[1] that a body circulating in the chasm between the bright rings known as " Cassini's division," would have a period nearly commensurable with those of *four* out of the eight moons ; and Dr. Meyer of Geneva has recently calculated all such combinations with the result of bringing out coincidences between regions of maximum perturbation and the limiting and dividing lines of the system.[2] This is in itself a strong confirmation of the view that the rings are made up of independently revolving small bodies.

On December 7, 1876, Professor Asaph Hall discovered at Washington a bright equatorial spot on Saturn, which he followed and measured through above sixty rotations, each performed in ten hours fourteen minutes twenty-four seconds.[3] He is careful to add that this represents the period, not necessarily of the *planet*, but only of the individual spot. The only previous determination of Saturn's axial movement (setting aside some insecure estimates by Schröter) was Herschel's in 1794, giving a period of ten hours sixteen minutes.

Saturn's outermost satellite, Japetus, is markedly variable— so variable that it sends us, when brightest, just $4\frac{1}{2}$ times as much light as when faintest. Moreover, its fluctuations depend upon its orbital position in such a way as to make it a conspicuous telescopic object when west, a scarcely discernible one when east of the planet. Herschel's inference [4] of a partially obscured globe turning always the same face towards

---

[1] *Meteoric Astronomy*, chap. xii. He carried the subject somewhat farther in 1871. See *Observatory*, vol. vi. p. 335.

[2] *Astr. Nach.*, No. 2527.   [3] *Am. Jour of Sc.*, vol. xiv. p. 325.

[4] *Phil. Trans.*, vol. lxxxii. p. 14.

its primary, seems the only admissible one, and is confirmed by Pickering's measurements of the varying intensity of its light. He remarks further that the dusky and brilliant hemispheres must be so posited as to divide the disc, viewed from Saturn, into nearly equal parts; so that this Saturnian moon, even when "full," appears very imperfectly illuminated over one-half of its surface.[1]

The spectrum of Saturn is closely similar to that of Jupiter. It shows the distinctive dark line in the red, which we may call the "red-star line;" and Janssen, examining it from the summit of Etna in 1867,[2] found unmistakable traces of aqueous absorption. The light from the ring is much less modified by original atmospheric action.

Uranus can now easily be seen with the naked eye as a star somewhat below the fifth magnitude. He thus appears considerably brighter than when discovered 105 years ago. Not, however, through any intrinsic change. He is at present conspicuous simply because he has but lately passed perihelion.[3] This circumstance has enabled astronomers, provided with the powerful telescopes of modern times, to make some highly interesting observations on this remote planet.

It will be remembered that Uranus presents the unusual spectacle of a system of satellites travelling nearly at right angles to the plane of the ecliptic. The existence of this anomaly gives a special interest to investigations of his axial movement, which the analogy of the other planets might lead us to presume to be executed in the same tilted plane. Yet this, strange to say, does not seem to be the case.

Mr. Buffham in 1870–72 caught traces of bright markings on the Uranian disc, suggesting, with much uncertainty, a rotation in about twelve hours in a plane *not* coincident with that in which his satellites circulate.[4] Dusky bands resembling

[1] *Smiths. Report,* 1880.   [2] *Comptes Rendus,* t. lxiv. p. 1304.
[3] Tebbutt, *Trans. Roy. Soc. N. S. Wales,* vol. xiv. p. 23.
[4] *Month. Not.,* vol. xxxiii. p. 164.

those of Jupiter, but very faint, were barely perceptible to Professor Young at Princeton in 1883. Yet, though inevitably inferred to be equatorial, they made a considerable angle with the trend of the satellites' orbits.[1] More distinctly by the brothers Henry, with the aid of their fine refractor, two grey parallel rulings, separated by a brilliant zone, were discerned every clear night at Paris from January to June 1884.[2] What were taken to be the polar regions appeared comparatively dusky. The direction of the equatorial rulings (for so we may safely call them) made an angle of 40° with the satellites' line of travel. Similar observations were made at Nice by MM. Perrotin and Thollon, March to June 1884, a lucid spot near the equator, in addition, indicating rotation in a period of about ten hours.[3]

Measurements of the little sea-green disc which represents to us the massive bulk of Uranus, give, however, a different result. Young, Schiaparelli,[4] and Schafarik have each found it to be quite distinctly *bulged ;* and all agree that the bulging lies just in the plane of the satellites' orbits. If this be so, there can be no question but that the same plane is that of the planet's rotation, the spheroidal shape of a rotating globe being the necessary consequence of the greater equatorial velocity of its particles. But the "equatorial" markings visibly assert a rapid whirling in a direction removed by nearly half a right angle from that plane. Which are we to believe? Where such minute quantities are concerned as in the *differences* between the various diameters of a disc about four seconds across, conclusions are of necessity highly precarious. They cannot weigh against the positive assurance conveyed by the parallel bands seen at Nice and Paris that Uranus now rotates in a plane widely removed from that in which the bodies dependent upon him circulate. This discrepancy may possibly be the result of a violent change in the axis of rotation ; and we might conjecture that the planet still retains the shape im-

[1] *Astr. Nach.*, No. 2545. [2] *Comptes Rendus*, t. xcviii. p. 1419.
[3] *Ibid.*, pp. 718, 967. [4] *Astr. Nach.*, No. 2526.

pressed by former conditions of movement, were it not that a globe almost certainly plastic, if not largely vaporous, would at once accommodate its form to their change.

The spectrum of Uranus was first examined by Father Secchi in 1869, and later, though with more advantages for accuracy, by Huggins and Vogel. It is a very remarkable one. In lieu of the reflected Fraunhofer lines, imperceptible perhaps through feebleness of light, six broad bands of original absorption appear,[1] one corresponding to the blue-green ray of hydrogen (F), another to the " red-star line " of Jupiter and Saturn, the rest as yet unidentified. The hydrogen band seems much too strong and diffused to be the mere echo of a solar line, and implies accordingly the presence of free hydrogen in the Uranian atmosphere, where a temperature must thus prevail sufficiently high to reduce water to its constituent elements.

Judging from the indications of an almost evanescent spectrum, Neptune, as regards physical condition, is the twin of Uranus, as Saturn of Jupiter. Of the circumstances of his rotation we are as good as completely ignorant. Mr. Maxwell Hall, indeed, noticed at Jamaica, in November and December 1883, certain rhythmical fluctuations of brightness, suggesting revolution on an axis in slightly less than eight hours ;[2] but Professor Pickering reduces the supposed variability to an amount altogether too small for certain perception, and Dr. G. Müller denies its existence *in toto*. It is true their observations were not precisely contemporaneous with those of Mr. Hall,[3] who believes the partial obscurations recorded by himself to have been of a passing kind, and to have suddenly ceased after a fortnight of prevalence. Their less conspicuous renewal was visible to him in November 1884, confirming a rotation period of 7.92 hours.

[1] Vogel, *Annalen der Phys.*, vol. clviii. p. 470.
[2] *Month. Not.*, vol. xliv. p. 257.
[3] *Observatory*, vol. vii. pp. 134, 221, 264.

The possibility that Neptune may not be the most remote body circling round the sun has been contemplated ever since he has been known to exist. Within the last few years the position at a given epoch of a planet far beyond him has been approximately fixed by two separate investigators. Its actual discovery is perhaps one of the prizes reserved for the astronomers of the future.

Professor George Forbes of Edinburgh hit upon in 1880 a novel plan of search for unknown members of the solar system. It depends upon the movements of comets. It is well known that those of moderately short periods are, for some reason, connected with the larger planets in such a way that the cometary aphelia fall near some planetary orbit. Jupiter claims above a dozen of such partial dependants, Neptune owns six, and there are two considerable groups, the farthest distances of which from the sun lie respectively near 100 and 300 times that of the earth. At each of these vast intervals, one involving a period of 1000, the other of 5000 years, Professor Forbes maintains that an unseen planet circulates. He has even computed elements for the nearer of the two, and fixed its place on the celestial sphere.[1]

In the meantime, Mr. D. P. Todd of Washington had been groping for the same object by the help of a totally different set of indications. The old approved method of perturbations was that adopted by him ; but those of Neptune have scarcely yet had time to develop, so that he was thrown back upon the "residual errors" of Uranus. They gave him a virtually identical situation for the new planet with that arrived at by Professor Forbes.[2] If this be a coincidence, it is a very remarkable one, the more so as each inquirer worked in complete ignorance of the results of the other.

[1] *Proc. Roy. Soc. Edinb.*, vol. x. p. 429 ; *Observatory*, vol. iii. p. 439.
[2] *Am. Jour. of Sc.*, vol. xx. p. 225.

CHAPTER IX.

THEORIES OF PLANETARY EVOLUTION.

WE cannot doubt that the solar system, as we see it, is the result of some process of growth—that, during innumerable ages, the forces of Nature were at work upon its materials, blindly modelling them into the shape appointed for them from the beginning by Omnipotent Wisdom. To set ourselves to inquire what that process was, may be an audacity, but it is a legitimate, nay, an inevitable one. For man's implanted instinct to "look before and after" does not apply to his own little life alone, but regards the whole history of creation, from the highest to the lowest—from the microscopic germ of an alga or a fungus to the visible frame and furniture of the heavens.

Kant considered that the inquiry into the mode of origin of the world was one of the easiest problems set by Nature; but it cannot be said that his own solution of it was a satisfactory one. He, however, struck out in 1755 a track which thought still pursues. In his *Allgemeine Naturgeschichte* the growth of sun and planets was traced from the cradle of a vast and formless mass of evenly diffused particles, and the uniformity of their movements was sought to be accounted for by the uniform action of attractive and repulsive forces, under the dominion of which their development was carried forwards.

In its modern form, the "Nebular Hypothesis" made its appearance in 1796.[1] It was presented by Laplace with diffidence, as a speculation unfortified by numerical buttresses of any kind, yet with visible exultation in having, as he thought,

---

[1] *Exposition du Système du Monde*, t. ii. p. 295.

penetrated the birth-secret of our system. He demanded, indeed, more in the way of postulates than Kant had done. He started with a sun ready-made,[1] and surrounded with a vast glowing atmosphere, extending into space out beyond the orbit of the farthest planet, and endowed with a slow rotatory motion. As this atmosphere or nebula cooled, it contracted; and as it contracted, its rotation, by a well-known mechanical law, became accelerated. At last, a point arrived when centrifugal force at the equator increased beyond the power of gravity to control, and equilibrium was restored by the separation of a nebulous ring revolving in the same period as the generating mass. After a time, the ring broke up into fragments, all eventually reunited in a single revolving and rotating body. This was the first and farthest planet.

Meanwhile the parent nebula continued to shrink and whirl quicker and quicker, passing, as it did so, through successive crises of instability, each resulting in, and terminated by, the formation of a planet, at a smaller distance from the centre, and with a shorter period of revolution than its predecessor. In these secondary bodies the same process was repeated on a reduced scale, the birth of satellites ensuing upon their contraction, or not, according to circumstances. Saturn's ring, it was added, afforded a striking confirmation of the theory of annular separation,[2] and appeared to have survived in its original form in order to throw light on the genesis of the whole solar system; while the four first discovered asteroids offered an example in which the *débris* of a shattered ring had failed to coalesce into a single globe.

This scheme of cosmical evolution was a characteristic bequest from the eighteenth century to the nineteenth. It possessed the self-sufficing symmetry and entireness appropriate to the ideas of a time of renovation, when the complexity of nature was little accounted of in comparison with the imperious orderliness of the thoughts of man. Since it was

---

[1] In later editions a retrospective clause was added admitting a prior condition of all but evanescent nebulosity.　[2] *Méc. Cél.*, lib. xiv. ch. iii.

propounded, however, knowledge has transgressed many
boundaries, and set at naught much ingenious theorising.
How has it fared with Laplace's sketch of the origin of the
world? It has at least not been discarded as effete. The
groundwork of speculation on the subject is still furnished by
it. It is, nevertheless, admittedly inadequate. Of much that
exists it gives no account, or an erroneous one. It is certain
that the march of events did not everywhere—it is doubtful
whether it anywhere—followed the exact path prescribed for it.
Yet modern science attempts to supplement, but scarcely ven-
tures to supersede it.

Thought has, in many directions, been profoundly modified
by Mayer's and Joule's discovery, in 1842, of the equivalence
between heat and motion. Its corollary was the grand idea of
the "conservation of energy," now one of the cardinal principles
of science. This means that, under the ordinary circumstances of observation, the old maxim *ex nihilo nihil fit* applies
to force as well as to matter. The supplies of heat, light,
electricity, must be kept up, or the stream will cease to flow.
The question of the maintenance of the sun's heat was thus
inevitably raised; and with the question of maintenance that
of origin is indissolubly connected.

Dr. Julius Robert Mayer, a physician residing at Heilbronn,
was the first to apply the new light to the investigation of what
Sir John Herschel had termed the "great secret." He showed
that if the sun were a body either simply cooling or in a state
of combustion, it must long since have "gone out." Had an
equal mass of coal been set alight, four or five centuries
after the building of the Pyramid of Cheops, and kept burning at such a rate as to supply solar light and heat during
the interim, only a few cinders would now remain in lieu of
our undiminished glorious orb. Mayer looked round for an
alternative. He found it in the "meteoric hypothesis" of
solar conservation.[1] The importance in the economy of our
system of the bodies known as falling stars was then (in 1848)

[1] *Beiträge zur Dynamik des Himmels*, p. 12.

beginning to be recognised. It was known that they revolved in countless swarms round the sun; that the earth daily encountered millions of them; and it was surmised that the cone of the zodiacal light represented their visible condensation towards the attractive centre. From the zodiacal light, then, Mayer derived the store needed for maintaining the sun's radiations. He proved that, by the stoppage of their motion through falling into the sun, bodies would evolve from 4600 to 9200 times as much heat (according to their ultimate velocity) as would result from the burning of equal masses of coal, their precipitation upon the sun's surface being brought about by the resisting medium observed to affect the revolutions of Encke's comet. There was, however, a difficulty. The quantity of matter needed to keep, by the sacrifice of its movement, the hearth of our system warm and bright, would be very considerable. Mayer's lowest estimate put it at 94,000 billion kilogrammes per second, or a mass equal to that of our moon bi-annually. But so large an addition to the gravitating power of the sun would quickly become sensible in the movement of the bodies dependent upon him. Their revolutions would be notably accelerated. Mayer admitted that each year would be shorter than the previous one by a not insignificant fraction of a second, and postulated an unceasing waste of substance, such as Newton had supposed must accompany emission of the material corpuscles of light, to neutralise continual reinforcement.

Mayer's views obtained a very small share of publicity, and owned Mr. Waterston as their independent author in this country. The meteoric, or "dynamical" theory of solar sustentation was expounded by him before the British Association in 1853. It was developed with his usual ability by Sir William Thomson in the following year. The inflow of meteorites, he remarked, "is the only one of all conceivable causes of solar heat which we know to exist from independent evidence."[1] We know it to exist, but we now also know it to

<hr>

[1] *Trans. Roy. Soc. of Edinburgh*, vol. xxi. p. 66.

be entirely insufficient. The supplies presumed to be contained in the zodiacal light would be quickly exhausted; a
constant inflow from space would be needed to meet the
demand. But if moving bodies were drawn into the sun at
anything like the required rate, the air, even out here at
ninety-three millions of miles distance, would be thick with
them; the earth would be red-hot from their impacts;[1] geological deposits would be largely meteoric;[2] to say nothing
of the effects on the mechanism of the heavens. Sir William
Thomson himself urged the inadmissibility of the "extra-
planetary" theory of meteoric supply on the very tangible
ground that, if it were true, the year would be shorter now,
actually *by six weeks*, than at the opening of the Christian era.
The "intra-planetary" supply, however, is too scanty to be
anything more than a temporary makeshift.

The meteoric hypothesis was naturally extended from the
maintenance of the sun's heat to the formation of the bodies
circling round him. The earth—no less doubtless than the
other planets—is still growing. Cosmical matter in the shape
of falling stars and aerolites, to the amount, adopting Professor
Newton's estimate, of 100 tons daily, is swept up by it as it
pursues its orbital round. Inevitably the idea suggested itself
that this process of appropriation gives the key to the life-
history of our globe, and that the momentary streak of fire in
the summer sky represents a feeble survival of the glowing
hail-storm by which, in old times, it was fashioned and warmed.
Mr. E. W. Brayley supported this view of planetary production in 1864,[3] and it has recommended itself to Haidinger,
Helmholtz, Proctor, and Faye. But the negative evidence of
geological deposits appears fatal to it.

The theory of solar energy now generally regarded as the
true one, was enounced by Helmholtz in a popular lecture
in 1854. It depends upon the same principle of the equi-

---

[1] Newcomb, *Pop. Astr.*, p. 521 (2d ed.)

[2] M. Williams, *Nature*, vol. iii. p. 26.

[3] *Comp. Brit. Almanac*, p. 94.

valence of heat and motion which had suggested the meteoric hypothesis. But here the movement surrendered and transformed belongs to the particles, not of any foreign bodies, but of the sun itself. Drawn together by the force of their own gravity from a wide ambit, their fall towards the sun's centre must have engendered a vast thermal store, of which $\frac{453}{454}$ are computed to be already spent. Presumably, however, this stream of reinforcement is still flowing. In the very act of parting with heat, the sun develops a fresh stock. His radiations, in short, are the direct result of shrinkage through cooling. A diminution of the solar diameter by 300 feet yearly (Langley) would just suffice to cover the present rate of emission. But the process, though not terminated, is strictly a terminable one. In five million years, the sun will have contracted to half its present bulk. In seven million more, it will be as dense as the earth. It is difficult to believe that it will then be a luminous body.[1] Nor can an unlimited past duration be admitted. Helmholtz considered that radiation might have gone on with its actual intensity for twenty-two, Langley allows only eighteen million years. The period can scarcely be stretched, by the most generous allowances, to double the latter figure. But this is far from meeting the demands of geologists and biologists.

An ingenious attempt has lately been made to supply the sun with machinery analogous to that of a regenerative furnace, enabling it to consume the same fuel over and over again, and so to prolong indefinitely its beneficent existence. The inordinate "waste" of energy, which shocks our thrifty ideas, was simultaneously abolished. The earth stops and turns variously to account one 2250-millionth part of the solar radiations; each of the other planets and satellites takes a proportionate share; the rest, being all but an infinitesimal fraction of the whole, is dissipated through endless space, to serve what purpose we know not. Now, on the late Sir William Siemens's plan, this reckless expenditure would cease; the solar incomings

[1] Newcomb, *Pop. Astr.*, pp. 521–525.

z

and outgoings would be regulated on approved economic principles, and the inevitable final bankruptcy would be staved off to remote ages.   Let us see how it is to be done.

We must first imagine space to be filled with combustible substances—hydrogen, hydro-carbons, and oxygen—in an excessively rarefied state. Next, that the sun keeps up, by its rotation, a fan-like action on this floating matter, drawing it inwards at the polar surfaces, and projecting it outwards at the equator "in a continuous disc-like stream." [1]   But it will not travel from the sun unchanged.   Combustion will have intervened.   In other words, the particles sucked in will have surrendered their stored-up energy in the shape of heat and light, and they will depart, no longer combustible, but the mere inert products of combustion.   By the very power of the radiations they had contributed to supply, however, they may be restored to activity.   Sir W. Siemens obtained some experimental evidence that carbonic acid and water may possibly be dissociated in space, as they undoubtedly are in the leaves of plants, by the power of direct sunshine.   Their particles, thus compulsorily separated, and by the act restocked with energy, are ready to rush together again with fresh evolution of heat and light.   A mechanical circulation is, in this way, combined with a pendulum-swing of chemical change, and the round might go on for ever, if only one condition were granted.   That one condition is an unlimited supply of motive power.   It is, however, an inexorable law of nature that there is no work without waste.   *Ex nihilo nihil fit.*

In this case, the heart-throb of the circulating system resides in the rotation of the sun.   Therein is contained a certain definite amount of mechanical power—enough, according to Sir W. Thomson, if directly converted into heat, to keep up the sun's emission during 116 years and six days—a mere moment in cosmical time.   More economically applied, it would no doubt go farther.   Its exhaustion would nevertheless, under the most favourable circumstances, ensue in a

[1] *Proc. Roy. Soc.*, vol. xxxiii. p. 393.

comparatively short period,[1] Many other objections equally unanswerable have been urged to the " regenerative" hypothesis, but this one suffices.

There remains, then, as the only intelligible rationale of solar sustentation, Helmholtz's shrinkage theory. And this has a very important bearing upon the nebular view of planetary formation ; it may, in fact, be termed its complement. For it involves the idea that the sun's materials, once enormously diffused, gradually condensed to their present volume with development of heat and light, and, it may plausibly be added, the separation of dependent globes. The data furnished by spectrum analysis, too, favour the supposition of a common origin for sun and planets by showing their community of substance; while gaseous nebulæ present examples of vast masses of tenuous vapour, such as our system may plausibly be conjectured to have primitively sprung from.

But recent science raises many objections to the details, if it supplies some degree of confirmation to the fundamental idea of Laplace's cosmogony. The detection of the retrograde movement of Neptune's satellite made it plain that the anomalous conditions of the Uranian world were due to no extraordinary disturbance, but to a systematic variety of arrangement at the outskirts of the solar domain. So that, were a trans-Neptunian planet discovered, we should be fully prepared to find it rotating, and surrounded by satellites circulating from east to west. The uniformity of movement, upon the probabilities connected with which the French geometer mainly based his scheme, thus at once vanishes.

The excessively rapid revolution of the inner Martian moon is a further stumbling-block. On the nebular view, *no* satellite can revolve in a shorter time than its primary rotates ; for in its period of circulation survives the period of rotation of the

---

[1] To this hostile argument, as urged by Mr. E. Douglas Archibald, Sir W. Siemens opposed the increase of rotative velocity through contraction (*Nature*, vol. xxv. p. 505). But contraction cannot restore lost momentum.

parent mass which filled the sphere of its orbit at the time of giving it birth. And rotation quickens as contraction goes on; therefore, the older time of axial rotation should invariably be the longer. There is, however, a way out of this difficulty, presently to be adverted to.

More serious is one connected with the planetary periods, pointed out by Babinet in 1861.[1] In order to make them fit in with the hypothesis of successive separation from a rotating and contracting body, certain arbitrary assumptions have to be made of fluctuations in the distribution of the matter forming that body at the various epochs of separation. Such expedients usually merit the distrust which they inspire.

Again, it was objected by Professor Kirkwood in 1869[2] that there could be no sufficient cohesion in such an enormously diffused mass as the planets are supposed to have sprung from, to account for the wide intervals between them. The matter separated, through the growing excess of centrifugal speed, would have been cast off, not by rarely recurring efforts, but continually, fragmentarily, *pari passu* with condensation and acceleration. Each wisp of nebula, as it found itself unduly hurried, would have declared its independence, and set about revolving and condensing on its own account. The result would have been a meteoric, not a planetary system.

Moreover, it is a question whether the relative ages of the planets do not follow an order just the reverse of that concluded by Laplace. Professor Newcomb holds the opinion that the rings which eventually constituted the planets, divided from the main body of the nebula almost simultaneously, priority, if there were any, being on the side of the inner and smaller ones;[3] while, in M. Faye's ingenious supplement to the nebular cosmogony,[4] the retrograde motion of the systems formed by the two outer planets is ascribed—on grounds, it is true, of dubious validity—to their comparatively late origin.

[1] *Comptes Rendus*, t. lii. p. 481. See also Kirkwood, *Observatory*, vol. iii. p. 409.   [2] *Month. Not.*, vol. xxix. p. 96.

[3] *Pop. Astr.*, p. 527.   [4] *Nature*, vol. xxxi. p. 194.

We now come to a most remarkable investigation—one, indeed, unique in its profession to lead us back with mathematical certainty towards the origin of a heavenly body. We refer to Mr. G. H. Darwin's recent inquiries into the former relations of the earth and moon.[1]

They deal exclusively with the effects of tidal friction, and primarily with those resulting, not from oceanic, but from "bodily" tides, such as the sun and moon must have raised in past ages on a liquid or viscous earth. The immediate effect of either is, as already explained, to destroy the rotation of the body on which the tide is raised, as regards the tide-raising body, bringing it to turn always the same face towards its disturber. This, we can see, has been completely brought about in the case of the moon. There is, however, a secondary or reactive effect. Action is always mutual. Precisely as much as the moon pulls the terrestrial tidal wave backward, the tidal wave pulls the moon forward. But pulling a body forward in its orbit implies the enlargement of that orbit; that is to say, the moon is, as a consequence of tidal friction, very slowly receding from the earth. This will go on (other circumstances remaining unchanged) until the lengthening day overtakes the more tardily lengthening month, when each will be of about 1400 hours. A position of what we may call tidal equilibrium between earth and moon will (apart from disturbance by other bodies) then be attained.

If, however, it be true that, in the time to come, the moon will be much farther from us, it follows that in the time past she was much nearer to us than she now is. Tracing back her history by the aid of Mr. Darwin's clue, we at length find her revolving in a period of somewhere between two and four hours, almost in contact with an earth rotating just at the same rate. This was before tidal friction had begun its work of grinding down axial velocity and expanding orbital range. But the position was not one of stable equilibrium. The slightest inequality must have set on foot a series of uncom-

[1] *Phil. Trans.*, vol. clxxi. p. 713.

pensated changes. If the moon had whirled the least iota
faster than the earth spun, she must have been precipitated
upon it. Her actual existence shows that the trembling
balance inclined the other way. By a second or two to begin
with, the month exceeded the day; the tidal wave crept ahead
of the moon; tidal friction came into play, and our satellite
started on its long spiral journey outward from the parent
globe. This must have occurred, it is computed, *at least* fifty-
four million years ago.

Assuming the exactness of the physical data involved—a
proviso which may cover a good deal of doubt—these conclusions are, in the opinion of those most competent to judge,
mathematically certain. An irresistible conjecture carries us
one step beyond them. The moon's time of revolution, when
so near the earth as barely to escape contact with it, must
have been, by Kepler's Law, more than two, and less than two
and a half hours. Now it happens that the most rapid rate of
rotation of a fluid mass of the earth's average density, consistent with spheroidal equilibrium, is two hours and twenty
minutes. Quicken the movement but by one second, and, the
globe must fly asunder. Hence the inference that the earth
actually *did* fly asunder through over-fast spinning, the ensuing
disruption representing the birth-throes of the moon. It is
likely that the event was hastened or helped by solar tidal
disturbance.

To recapitulate. Analysis tracks backward the two bodies
until it leaves them in very close contiguity, one rotating and
the other revolving in approximately the same time, and that
time certainly not far different from, and quite possibly
identical with, the critical period of instability for the terrestrial
spheroid. "Is this," Mr. Darwin asks, "a mere coincidence,
or does it not rather point to the break-up of the primeval
planet into two masses in consequence of a too rapid rotation?"[1] Few will hesitate as to the answer.

---

[1] *Phil. Trans.*, vol. clxxi. p. 835.

This investigation was communicated to the Royal Society, December 18, 1879. It was followed, January 20, 1881,[1] by an inquiry on the same principles into the earlier condition of the entire solar system. The results were a warning against hasty generalisation. They showed that the lunar terrestrial system, far from being a pattern for their development, was a singular exception amongst the bodies swayed by the sun. Its peculiarity resides in the fact that the moon is *proportionately* by far the most massive attendant upon any known planet. Its disturbing power over its primary is thus abnormally great, and tidal friction has, in consequence, played a predominant part in bringing their mutual relations into their present state.

The comparatively late birth of the moon tends to ratify this inference. The dimensions of the earth did not differ (according to Mr. Darwin) very greatly from what they now are when her solitary offspring came, somehow, into existence. This is found not to have been the case with any other of the planets. It is unlikely that the satellites of Jupiter, Saturn, or Mars (we may safely add of Uranus or Neptune) ever revolved in much narrower orbits than those they now traverse; it is practically certain that they did not originate close to the *present* surfaces of their primaries, like our moon.[2] What follows? The tide-raising power of a body grows with nearness in a rapidly accelerated ratio. Lunar tides must then have been on an enormous scale when the moon swung round just outside the earth's atmosphere. But no other satellite with which we are acquainted occupied at any time a corresponding position. Hence no other satellite ever possessed tide-raising capabilities in the least comparable to those of the moon. We conclude once more that tidal friction had an influence here quite different from its influence elsewhere.

There is, however, another branch of the same subject. We know that the sun as well as the moon causes tides in our oceans. There must then be solar, no less than lunar tidal

[1] *Phil. Trans.*, vol. clxxii. p. 491.   [2] *Ibid.*, p. 530.

friction. The question at once arises : What part has it played
in the development of the solar system? Has it ever been
one of leading importance, or has its influence always been, as
it now is, subordinate, almost negligeable? To this, too, Mr.
Darwin supplies an answer.

It can be stated without hesitation that the sun did *not* give
birth to the planets, as the earth may have given birth to the
moon, by the disruption of its already condensed, though
plastic and glowing mass, pushing them then gradually backward from its surface into their present places. For the utmost
possible increase in the length of the year through tidal friction
is one hour ; and five minutes is a more probable estimate.[1]
So far as the pull of tide-waves raised on the sun by the
planets is concerned, then, the distances of the latter have
never been notably different from what they now are ; though
that cause may have converted the paths traversed by them
from circles into ellipses.

Over their *physical* history, however, it was probably in a
large measure influential. The first vital issue for each of
them was—satellites or no satellites? Were they to be gover-
nors as well as governed, or should they revolve in sterile
isolation throughout the æons of their future existence? Here
there is strong reason to believe that solar tidal friction
was the overruling power. It is remarkable that planetary
fecundity increases—at least so far outward as Saturn—with
distance from the sun. Can these two facts be in any way
related? In other words, is there any conceivable way by
which tidal influence could prevent or impede the throwing-
off of secondary bodies? We have only to think for a moment
in order to see that this is precisely one of its direct results.

Tidal friction, whether solar or lunar, tends to reduce the
axial movement of the body it acts upon. But the separation
of satellites depends—according to the received view—upon
the attainment of a disruptive rate of rotation. Hence, if
solar tidal friction were strong enough to keep down the pace

[1] *Phil. Trans.*, vol. clxxii. p. 533.

below this critical point, the contracting mass would remain intact—there would be no satellite-production. This, in all probability, actually occurred in the case both of Mercury and Venus. They cooled without dividing, because the solar friction-brake applied to them was too strong to permit acceleration to pass the limit of equilibrium. The earth barely escaped the same fate of loneliness. Her first and only epoch of instability was retarded until she had nearly reached maturity. The late appearance of the moon accounts for its large relative size—through the increased cohesion of an already strongly condensed parent mass—and for the distinctive peculiarities of its history and influence on the producing globe.

Solar tidal friction is still considerably effective at the distance of Mars. It did not, indeed, hinder the formation of two minute dependants, but it explains the anomalously rapid revolution of one of them. Phobos, we have seen, completes more than three revolutions while Mars rotates once. But this was probably not always so. The two periods were originally nearly equal. The difference was brought about by tidal waves raised on the viscous spheroid of Mars by the sun. Rotatory velocity was thereby destroyed, the Martian day slowly lengthened, and, as a secondary consequence, the period of the inner satellite, become shorter than the augmented day, began progressively to diminish. So that Phobos, unlike our moon, was in the beginning farther from its primary than now. The attraction of the tidal wave raised by the sun on the globe of Mars is gradually drawing it inward, and threatens to effect its eventual precipitation upon his surface. The same destiny, it may be added, awaits our own satellite, should the present order of things endure long enough to enable solar tidal friction to bring about that indefinitely remote end.

Outside the orbit of Mars, this agency can scarcely be said to possess any sensible power. In the systems of Jupiter, Saturn, Uranus, and Neptune, tides are probably effective chiefly on the rotation of satellites, compelling them to turn always the same faces towards their primaries.

The general outcome of Mr. Darwin's researches has been to leave Laplace's cosmogony untouched. He concludes nothing against it, and, what perhaps tells with more weight in the long run, has nothing to substitute for it. In one form or the other, if we speculate at all on the development of the planetary system, our speculations are driven into conformity with the broad lines of the Nebular Hypothesis—so far, at least, as admitting an original material unity and motive uniformity. But we can see now, better than formerly, that these supply a bare and imperfect sketch of the truth. We should err gravely were we to suppose it possible to reconstruct, with the help of any knowledge our race is ever likely to possess, the real and complete history of our admirable system. "The subtlety of nature," Bacon says, "transcends in many ways the subtlety of the intellect and senses of man." By no mere barren formula of evolution, indiscriminately applied all round, the results we marvel at, and by a fragment of which our life is conditioned, were brought forth; but by the manifold play of interacting forces, variously modified and variously prevailing, according to the local requirements of the design they were appointed to execute.

CHAPTER X.

RECENT COMETS.

ON the 2d of June 1858, Giambattista Donati discovered at
Florence a feeble round nebulosity in the constellation Leo,
about one-tenth the diameter of the full moon. It proved to
be a comet approaching the sun. But it changed little in
apparent place or brightness for some weeks. The gradual
development of a central condensation of light was the first
symptom of coming splendour. At Harvard, in the middle of
July, a strong stellar nucleus was seen; on August 14 a tail
began to be thrown out. As the comet wanted still over six
weeks of the time of its perihelion-passage, it was obvious that
great things might be expected of it. They did not fail of
realisation.

Not before the early days of September was it generally
recognised with the naked eye, though it had been detected
without a glass at Pulkowa, August 19. But its growth was
thenceforward a surprisingly rapid one, as it swept with accelerated motion under the hindmost foot of the Great Bear,
and past the starry locks of Berenice. A sudden leap upward
in lustre was noticed on September 12, when the nucleus
shone with about the brightness of the pole-star, and the tail,
notwithstanding large fore-shortening, could be traced with
the lowest telescopic power over six degrees of the sphere.
The appendage, however, attained its full development only
after perihelion, September 30, by which time, too, it lay nearly
square to the line of sight from the earth. On October 10
it stretched in a magnificent scimitar-like curve over a third

and upwards of the visible hemisphere, representing a real extension in space of fifty-four million miles. But the most striking view was presented on October 5, when the brilliant star Arcturus became involved in the brightest part of the tail, and during many hours contributed, its lustre undiminished by the interposed nebulous screen, to heighten the grandeur of the most majestic celestial object of which living memories retain the impress. Donati's comet was, according to Admiral Smyth's testimony,[1] outdone " as a mere *sight*-object " by the great comet of 1811; but what it lacked in splendour, it surely made up in grace, and variety of what we may call "scenic" effects.

Some of these were no less interesting to the student than they were impressive to the spectator. At Pulkowa, on the 16th September, Winnecke[2] observed a faint outer envelope resembling a veil of almost evanescent texture flung somewhat widely over the head. Next evening, the first of the "secondary" tails appeared, possibly as part of the same phenomenon. This was a narrow, straight ray, forming a tangent to the strong curve of the primary tail, and reaching to a still greater distance from the nucleus. It continued faintly visible for about three weeks, during part of which time it was seen in duplicate. For from the chief train itself, at a point where its curvature abruptly changed, issued, as if through the rejection of part of its materials, a second beam nearly parallel to the first, the rigid line of which contrasted singularly with the softly diffused, and waving aspect of the plume of light from which it sprang. Olbers's theory of unequal repulsive forces was never more beautifully illustrated. The triple tail was a visible solar analysis of cometary matter.

The processes of luminous emanation going on in this body forcibly recalled the observations made on the comets of 1744 and 1835. From the middle of September, the nucleus, estimated by Bond to be under five hundred miles in diameter,

---

[1] *Month. Not.*, vol. xix. p. 27.

[2] *Mém. de l'Ac. Imp.*, t. ii. 1859, p. 46.

was the centre of action of the most energetic kind. Seven distinct "envelopes" were detached in succession from the nebulosity surrounding the head, and after rising towards the sun during periods of from four to seven days, finally cast their material backward to form the right and left branches of the great train. The separation of these by an obscure axis—apparently as black, quite close up to the nucleus, as the sky—indicated for the tail a hollow, cone-like structure;[1] while the repetition of certain spots and rays in the same corresponding situation on one envelope after another, served to show that the nucleus—to some local peculiarity of which they were doubtless due—had no proper rotation, but merely shifted sufficiently on an axis to preserve the same aspect towards the sun as it moved round it.[2] This observation of Bond's was strongly confirmatory of Bessel's hypothesis of opposite polarities in such bodies' opposite sides.

The protrusion towards the sun, on September 25, of a brilliantly luminous, fan-shaped sector completed the resemblance to Halley's comet. The appearance of the head was now somewhat that of a "bat's-wing" gaslight. There were, however, no oscillations to and fro, such as Bessel had seen and speculated upon in 1835. As the size of the nucleus contracted with approach to perihelion, its intensity augmented. On October 2, it outshone Arcturus, and for a week or ten days was a conspicuous object half an hour after sunset. Its lustre—setting aside the light emitted from the tail—was, at that date, 6300 times what it had been in June 15, though *theoretically*—taking into account, that is, only the differences of distance from sun and earth—it should have been only $\frac{1}{33}$ of that amount. Here, it might be thought, was convincing evidence of the comet itself becoming ignited under the growing intensity of the solar radiations. Experiments with the polariscope were, however, interpreted in an adverse sense, and Bond's conclusion that the comet sent us virtually unmixed reflected sunshine was generally acquiesced in. It did

---

[1] *Harvard Annals*, vol. iii. p. 368.   [2] *Ibid.*, p. 371.

not, nevertheless, survive the first application of the spectroscope to these bodies.

Very few comets have been so well or so long observed as
Donati's. It was visible to the naked eye during 112 days;
it was telescopically discernible for 275, the last observation
having been made by Dr. Mann at the Cape of Good Hope,
March 4, 1859. Its course through the heavens combined
singularly with the orbital place of the earth to favour curious
inspection. The tail, when near its greatest development,
lost next to nothing by the effects of perspective, and at the
same time lay in a plane sufficiently inclined to the line of
sight to enable it to display its exquisite curves to the greatest
advantage. Even the weather was, on both sides of the
Atlantic, propitious during the period of greatest interest, and
the moon as little troublesome as possible. The splendid
volume compiled by the younger Bond is a monument to the
care and skill with which these advantages were turned to
account. Yet this stately apparition marked no turning-point
in the history of cometary science. By its study knowledge
was indeed materially advanced, but along the old lines.
No quick and vivid illumination broke upon its path. Quite
insignificant objects—as we have already partly seen—have
often proved more vitally instructive.

Donati's comet has been identified with no other. Its path
is an immensely elongated ellipse, lying in a plane far apart
from that of the planetary movements, carrying it at perihelion
considerably within the orbit of Venus, and at aphelion out into
space to $5\frac{1}{2}$ times the distance from the sun of Neptune. The
entire circuit occupies over 2000 years, and is performed
in a retrograde direction, or against the order of the Signs.
Before its next return, about the year 4000 A.D., the enigma
of its presence and its purpose may have been to some extent
—though we may be sure not completely—penetrated.

On June 30, 1861, the earth passed, for the second time in
this century, through the tail of a great comet. Many of our
readers must remember the unexpected disclosure, on the

withdrawal of the sun below the horizon on that evening, of an object so remarkable as to challenge universal attention. A golden-yellow planetary disc, wrapt in dense nebulosity, shone out while the June twilight of these latitudes was still in its first strength. The number and complexity of the envelopes surrounding the head produced, according to the late Mr. Webb,[1] a magnificent effect. Portions of six distinct emanations were traceable. " It was as though a number of light, hazy clouds were floating round a miniature full moon." As the light faded, the tail came out.[2] Although in brightness and sharpness of definition it could not compete with the display of 1858, its dimensions proved to be extraordinary. It reached upwards beyond the zenith when the head had already set. By some authorities its extreme length was stated at 118°, and it showed no trace of curvature. Most remarkable, however, was the appearance of two widely divergent rays, each pointing towards the head, though cut off from it by sky-illumination, of which one was seen by Mr. Webb, and both by Mr. Williams at Liverpool, a quarter of an hour before midnight. There seems no doubt that Mr. Webb's interpretation was the true one, and that these beams were, in fact, " the perspective representation of a conical or cylindrical tail, hanging closely above our heads, and probably just being lifted up out of our atmosphere." [3] The cometary train was then rapidly receding from the earth, so that the sides of the " outspread fan " of light which it showed when we were right in the line of its axis, must have appeared (as they did) to close up in departure. The swiftness with which the visually opened fan shut, proved its vicinity ; and indeed Mr. Hind's calculations showed that we were not so much near, as actually within its folds at that very time.

Already M. Liais, from his observations at Rio de Janeiro, June 11 to 14, had anticipated, as a probability, such an encounter, and had subsequently proved that it must have

---

[1] *Month. Not.*, vol. xxii. p. 306.   [2] Stothard in *ibid.*, vol. xxi. p. 243.
[3] *Intell. Observer*, vol. i. p. 65.

occurred in such a way as to cause an immersion of the earth in cometary matter to a depth of 300,000 miles.[1] The comet then lay between the earth and the sun at a distance of about fourteen million miles from the former; its tail stretched outward just along the line of intersection of its own with the terrestrial orbit to an extent of fifteen million miles; so that our globe, happening to pass at the time, found itself during some hours involved in the flimsy appendage.

No perceptible effects were produced by the meeting; it was known to have occurred by theory alone. A peculiar glare in the sky, thought by some to have distinguished the evening of June 30, was, at best, inconspicuous. Nor were there any symptoms of unusual electric excitement. The Greenwich instruments were, indeed, disturbed on the following night; but it would be rash to infer that the comet had art or part in their agitation.

The perihelion-passage of this body occurred June 11, 1861; and its orbit has been shown by M. Kreutz of Bonn, from a very complete investigation founded on observations extending over nearly a year, to be an ellipse traversed in a period of 409½ years.[2]

Towards the end of August 1862, a comet became visible to the naked eye high up in the northern hemisphere, with a nucleus equalling in brightness the lesser stars of the Plough and a feeble tail 20° in length. It thus occupied quite a secondary position among the members of its class. It was, nevertheless, a splendid object in comparison with a telescopic nebulosity discovered by Tempel at Marseilles, December 19, 1865. This, the sole comet of 1866, slipped past perihelion January 11, without pomp of train or other appendages, and might have seemed hardly worth the trouble of pursuing. Fortunately, however, this was not the view entertained by observers and computers; since upon the knowledge acquired of the movements of these two bodies has been founded one of the most significant discoveries of modern times. The first

[1] *Comptes Rendus*, t. lxi. p. 953.   [2] *Smiths. Report*, 1881 (Holden).

of them is now styled the comet (1862 iii.) of the August meteors, the second (1866 i.) that of the November meteors. The steps by which this curious connection came to be ascertained were many, and were taken in succession by a number of individuals. But the final result was reached by Schiaparelli of Milan, and remains deservedly associated with his name.

The idea prevalent in the last century as to the nature of shooting stars was that they were mere aerial *ignes fatui*—inflammable vapours accidentally kindled in our atmosphere. But Halley had already entertained the opinion of their cosmical origin; and Chladni in 1794 formally broached the theory that space is filled with minute circulating atoms, which, drawn by the earth's attraction, and ignited by friction in its gaseous envelope, produce the luminous effects so frequently witnessed.[1] Acting on his suggestion, Brandès and Benzenberg, two students at the University of Göttingen, began in 1798 to determine the heights of falling stars by simultaneous observations at a distance. They soon found that they move with planetary velocities in the most elevated regions of our atmosphere, and by the ascertainment of this fact laid a foundation of distinct knowledge regarding them. Some of the data collected, however, served only to perplex opinion, and even caused Chladni temporarily to renounce his. Many high authorities, headed by Laplace in 1802, declared for the lunar-volcanic origin of meteorites; but thought on the subject was turbid, and inquiry seemed only to stir up the mud of ignorance. It needed one of those amazing spectacles, at which man assists, no longer in abject terror for his own frail fortunes, but with keen curiosity and the vivid expectation of new knowledge, to bring about a clarification.

On the night of November 12–13, 1833, a tempest of falling stars broke over the earth. North America bore the brunt of its pelting. From the Gulf of Mexico to Halifax, until daylight with some difficulty put an end to the display, the sky was scored in every direction with shining tracks and

[1] *Ueber den Ursprung der von Pallas gefundenen Eisenmassen*, p. 24.

2 A

illuminated with majestic fireballs.   At Boston, the frequency
of meteors was estimated to be about half that of flakes of
snow in an average snowstorm.   Their numbers, while the
first fury of their coming lasted, were quite beyond counting;
but as it waned, a reckoning was attempted, from which it was
computed, on the basis of that much-diminished rate, that
240,000 must have been visible during the nine hours they
continued to fall.[1]

Now there was one very remarkable feature common to the
innumerable small bodies which traversed, or were consumed
in our atmosphere that night.   *They all seemed to come from
the same part of the sky.*   Traced backwards, their paths were
invariably found to converge to a point in the constellation
Leo.   Moreover, that point travelled with the stars in their
nightly round.   In other words, it was entirely independent of
the earth and its rotation.   It was a point in inter-planetary
space.

The *effective* perception of this fact [2] amounted to a discovery,
as Olmsted and Twining, who had "simultaneous ideas" on
the subject, were the first to realise.   Denison Olmsted was
then professor of mathematics in Yale College.   He showed
early in 1834 [3] that the emanation of the showering meteors
from a fixed "radiant" proved their approach to the earth
along nearly parallel lines, appearing to diverge by an effect of
perspective; and that those parallel lines must be sections of
orbits described by them round the sun and intersecting that
of the earth.   For the November phenomenon was now seen
to be a periodical one.   On the same night of the year 1832,
although with less dazzling and universal splendour than in
America in 1833, it had been witnessed over great part of
Europe and in Arabia.   Olmsted accordingly assigned to the
cloud of cosmical particles (or " comet," as he chose to call

[1] Arago, *Annuaire*, 1836, p. 294.

[2] Humboldt had noticed the emanation of the shooting stars of 1799
from a single point, or "radiant," as Greg long afterwards termed it ; but
no reasoning was founded on the observation.

[3] *Am. Jour. of Sc.*, vol. xxvi. p. 132.

it), by terrestrial encounters with which he supposed the appearances in question to be produced, a period of about 182 days; its path a narrow ellipse, meeting, near its farthest end from the sun, the place occupied by the earth on November 12.

Once for all, then, as the result of the star-drift of 1833, the study of luminous meteors became an integral part of astronomy. Their membership of the solar system was no longer a theory or a conjecture—it was an established fact. The discovery might be compared to, if it did not transcend in importance, that of the asteroidal group. " C'est un nouveau monde planétaire," Arago wrote,[1] " qui commence à se révéler à nous."

Evidences of periodicity continued to accumulate. It was remembered that Humboldt and Bonpland had been the spectators, at Cumana, after midnight of November 12, 1799, of a fiery shower little inferior to that of 1833, and reported to have been visible from the equator to Greenland. Moreover, in 1834 and some subsequent years, there were waning repetitions of the display, as if through the gradual thinning-out of the meteoric supply. The extreme irregularity of its distribution was noted by Olbers in 1837, who conjectured that we might have to wait until 1867 to see the phenomenon renewed on its former scale of magnificence.[2] This was the first hint of a thirty-three or thirty-four year period.

The falling stars of November did not alone attract the attention of the learned. Similar appearances were traditionally associated with August 10 by the popular phrase in which they figured as " the tears of St. Lawrence." But the association could not be taken on trust from mediæval authority. It had to be proved scientifically, and this Quetelet of Brussels succeeded in doing in December 1836.[3]

A second meteoric revolving system was thus shown to exist. But its establishment was at once perceived to be fatal to the "cosmical cloud" hypothesis of Olmsted. For if it be a violation of probability to attribute to *one* such agglomera-

---

[1] *Annuaire*, 1836, p. 297.
[2] *Ann. de l'Observ.*, Bruxelles, 1839, p. 248   [3] *Ibid.*, 1837, p. 272.

tion a period of an exact year, or sub-multiple of a year, it
would be plainly absurd to suppose the movements of *two*
or more regulated by such highly artificial conditions. An
alternative was proposed by Adolf Erman of Berlin in 1839.[1]
No longer in *clouds*, but in closed *rings*, he supposed meteoric
matter to revolve round the sun. Thus the mere circumstance
of intersection by a meteoric, of the terrestrial orbit, without
any coincidence of period, would account for the earth meeting some members of the system at each annual passage
through the "node" or point of intersection. This was an
important step in advance, yet it decided nothing as to the
forms of the orbits of such annular assemblages ; nor was it
followed up in any direction for a quarter of a century.

Professor Hubert A. Newton of Yale College took up,
however, the dropped thread of inquiry in 1864.[2] A search
through old records carried the November phenomenon back
to the year 902 A.D., long distinguished as "the year of the
stars." For in the same night in which Taormina was captured
by the Saracens, and the cruel Aghlabite tyrant Ibrahim ibn
Ahmed died "by the judgment of God" before Cosenza, stars
fell from heaven in such abundance as to amaze and terrify be-
holders far and near. This was on October 13, and recurrences
were traced down through the subsequent centuries, always
with a day's delay in about seventy years. It was easy, too,
to derive from the dates a cycle of $33\frac{1}{4}$ years, so that Professor
Newton did not hesitate to predict the exhibition of an unusually striking meteoric spectacle for November 13–14, 1866.[3]

For the astronomical explanation of the phenomena, recourse
was had to a method introduced by Erman of computing
meteoric orbits. It was found, however, that conspicuous
recurrences every thirty-three or thirty-four years could be
explained on the supposition of five widely different periods,
combined with varying degrees of extension in the revolving

---

[1] *Astr. Nach.*, Nos. 385, 390.
[2] *Am. Jour. of Sc.*, vol. xxxvii. (2d ser.), p. 377.
[3] *Ibid.*, vol. xxxviii. p. 61.

group.   Professor Newton himself gave the preference to the
shortest of the five—of 354½ days—but indicated the means
of deciding with certainty upon the true one.   It was furnished
by the advancing motion of the node, or that day's delay of the
November shower every seventy years, which the old chronicles had supplied data for detecting.   For this is a strictly
measurable effect of gravitational disturbance by the various
planets, the amount of which naturally depends upon the
course pursued by the disturbed bodies.   Here the great
mathematical resources of Professor Adams were brought to
bear.   By laborious processes of calculation, he ascertained
that four out of Newton's five possible periods were entirely
incompatible with the observed nodal displacement, while for
the fifth—that of 33¼ years—a perfectly harmonious result was
obtained.[1]   This was the last link in the chain of evidence
proving that the November meteors—or "Leonids," as they
had by that time come to be called—revolve round the sun in
a period of 33.27 years, in an ellipse spanning the vast gulf
between the orbits of the earth and Uranus, the group
being so extended as to occupy six or eight years in defiling
past the scene of terrestrial encounters.   But before it was
completed in March 1867, the subject had assumed a new
aspect and importance.

Professor Newton's prediction of a remarkable star-shower
in November 1866 was punctually fulfilled.   This time,
Europe served as the main target of the celestial projectiles,
and observers were numerous and forewarned.   The display,
although, according to Mr. Baxendell's memory,[2] inferior to
that of 1833, was of extraordinary impressiveness.   Dense
crowds of meteors, equal in lustre to the brightest stars, and
some rivalling Venus at her best,[3] darted from east to west
across the sky with enormous apparent velocities, and with a
certain determinateness of aim, as if let fly with a purpose, and

---

[1] *Month. Not.*, vol. xxvii. p. 247.

[2] *Am. Jour. of Sc.*, vol. xliii. (2d ser.), p. 87.

[3] Grant, *Month. Not.*, vol. xxvii. p. 29.

at some definite object.[1] Nearly all left behind them trains of emerald-green or clear blue light, which occasionally lasted many minutes, before they shrivelled, and curled up out of sight. The maximum rush occurred a little after one o'clock on the morning of November 14, when attempts to count were overpowered by frequency. But during a previous interval of seven minutes five seconds, four observers at Mr. Bishop's observatory at Twickenham reckoned 514, and during an hour 1120.[2] Before daylight the earth had fairly cut her way through the star-bearing stratum ; the "ethereal rockets" had ceased to fly.

This event brought the subject of shooting stars once more vividly to the notice of astronomers. Schiaparelli had, indeed, been already attracted by it. The results of his studies were made known in four remarkable letters, addressed, before the close of the year 1866, to Father Secchi, and published in the *Bullettino* of the Roman Observatory.[3] Their upshot was to show, in the first place, that meteors possess a real velocity considerably greater than that of the earth, and travel, accordingly, to enormously greater distances from the sun, along tracks resembling those of comets in being very eccentric, in lying at all levels indifferently, and in being pursued in either direction. It was next inferred that comets and meteors equally have an origin foreign to the solar system, but are drawn into it temporarily by the sun's attraction, and occasionally fixed in it by the backward pull of some planet. But the crowning fact was reserved for the last. It was the astonishing one that the August meteors move in the same orbit with the bright comet of 1862—that the comet, in fact, is but a larger member of the family of Perseids (so named because their radiant point is situated in the constellation Perseus).

This discovery was quickly capped by others of the same kind. Leverrier published, January 21, 1867,[4] elements for the November swarm, founded on the most recent and authen-

---

[1] P. Smyth, *Month. Not.*, vol. xxvii. p. 256. [2] Hind, *ibid.*, p. 49.

[3] Reproduced in *Les Mondes*, t. xiii.

[4] *Comptes Rendus*, t. lxiv. p. 96.

tic observations; at once identified by Dr. Peters of Altona, the late distinguished editor of the *Astronomische Nach richten*, with Oppolzer's elements for Tempel's comet of 1866.[1] A few days later, Schiaparelli, having re-calculated the orbit of the meteors from improved data, arrived at the same conclusion; while Professor Weiss of Vienna pointed to the agreement between the orbits of a comet which had appeared in 1861 and of a star-shower found to recur on April 20 (Lyraïds), as well as between those of Biela's comet and certain conspicuous meteors of November 28.[2]

These instances do not seem to be exceptional. The number of known or suspected accordances of cometary tracks with meteor streams contained in a list drawn up in 1878 [3] by Professor Alexander S. Herschel (who has made the subject peculiarly his own), amounts to seventy-six; although the four first detected still remain the most conspicuous examples of a relation as significant as it was, to most astronomers, unexpected.

There had, indeed, been anticipatory ideas. Not that Kepler's comparison of shooting stars to "minute comets," or Maskelyne's "forse risulterà che essi sono comete," in a letter to the Abate Cesaris, December 12, 1783,[4] need count for much. But Chladni, in 1819,[5] considered both to be fragments or particles of the same primitive matter, irregularly dispersed through space as nebulæ; and Morstadt of Prague suggested about 1837 [6] that the November meteors might be dispersed atoms from the tail of Biela's comet, the path of which is cut across by the earth near that epoch. Professor Kirkwood, however, by a luminous intuition, penetrated the whole secret, so far as it has yet been made known. In an article published, or rather buried, in the *Danville Quarterly Review* for December 1861, he argued from the observed division of Biela, and other less noted instances of the same

---

[1] *Astr. Nach.*, No. 1626. [2] *Ibid.*, No. 1632.
[3] *Month. Not.*, vol. xxxviii. p. 369.
[4] Schiaparelli, *Le Stelle Cadenti*, p. 54.
[5] *Ueber Feuer-Meteore*, p. 406. [6] *Astr. Nach.*, No. 347 (Mädler).

kind, that the sun exercises a "divellent influence" on the nuclei of comets, which may be presumed to continue its action until their corporate existence (so to speak) ends in complete pulverisation. "May not," he continued, "our periodic meteors be the débris of ancient, but now disintegrated comets, whose matter has become distributed round their orbits?"[1]

The gist of Schiaparelli's discovery could not be more clearly conveyed. For it must be borne in mind that with the ultimate destiny of comets' *tails* this had nothing to do. The tenuous matter composing them is, no doubt, permanently lost to the body from which it emanated; but science does not pretend to track its further wanderings through space. It can, however, state categorically that these will no longer be conducted along the path forsaken under solar compulsion. From the central, and probably solid parts of comets, on the other hand, are derived the granules by the swift passage of which our skies are seamed with periodic fires. It is certain that a loosely agglomerated mass (such as there is every reason to believe cometary nuclei to be) must gradually separate through the unequal action of gravity on its various parts—through, in short, solar tidal influence. Thenceforward its fragments will revolve independently in parallel orbits, at first as a swarm, finally—when time has been given for the full effects of the lagging of the slower moving particles to develop—as a closed ring. The first condition is still, more or less, that of the November meteors; those of August have already arrived at the second. For this reason, Leverrier pronounced, in 1867, the Perseid to be of older formation than the Leonid system. He even assigned a date at which the introduction of the last-named bodies into their present orbit was probably effected through the influence of Uranus.[2] In 126 A.D. a close approach must have taken place between

---

[1] *Nature*, vol. vi. p. 148.

[2] Mr. Proctor's recent inquiries have shown that the effect of no single planetary encounter can suffice (as the "capture theory" of comets requires that it should) to compel a body approaching the sun from an

the planet and the parent comet of the November stars, after which its regular returns to perihelion, and the consequent process of its disintegration, set in. Though not complete, it is already far advanced.

The view that meteorites are the dust of decaying comets, was now to be put to a definite test of prediction. Biela's comet had not been seen since its duplicate return in 1852. Yet it had been carefully watched for with the best telescopes; its path was accurately known; every perturbation it could suffer was scrupulously taken into account. Under these circumstances, its repeated failure to come up to time might fairly be thought to imply a cessation from visible existence. Might it not, however, be possible that it would appear under another form—that a star-shower might have sprung from, and would commemorate its dissolution?

An unusually large number of falling stars was seen by Brandès, December 7, 1798. Similar displays were noticed in the years 1830, 1838, and 1847 (a day earlier on the two latter occasions), and the point from which they emanated was shown by Heis at Aix-la-Chapelle to be situated near the bright star γ Andromedae.[1] Now this is precisely the direction in which the orbit of Biela's comet would seem to lie, as it runs down to cut the terrestrial track very near the place of the earth at the above dates. The inference was then an easy one that the meteors were pursuing the same path with the comet; and it was separately arrived at, early in 1867, by Weiss, D'Arrest, and Galle.[2] But Biela travels in the opposite direction to Tempel's comet and its attendant "Leonids;" its motion is direct, or from west to east, while theirs is retrograde. Consequently, the motion of its node is in the opposite direction too. In other words, the meeting-place of its orbit with that of the earth retreats (and very rapidly) along

indefinite distance to revolve thenceforth in an orbit having its aphelion near the meeting-place. Several successive encounters, however, may have done the work.

[1] A. S. Herschel, *Month. Not.*, vol. xxxii. p. 355.

[2] *Astr. Nach.*, Nos. 1632, 1633, 1635.

the ecliptic, instead of advancing.   So that if the "Andromeds"
possessed the intimate relation supposed to Biela's comet,
they might be expected to anticipate the times of their recurrence by as much as a week (or thereabouts) in half a century.
All doubt as to the fact may be said to have been removed by
Signor Zezioli's observation of the annual shower in more than
usual abundance, at Bergamo, November 30, 1867.

The missing comet was next due at perihelion in the year
1872, and the probability was contemplated by both Weiss and
Galle of its being replaced by a somewhat dense drift of falling
stars.   The precise date of the occurrence was not easily
determinable, but Galle thought the chances in favour of
November 28.   The event anticipated the prediction by
twenty-four hours.   Scarcely had the sun set in Western
Europe on November 27, when it became evident that Biela's
comet was shedding over us the pulverised products of its
disintegration.   The meteors came in volleys from the foot of
the Chained Lady, their numbers at times baffling the attempt
to keep a reckoning.   At Moncalieri, about 8 P.M., they constituted (as Father Denza said [1]) a "real rain of fire."   Four
observers counted, on an average, four hundred each minute
and a half; and not a few fireballs equalling the moon in diameter
traversed the sky.   On the whole, however, the stars of 1872,
though about equally numerous, were less brilliant than those
of 1866 ; the phosphorescent tracks marking their passage were
comparatively evanescent, and their movements sluggish.   This
is easily understood when we remember that the Andromeds
overtake the earth, while the Leonids rush to meet it ; the
velocity of encounter for the first class of bodies being under
twelve, for the second above forty-four miles a second.   The
spectacle was, nevertheless, magnificent.   It presented itself
successively to various parts of the earth, from Bombay and
the Mauritius to New Brunswick and Venezuela, and was most
diligently and extensively observed.   Here it had well-nigh
terminated by midnight.[2]

[1] *Nature*, vol. vii. p. 122.
[2] A. S. Herschel, *Report Brit. Ass.*, 1873, p. 390.

It was attended by a slight aurora, and although Tacchini had telegraphed that the state of the sun rendered some show of polar lights probable, it has too often figured as an accompaniment of star-showers to permit the coincidence to rank as fortuitous. Admiral Wrangel was accustomed to describe how, during the prevalence of an aurora on the Siberian coast, the passage of a meteor never failed to extend the luminosity to parts of the sky previously dark;[1] and the power of exciting electrical disturbance seems to belong to all such flitting cosmical atoms.

A singular incident connected with the meteors of 1872 has now to be recounted. The late Professor Klinkerfues, who had observed them very completely at Göttingen, was led to believe that not merely the débris strewn along its path, but the comet itself, must have been in the closest proximity to the earth during their appearance.[2] If so, it might be possible, he thought, to descry it as it retreated in the diametrically opposite direction from that in which it had approached. On November 30, accordingly, he telegraphed to Mr. Pogson, the Madras astronomer, " Biela touched earth November 27; search near Theta Centauri "—the " anti-radiant," as it is called, being situated close to that star. Bad weather prohibited observation during thirty-six hours, but when the rain-clouds broke on the morning of December 2, there a comet was, just in the indicated position. In appearance it might have passed well enough for one of the Biela twins. It had no tail, but a decided nucleus, and was about 45 seconds across, being thus altogether below the range of naked-eye discernment. It was again observed December 3, when a short tail was perceptible; but overcast skies supervened, and it has never since been seen. Its identity accordingly remained in doubt. It seems tolerably certain, however, that it was *not* the lost comet, which ought to have passed that spot twelve weeks earlier, and was subject to no conceivable disturbance capable of delaying to that extent

[1] Humboldt, *Cosmos*, vol. i. p. 114 (Otte's trans.)
[2] *Month. Not.*, vol. xxxiii. p. 128.

its revolution. On the other hand, there is the strongest likelihood that it belonged to the same system [1]—that it was a third
fragment, torn from the parent-body of the Andromeds at a
period anterior to our first observations of it. Nor did the
meteors of November 27 directly replace the vanished comet.
They too must have separated from it at a much earlier stage
of its history.

Biela does not offer the only example of cometary disruption.
Setting aside the unauthentic reports of early chroniclers, we
meet the "double comet" discovered by Liais at Olinda
(Brazil), February 27, 1860, of which the division appeared
recent, and about to be carried farther.[2] But a division once
established, separation must continually progress. The periodic
times of the fragments will never be identical; one must drop
a little behind the other at each revolution, until at length
they come to travel in remote parts of nearly the same orbit.
Thus the comet predicted by Klinkerfues and discovered by
Pogson had already lagged to the extent of twelve weeks, and
we shall meet instances farther on where the retardation is
counted, not by weeks, but by years. Here, original identity
emerges only from calculation and comparison of orbits.

Comets then die, as Kepler wrote long ago, *sicut bombyces
filo fundendo.* This certainty, anticipated by Kirkwood in
1861, we have at least acquired from the discovery of their
generative connection with meteors. Nay, their actual materials become, in smaller or larger proportions, incorporated
with our globe. Whether, indeed, the ponderous masses of
which, according to Daubrée's estimate,[3] 600 or 700 fall annually from space upon the earth, ever formed part of the
bodies known to us as comets, is a question. Some follow
Tschermak in attributing to aerolites a totally different origin
from that of periodical shooting-stars. That no clear line of

[1] Even this was denied by Bruhns, *Astr. Nach.*, No. 2054.
[2] *Month. Not.*, vol. xx. p. 336.
[3] Newton, *Ency. Brit.*, vol. xvi. p. 109.

demarcation can be drawn is no valid reason for asserting that
no real distinction exists; and it is certainly remarkable that a
meteoric fusillade may be kept up for hours without a single
solid projectile reaching its destination. It would seem as if
the celestial army had been supplied with blank cartridges.
There is, indeed, much probability that few of the components
of the recent brilliant showers attained the dimensions of a
canary-seed.

It would gratify curiosity to think that we might, by actual
inspection and analysis, ascertain the composition of those
mysterious visitors, the "brandishing" of whose "crystal
tresses" in our skies was wont, in times past, to "import
change of times and states." But if this be denied us, another
way has been laid open towards the same end.

The first successful application of the spectroscope to such
bodies was by Donati in 1864.[1] A comet discovered by
Tempel, July 4, brightened until it appeared like a star somewhat below the second magnitude, with a feeble tail 30° in
length. It was remarkable as having, on August 7, almost
totally eclipsed a small star—a very rare occurrence.[2] On
August 5, Donati admitted its light through his train of prisms,
and found it, thus analysed, to consist of three bright bands—
yellow, green and blue—separated by wider dark intervals.
This implied a good deal. Comets had previously been considered, as we have seen, to shine mainly, if not wholly, by reflected sunlight. They were now perceived to be self-luminous,
and to be formed, to a large extent, of glowing gas. The next
step was to determine what *kind* of gas it was that was thus
glowing in them; and this was taken by Dr. Huggins in 1868.[3]

A comet of subordinate brilliancy, known as comet 1868 ii.,
or sometimes as Winnecke's, was the subject of his experiment. On comparing its spectrum with that of an olefiant-
gas "vacuum tube" rendered luminous by electricity, he
found the agreement exact. It has since been abundantly

[1] *Astr. Nach.*, No. 1488.   [2] *Annuaire*, Paris, 1883, p. 185.
[3] *Phil. Trans.*, vol. clviii. p. 556.

confirmed. All the eighteen comets of which the light had
been analysed down to 1880, showed the typical hydro-carbon
spectrum [1] common to the whole group of those compounds,
but probably due immediately to the presence of acetylene.
Some slight apparent anomalies have been almost certainly
caused, not by any real differences of constitution, but by
deficient light-power, rendering observations difficult and insecure. The brighter the comet, the more perfect proved
its conformity to the type.

The earliest comet of first-class lustre to present itself for
spectroscopic examination, was that discovered by Coggia at
Marseilles, April 17, 1874. Invisible to the naked eye until
June, it blazed out in July a splendid ornament of our northern
skies, with a just perceptibly curved tail, reaching more than
half way from the horizon to the zenith, and a nucleus surpassing in brilliancy the brightest stars in the Swan. Bredichin, Vogel, and Huggins [2] were unanimous in pronouncing
its spectrum to be that of marsh, or olefiant gas. Father Secchi,
in the clear sky of Rome, was able to push the identification
even closer than had heretofore been done. The *complete*
hydro-carbon spectrum consists of five zones of variously
coloured light. Three of these only—the three central ones
—had till then been obtained from comets; it was supposed,
because their temperature was not high enough to develop
the others. The light of Coggia's comet, however, was found
to contain all five, traces of the violet band emerging June 4,
of the red, July 2.[3] Presumably, all five would show universally
in cometary spectra, were the dispersed rays strong enough to
enable them to be seen.

The gaseous surroundings of comets are then made up of
a compound of hydrogen with carbon. Other materials are
also present, as will be seen by and by; but the hydro-carbon
element is probably unfailing and predominant. Its luminosity

---

[1] Hasselberg, *Mém. de l'Ac. Imp. de St. Pétersbourg*, t. xxviii. (7th ser.),
No. 2, p. 66.   [2] *Proc. Roy. Soc.*, vol. xxiii. p. 154.
[3] Hasselberg, *loc. cit.*, p. 58.

is, there is little doubt, an effect of electrical excitement. Zöllner showed in 1872 [1] that, owing to evaporation, and other changes produced by rapid approach to the sun, electrical processes of considerable intensity must take place in comets; and that their original light is immediately connected with these, and is only an indirect result of solar radiation, may be considered a truth permanently acquired to science.[2] They are not, it thus seems, bodies incandescent through heat, but glowing by electricity; and this is compatible with a relatively low temperature.

The gaseous spectrum of comets is accompanied, in varying degrees, by a continuous spectrum. This is usually derived most strongly from the nucleus, but extends, more or less, to the nebulous appendages. In part, it is certainly due to reflected sunlight; in part, it is likely, to the ignition of minute solid particles.

[1] *Ueber die Natur. der Cometen*, p. 112.
[2] Hasselberg, *loc. cit.*, p. 38.

CHAPTER XI.

RECENT COMETS (*continued*).

THE mystery of comets' tails has been to some extent penetrated ; so far, at least, that, by making certain assumptions
strongly recommended by the facts of the case, their forms can
be, with very approximate precision, calculated beforehand.
We have, then, the assurance that these extraordinary appendages are composed of no ethereal or super-sensual stuff, but
of matter such as we know it, and subject to the ordinary laws
of motion, though in a state of extreme tenuity. This is
unquestionably one of the most remarkable discoveries of our
time.

Olbers, as already stated, originated in 1812 the view that
the tails of comets are made up of particles subject to a force
of electrical repulsion proceeding from the sun. It was
developed and enforced by Bessel's discussion of the appearances presented by Halley's comet in 1835. He, moreover,
provided a formula for computing the movement of a particle
under the influence of a repulsive force of any given intensity,
and thus laid firmly the foundation of a mathematical theory
of cometary emanations. Professor W. A. Norton of Yale College considerably improved this by inquiries begun in 1844,
and resumed on the apparition of Donati's comet ; and Dr.
C. F. Pape at Altona [1] gave numerical values for the impulses
outward from the sun, which must have actuated the materials

[1] *Astr. Nach.*, Nos. 1172-74.

respectively of the curved and straight tails adorning the same beautiful and surprising object.

The *physical* theory of repulsion, however, was, it might be said, still in the air. Nor did it assume an aspect of even moderate plausibility until Zöllner took it in hands in 1871.[1] It is perfectly well ascertained that the energy of the push or pull produced by electricity depends (other things being the same) upon the *surface* of the body acted on; that of gravity, upon its *mass*. The efficacy of solar electrical repulsion relatively to solar gravitational attraction grows, consequently, as the size of the particle diminishes. Make this small enough, and it will virtually cease to gravitate, and will unconditionally obey the impulse to recession.

This principle Zöllner was the first to realise in its application to comets. It gives the key to their constitution. Admitting (as we seem bound to do) that the sun and they are similarly electrified, their more substantially aggregated parts will still follow the solicitations of his gravity, while the finely-divided particles escaping from them will, simply by reason of their minuteness, fall under the sway of his repellent electric power. They will, in other words, form "tails." Nor is any extravagant assumption called for as to the intensity of the electrical charge concerned in producing these effects. Zöllner, in fact, showed[2] that it need not be higher than that attributed by the best authorities to the terrestrial surface.

It is now nearly a quarter of a century since M. Bredichin, late director of the Moscow Observatory, directed his attention to these curious phenomena. His persistent inquiries on the subject, however, date from the appearance of Coggia's comet in 1874. On computing the value of the repulsive force exerted in the formation of its tail, and comparing it with values of the same force arrived at by him in 1862 for some other conspicuous comets, it struck him that the numbers representing them fell into three well-defined classes. " I

[1] *Berichte Sächs. Ges.*, 1871, p. 174.
[2] *Natur der Cometen*, p. 124; *Astr. Nach.*, No. 2086.

suspect," he wrote in 1877, "that comets are divisible into groups, for each of which the repulsive force is perhaps the same."[1] This idea was confirmed on fuller investigation. In 1882 the appendages of thirty-six well-observed comets had been re-constructed in the study, without a single exception being met with to the rule of the three types.

In the first of these, the repellent energy of the sun is twelve times as strong as his attractive energy ; the particles forming the enormously long, straight rays projected outwards from this kind of comet, leave the nucleus with a velocity of $4\frac{1}{2}$ kilometres per second, which becoming constantly accelerated, carries them in a few days to the limit of visibility. The great comets of 1811, 1843, and 1861, that of 1744 (so far as its principal tail was concerned), and Halley's comet at its various apparitions, belonged to this class. For the second type, the value of the repulsive force employed is less narrowly limited. It may range as high as $2\frac{1}{2}$ (2.6) times, or descend as low as $\frac{8}{10}$ the power of solar gravity ;[2] but, on an average, it is just equal to it. The corresponding initial velocity is 900 metres a second, and the resulting appendage a scimitar-like or plumy tail, such as Donati's and Coggia's comets furnished splendid examples of. Tails of the third type are constructed with a force of repulsion from the sun one-fifth (or, at the most, three-tenths) that of his gravity, producing an accelerated movement of attenuated matter from the nucleus, beginning at the leisurely rate of 300 metres a second. They are short, strongly bent, brush-like emanations, and in bright comets seem to be only found in combination with tails of the higher classes. Multiple tails, indeed—that is, tails of different types emitted simultaneously by one comet—are perceived, as experience advances and observation becomes closer, to be rather the rule than the exception.[3]

Now what is the meaning of these three types? Is any translation of them into physical fact possible? To this

[1] *Annales de l'Obs. de Moscou*, t. iii. pt. i. p. 37.
Ibid., t. vii. pt. ii. p. 56.   [3] Faye, *Comptes Rendus*, t. xciii. p. 13.

question Bredichin supplied in 1879 a plausible answer.[1] It was already a current conjecture that multiple tails are composed of different kinds of matter, differently acted on by the sun. Both Olbers and Bessel had suggested this explanation of the straight and curved emanations from the comet of 1807 ; Norton had applied it to the faint light-tracks proceeding from that of Donati ;[2] Winnecke, to the varying deviations of its more brilliant plumage. Bredichin went further. He undertook to determine (provisionally as yet) the several kinds of matter appropriated severally to the three classes of tails. These he found to be, hydrogen for the first, hydro-carbons for the second, and iron for the third. The ground of this apportionment is that the atomic weights of these substances bear to each other the same inverse proportion as the repulsive forces employed in producing the appendages they are supposed to form ; and Zöllner had pointed out in 1875 that the "heliofugal" power by which comets' tails are developed, would in fact be effective just in that ratio.[3] Hydrogen, as the lightest known element—that is, the least under the influence of gravity—was naturally selected as that which yielded most readily to the counter-persuasions of electricity. Hydro-carbons had been shown by the spectroscope to be present in comets, and were fitted by their specific weight, as compared with that of hydrogen, to form tails of the second type ; while the atoms of iron were just heavy enough to compose those of the third, and, from the plentifulness of their presence in meteorites, might be presumed to enter, in no inconsiderable proportion, into the mass of comets. These three substances, however, were by no means supposed to be the sole constituents of the appendages in question. On the contrary, the great breadth of what, for the present, were taken to be characteristically "iron" tails, was attributed to the presence of many kinds of matter of high, and slightly different specific weights ;[4] while

---

[1] *Annales,* t. v. pt. ii. p. 137.
[2] *Am. Jour. of Sc.,* vol. xxxii. (2d ser.), p. 57.
[3] *Astr. Nach.,* No. 2082.   [4] *Annales,* t. vi. pt. i. p. 60.

the expanded plume of Donati was shown to be, in reality, a whole system of tails, made up of many substances, each spreading into a separate hollow cone, more or less deviating from, and partially superposed upon the others.

Never was a theory more promptly or profusely illustrated than this of Bredichin. Within three years of its promulgation five bright comets made their appearance, each presenting some distinctive peculiarity by which knowledge of these curious objects was materially helped forward. The first of these is remembered as the " Great Southern Comet." It was never visible in these latitudes, but made a short, though stately progress through southern skies. Its earliest detection was at Cordoba on the last evening of January, 1880 ; and it was seen on February 1 as a luminous streak, extending just after sunset from the south-west horizon towards the pole, in New South Wales, at Monte Video, and the Cape of Good Hope. The head was lost in the solar rays until February 4, when Dr. Gould, director of the National Observatory of the Argentine Republic at Cordoba, caught a glimpse of it very low in the west ; and on the following evening, Mr. Eddie, at Graham's Town, discovered a faint nucleus, of a straw-coloured tinge, about the size of the annular nebula in Lyra. Its condensation, however, was very imperfect, and the whole apparition was of an exceedingly filmy texture. The tail was enormously long. On February 5 it extended—large perspective retrenchment notwithstanding—over an arc of 50° ; but its brightness nowhere exceeded that of the Milky Way in Taurus. There was little curvature perceptible ; the edges of the appendage ran parallel, forming a nebulous causeway from star to star ; and the comparison to an auroral beam was appropriately used. The aspect of the famous comet of 1843 was forcibly recalled to the memory of Mr. Janisch, governor of St. Helena ; and the resemblance proved not merely superficial. But the comet of 1880 was less brilliant, and even more evanescent. After only eight days of visibility, it had faded so much as no longer to strike, though still discoverable by, the unaided eye ; and on February

20 it was invisible with the great Cordoba equatoreal pointed to its known place.

But the most astonishing circumstance connected with this body is the identity of its path with that of its predecessor in 1843. This is undeniable. Dr. Gould,[1] Mr. Hind, and Dr. Copeland,[2] each computed a separate set of elements from the first rough observations, and each was struck with an agreement between the two orbits so close as to render them virtually indistinguishable. " Can it be possible," Mr. Hind wrote to Sir George Airy, " that there is such a comet in the system, almost grazing the sun's surface in perihelion, and revolving in less than thirty-seven years ? I confess I feel a difficulty in admitting it, notwithstanding the above extraordinary resemblance of orbits." [3]

Mr. Hind's difficulty was shared by other astronomers. It would, indeed, be a violation of common sense to suppose that a celestial visitant so striking in appearance had been for centuries back an unnoticed frequenter of our skies. Various expedients accordingly were resorted to for getting rid of the anomaly. The most promising at first sight was that of the " resisting medium." It was hard to believe that a body, largely vaporous, shooting past the sun at a distance of less than a hundred thousand miles from his surface, should have escaped powerful retardation. It must have passed through the very midst of the corona. It might easily have had an actual encounter with a prominence. Escape from such proximity might, indeed, very well have been judged beforehand to be impossible. Even admitting no other kind of opposition than that met with by Encke's and Winnecke's comets, the effect in shortening the period ought to be of the most marked kind. It was proved by Oppolzer[4] that if the comet of 1843 had entered our system from stellar space with parabolic velocity, it would, by the action of a medium such as Encke postulated (varying in density inversely as the square of the distance from the sun),

[1] *Astr. Nach.*, No. 2307.   [2] *Ibid.*, No. 2304.
[3] *Observatory*, vol. iii. p. 390.   [4] *Astr. Nach.*, No. 2319.

have been brought down, by its first perihelion passage, to elliptic movement in a period of twenty-four years, with such rapid diminution that its next return would be in about ten. But such restricted observations as were on either occasion of its visibility available, gave no sign of such a rapid progress towards engulfment.

Another form of the theory was advocated by Klinkerfues. He supposed that four returns of the same body had been witnessed within historical memory—the first in 371 B.C., the next in 1668, besides those of 1843 and 1880; an original period of 2039 years being successively reduced by the withdrawal at each perihelion passage of $\frac{1}{1320}$ of the velocity acquired by falling from the far extremity of its orbit towards the sun, to 175 and 37 years. A continuance of the process would bring the comet of 1880 back in 1897.

Unfortunately, the earliest of these apparitions cannot be identified with the recent ones unless by doing violence to the plain meaning of Aristotle's words in describing it. He states that the comet was first seen " during the frosts and in the clear skies of winter," setting due west nearly at the same time as the sun.[1] This implies some considerable north latitude. But the objects lately observed had practically *no* north latitude. They accomplished their entire course *above* the ecliptic in two hours and a quarter, during which space they were barely separated a hand's-breadth (one might say) from the sun's surface. For the purposes of the desired assimilation, Aristotle's comet should have appeared in March. It is not credible, however, that even a native of Thrace should have termed March " winter."

With the comet of 1668 the case is more dubious. The circumstances of its appearance are barely reconcilable with the identity attributed to it, although too vaguely known to render certainty one way or the other attainable. It might, however, be expected that recent observations would at least decide the questions whether the comet of 1843 could have

[1] *Meteor*, lib. i. cap. 6.

returned in less than thirty-seven years, and whether the comet of 1880 was to be looked for at the end of $17\frac{1}{2}$. But the truth is that both these objects were observed over so small an arc—8° and 3° respectively—that their periods remained virtually undetermined. For while the shape and position of their orbits could be and were fixed with a very close approach to accuracy, the length of those orbits might vary enormously without any very sensible difference being produced in the small part of the curves traced out near the sun. It is, however, remarkable that Dr. Wilhelm Meyer arrived, by an elaborate discussion, at a period of thirty-seven years for the comet of 1880,[1] while the observations of 1843 are admittedly best fitted by Hubbard's ellipse of 533 years; but these Dr. Meyer supposes to be affected by some constant source of error, such as would be produced by a mistaken estimate of the position of the comet's centre of gravity. He infers finally that, in spite of previous non-appearances, we really have to do with a regular denizen of our system, returning once in thirty-seven years along an orbit of such extreme eccentricity that its movement might be described as one of precipitation towards, and rapid escape from the sun, rather than of sedate circulation round it.

The *geometrical* test of identity has hitherto been the only one which it was possible to apply to comets, and in the case before us it may fairly be said to have broken down. We may, then, tentatively, and with much hesitation, try a *physical* test, though scarcely yet, properly speaking, available. We have seen that the comets of 1843 and 1880 were strikingly alike in general appearance, though the absence of a formed nucleus in the latter, and its inferior brilliancy, detracted from the convincing effect of the resemblance. Nor was it maintained when tried by exact methods of inquiry. M. Bredichin found that the gigantic ray emitted in 1843 belonged to his type No. 1; that of 1880 to type No. 2.[2] The particles forming the

[1] *Mém. Soc. Phys. de Genève*, t. xxviii. p. 23.
[2] *Annales*, t. vii. pt. i. p. 60.

one were actuated by a repulsive force ten times as powerful as those forming the other. It is true that a second noticeably curved tail was seen in Chili, March 1, and at Madras, March 11, 1843; and M. Bredichin, accordingly, thinks the conjecture justified that the materials composing on that occasion the principal appendage having become exhausted, those of the secondary one remained predominant, and reappeared alone in the "hydro-carbon" train of 1880. But the one known instance in point is against such a supposition. Halley's comet, the only *great* comet of which the returns have been securely authenticated and carefully observed, has preserved its "type" unchanged through many successive revolutions. The dilemma presented to astronomers by the Great Southern Comet of 1880 was unexpectedly renewed in the following year.

On the 22d of May 1881, Mr. John Tebbutt of Windsor, New South Wales, scanning the western sky, discerned a hazy-looking object which he felt sure was a strange one. A marine telescope at once resolved it into two small stars and a comet, the latter of which quickly attracted the keen attention of astronomers; for Dr. Gould, computing its orbit from his first observations at Cordoba, found it to agree so closely with that arrived at by Bessel for the comet of 1807, that he telegraphed to Europe, June 1, announcing the unexpected return of that body. So unexpected, that theoretically it was not possible before the year 3346; and Bessel's investigation was one which inspired, and eminently deserved confidence. Here then once more the perplexing choice had to be made between a premature and unaccountable reappearance, and the admission of a plurality of comets moving nearly in the same path. But in this case facts proved decisive.

Tebbutt's comet passed the sun June 16, at a distance of sixty-eight millions of miles, and became visible in Europe six days later. It was, in the opinion of some, the finest object of the kind since 1861. In traversing the constella-

tion Auriga, on its *début* in these latitudes, it outshone
Capella. On June 24 and some subsequent nights, it was
unmatched in brilliancy by any star in the heavens. In the
telescope, the "two interlacing arcs of light" which had
adorned the head of Coggia's comet were reproduced; while
a curious *dorsal spine* of strong illumination formed the axis
of the tail, which extended in clear skies over an arc of 20°.
It belonged to the same "type" as Donati's great plume;
the particles composing it being driven *from* the sun by a
force twice as powerful as that urging them *towards* it.[1] But
the appendage was, for a few nights, and by two observers,
perceived to be double. Tempel on June 27, and Lewis
Boss, at Albany (N.Y.), June 26 and 28, saw a long straight
ray corresponding to a far higher rate of emission than the
curved train, and shown by Bredichin to be a member of the
(so-called) hydrogen class. It had vanished by July 1, but
made a temporary reappearance July 22.[2]

The appendages of this comet were of remarkable transparency. Small stars shone wholly undimmed across the
tail, and a very nearly central transit of the head over one of
the seventh magnitude on the night of June 29, produced—if
any change—an increase of brilliancy in the object of this
spontaneous experiment.[3] Yet Dr. Meyer, at the Geneva
Observatory, found distinct evidence of refraction suffered by
stellar rays under these circumstances. Three times he pursued
with micrometric measurements the course of a star across
the cometary surroundings; and on each occasion the uniformity of its progress was disturbed in a manner corresponding
to the optical action of a gaseous mass increasing in density
and refractive power as the square of the distance from the
nucleus diminished. Supposing olefiant gas to be in question,

---

[1] Bredichin, *Annales*, t. viii. p. 68.

[2] *Am. Jour. of Sc.*, vol. xxii. p. 305.

[3] Messrs. Burton and Green observed a dilatation of the stellar image
into a nebulous patch by the transmission of its rays through a nuclear
jet of the comet. *Am. Jour. of Sc.*, vol. xxii. p. 163.

its density, 102,000 kilometres from the nucleus, was estimated to be $\frac{7}{1000}$ that of our atmosphere at the sea-level.[1]   This was the first successful attempt to measure the effects of cometary refraction, and will doubtless be renewed on a favourable opportunity.

The track pursued by this comet gave peculiar advantages for its observation.   Ascending from Auriga through Camelopardus, it stood, July 19, on a line between the Pointers and the Pole, within 8° degrees of the latter, thus remaining for a considerable period constantly above the horizon of northern observers.   Its brightness, too, was no transient blaze, but had a lasting quality which enabled it to be kept steadily in view during nearly nine months.   Visible to the naked eye until the end of August, the last telescopic observation of it was made February 14, 1882, when its distance from the earth considerably exceeded 300 million miles.   Under these circumstances, the knowledge acquired of its orbit was of more than usual accuracy, and showed conclusively that the comet was not a simple return of Bessel's ; for this would involve a period of seventy-four years, whereas Tebbutt's comet cannot revisit the sun until after the lapse of close upon three millenniums.   Nevertheless, the two bodies move so nearly in the same path that an original connection of some kind is obvious ; and the recent example of Biela readily suggested a conjecture as to what the nature of that connection might have been. The comets of 1807 and 1881 are then regarded with much probability as fragments of a primitive disrupted body, one following in the wake of the other at an interval of seventy-four years.

Tebbutt's comet was the first of which a satisfactory photograph was obtained.   The difficulties to be overcome were very great.   The chemical intensity of cometary light is, to begin with, extraordinarily small.   Janssen estimates it at $\frac{1}{300,000}$ of moonlight.[2]   So that, if the ordinary process by

---

[1] *Archives des Sciences,* t. viii. p. 535.   Meyer founded his conclusions on the theory of M. Gustave Cellérier.   [2] *Annuaire,* Paris, 1882, p. 781.

which lunar photographs are taken had been applied to the comet of 1881, an exposure of at least *three days* would have been required in order to get an impression of the head with about a tenth part of the tail.   But by that time a new method of vastly increased sensitiveness had been rendered available, by which dry gelatine-plates were substituted for the wet collodion-plates hitherto in use ; and this improvement alone reduced the necessary time of exposure to two hours. It was brought down to half an hour by Janssen's employment of a reflector specially adapted to give an image illuminated eight or ten times as strongly as that produced in the focus of an ordinary telescope.[1]

The photographic feebleness of cometary rays was not the only obstacle in the way of success.   The proper motion of these bodies is so rapid as to render the usual devices for keeping a heavenly body steadily in view quite inapplicable. The machinery by which the diurnal movement of the sphere is followed, must be specially modified to suit each eccentric career.   This too was done, and on June 30, 1881, Janssen secured a perfect photograph of the brilliant object then visible, showing the structure of the tail with beautiful distinctness to a distance of $2\frac{1}{2}°$ from the head.   An impression to nearly 10° was obtained about the same time by Dr. Henry Draper at New York, with an exposure of 162 minutes.[2]

Tebbutt's (or comet 1881 iii.) was also the first comet of which the spectrum was so much as attempted to be chemi-cally recorded.   Both Dr. Huggins and Dr. Draper were successful in this respect, but Dr. Huggins the more com-pletely so.[3]   The importance of the feat consisted in its throwing open to investigation a part of the spectrum invisible to the eye, and so affording an additional test of cometary constitution.   The result was fully to confirm the origin from carbon-compounds assigned to the visible rays, by disclosing additional bands belonging to the same series in the ultra-

[1] *Annuaire*, p. 776.   [2] *Am. Jour. of Sc.*, vol. xxii. p. 134.
[3] *Report Brit. Ass.*, 1881, p. 520.

violet; as well as to establish unmistakably the presence of a not inconsiderable proportion of reflected solar light by the clear impression of some of the principal Fraunhofer lines. Thus the polariscope was found to have told the truth, though not the whole truth.

The photograph so satisfactorily communicative was taken by Dr. Huggins on the night of June 24; and on the 29th, at Greenwich, the tell-tale Fraunhofer lines werè perceived to interrupt the visible range of the spectrum. This was at first so vividly continuous, that the characteristic cometary bands could scarcely be detached from their bright background. But as the nucleus faded towards the end of June, they came out strongly, and were more and more clearly seen, both at Greenwich and at Princeton, to agree, *not* with the spectrum of hydro-carbons lit up in a vacuum tube by an electric discharge, but with that of the same substances burning in a Bunsen flame.[1] Here we have an additional clue to the molecular condition of cometary materials. It need not, however, be inferred that they are really in a state of combustion. This, from all that we know, may be called an impossibility. The truth pointed to seems rather to be that the electricity by which comets are rendered luminous is of very low intensity.[2]

The spectrum of the tail was, in this comet, found to be not essentially different from that of the head. Professor Wright of Yale College ascertained a large, but probably variable percentage of its light to be polarised in a plane passing through the sun, and hence to be reflected sunlight.[3] A faint continuous spectrum corresponded to this portion of its radiance; but gaseous emissions were also present. At Potsdam, on June 30, the hydro-carbon bands were traced by Vogel to the very end of the tail;[4] and they were kept in sight by Young at a greater distance from the nucleus than the more equably dispersed light. There seems little doubt that, as in the solar

---

[1] *Month. Not.*, vol. xlii. p. 14; *Am. Jour. of Sc.*, vol. xxii. p. 136.
[2] Piazzi Smyth, *Nature*, vol. xxiv. p. 430.
[3] *Astr. Nach.*, No. 2395.　　　　[4] *Ibid.*

corona, the relative strength of the two orders of spectrum is subject to fluctuations.

The comet 1881 iii. was thus of signal service to science. It afforded, when compared with the comet of 1807, the first undeniable example of two such bodies travelling so nearly in the same orbit as to leave absolutely no doubt of the existence of a genetic tie between them. Cometary photography came to its earliest fruition with it; and cometary spectroscopy made a notable advance by means of it. Before it was yet out of sight, it was provided with a successor.

At Ann Arbor Observatory, Michigan, on July 14, a comet was discovered by Dr. Schäberle, which, as his claim to priority is undisputed, is often allowed to bear his name. In strict scientific parlance, however, it is designated comet 1881 iv. It was observed in Europe after three days, became just discernible by the naked eye at the end of July, and brightened consistently up to its perihelion passage, August 22, when it was still about fifty million miles from the sun. During many days of that month, the uncommon spectacle was presented of two bright comets circling together, though at widely different distances, round the north pole of the heavens. The new-comer, however, never approached the pristine lustre of its predecessor. Its nucleus, when brightest, was comparable to the star Cor Caroli, a narrow, perfectly straight ray proceeding from it to a distance of 10°. This was easily shown by Bredichin to belong to the hydrogen type of tails;[1] while a "strange, faint second tail, or bifurcation of the first one," observed by Captain Noble, August 24,[2] fell into the hydro-carbon class of emanations. It was seen, August 22 and 24, by Dr. F. Terby of Louvain,[3] as a short nebulous brush, like the abortive beginnings of a congeries of curving trains; but appeared no more. Its well-attested presence was, however, significant of the complex constitution of such bodies, and the manifold kinds of action progressing in them.

[1] *Astr. Nach.*, No. 2411.   [2] *Month. Not.*, vol. xlii. p. 49.
[3] *Astr. Nach.*, No. 2414.

The only peculiarity in the spectrum of Schäberle's comet consisted in the almost total absence of continuous light. The carbon-bands were nearly isolated and very bright. Barely from the nucleus proceeded a rainbow-tinted streak, indicative of solid or liquid matter, which, in this comet, must have been of very scanty amount. Its visit to the sun in 1881 was, so far as is known, the first. The elements of its orbit showed no resemblance to those of any previous comet, nor any marked signs of periodicity. So that, although it may be considered probable, we do not *know* that it is moving in a closed curve, or will ever again penetrate the precincts of the solar system. It was last seen from the southern hemisphere, October 19, 1881.

The third of a quartette of lucid comets visible within sixteen months, was discovered by Mr. C. S. Wells at the Dudley Observatory, Albany, March 17, 1882. Two days later it was described by Mr. Lewis Boss as "a great comet in miniature," so well defined and regularly developed were its various parts and appendages. It was discernible without optical aid early in May; and on June 5 it was observed on the meridian at Albany just before noon—an astronomical event of extreme rarity. Comet Wells, however, never became an object so conspicuous as to attract general attention, owing to its immersion in the evening twilight of our northern June.

But the study of its spectrum revealed new facts of the utmost interest. All the comets till then examined had been found to conform to one invariable type of luminous emission. Individual distinctions there had been, but no specific differences. Now all these bodies had kept at a respectful distance from the sun; for of the great comet of 1880 no spectroscopic inquiries had been made. Comet Wells, on the other hand, approached his surface within little more than five million miles on June 10, 1882; and it is not doubtful that to this circumstance the novel feature in its incandescence was due.

During the first half of April its spectrum was of the normal type, though the carbon bands were unusually weak; but with increasing vicinity to the sun they died out, and the entire light seemed to become concentrated into a narrow, unbroken, brilliant streak, hardly to be distinguished from the spectrum of a star. This unusual behaviour excited attention, and a strict watch was kept. It was rewarded at the Dunecht Observatory (Lord Crawford's), May 27, by the discernment of what had never before been seen in a comet—the yellow ray of sodium.[1] By June 1, this had kindled into a blaze overpowering all other emissions. The light of the comet was practically monochromatic; and the image of the entire head, with the root of the tail, could be observed, like a solar prominence, depicted, in its new saffron vesture of vivid illumination, within the jaws of an open slit.

At Potsdam, the bright yellow line was perceived with astonishment by Vogel on May 31, and was next evening identified with Fraunhofer's " D." Its character led him to infer a very considerable density in the glowing vapour emitting it.[2] Hasselberg founded an additional argument in favour of the electrical origin of cometary light on the changes in the spectrum of comet Wells.[3] For they were closely paralleled by some earlier experiments of Wiedemann, in which the gaseous spectra of vacuum tubes were at once effaced on the introduction of metallic vapours. It seemed as if the metal had no sooner been rendered volatile by heat, than it usurped the entire office of carrying the discharge, the resulting light being thus exclusively of its production. Had simple incandescence by heat been in question, the effect would have been different; the two spectra would have been superposed without prejudice to either. Similarly, the replacement of the hydro-carbon bands in the spectrum of the comet by the sodium line, proved electricity to be the exciting agent. For the increasing thermal power of the sun might, indeed, have ignited the sodium, but it could not have extinguished the hydro-carbons.

Dr. Huggins succeeded in photographing the spectrum of

[1] *Copernicus*, vol. ii. p. 229.
[2] *Astr. Nach.*, Nos. 2434, 2437.   [3] *Ibid.*, No. 2441.

comet Wells by an exposure of one hour and a quarter.[1] The
result was to confirm the novelty of its character. None of
the ultra-violet carbon-groups were apparent; but certain
bright rays, as yet unidentified, had imprinted themselves.
Otherwise the spectrum was strongly continuous, uninterrupted
even by the Fraunhofer lines detected in the spectrum of
Tebbutt's comet. Hence it was concluded that a smaller
proportion of reflected light was mingled with the native
emissions of the later arrival.

All that is certainly known about the *extent* of the orbit
traversed by the first comet of 1882 is that it came from, and
is now retreating towards, vastly remote depths of space. An
American computer [2] found a period indicated for it of no
less than 400,000 years ; A. Thraen of Dingelstädt arrived at
one of 3617.[3] Both are perhaps equally insecure.

We have now to give some brief account of one of the
most remarkable cometary apparitions on record, and—with
the single exception of that identified with the name of Halley
—the most instructive to astronomers. The lessons learned
from it were as varied and significant as its aspect was splendid ;
although from the circumstance of its being visible in general
only before sunrise, the spectators of its splendour were comparatively few.

The discovery of a great comet at Rio Janeiro, September
11, 1882, became known in Europe through a telegram from
M. Cruls, director of the observatory at that place. It had,
however (as appeared subsequently), been already seen on the
8th by Mr. Finlay, assistant at the Cape Observatory, and at
Auckland as early as September 3. A later, but very singularly conditioned detection, quite unconnected with any of the
preceding, was effected by Mr. Common at Ealing. Since the
eclipse of May 17, when a comet—named "Tewfik" in honour
of the Khedive of Egypt—was caught on Dr. Schuster's photo-

---

[1] *Report Brit. Ass.*, 1882, p. 442.
[2] J. J. Parsons, *Am. Jour. of Science*, vol. xxvii. p. 34.
[3] *Astr. Nach.*, No. 2441.

graphs, entangled, one might almost say, in the outer rays of the corona, he had scrutinised the neighbourhood of the sun on the infinitesimal chance of intercepting another such body on its rapid journey thence or thither. We record with wonder that, after an interval of exactly four months, that infinitesimal chance turned up in his favour.

On the forenoon of Sunday, September 17, he saw a great comet close to, and rapidly approaching the sun. It was, in fact, then within a few hours of perihelion. Some measures of position were promptly taken; but a cloud-veil covered the interesting spectacle before midday was long past. Mr. Finlay at the Cape was more completely fortunate. Divided from his fellow-observer by half the world, he unconsciously finished, under a clearer sky, his interrupted observation. The comet, of which the silvery radiance contrasted strikingly with the reddish-yellow glare of the sun's margin it drew near to, was followed "continuously right into the boiling at the limb"— a circumstance without precedent in cometary history.[1] Dr. Elkin, who watched the progress of the event with another instrument, thought the intrinsic brilliancy of the nucleus scarcely surpassed by that of the sun's surface. Nevertheless it had no sooner touched it than it vanished as if annihilated. So sudden was the disappearance (at 4h. 50m. 58s. Cape mean time), that it was at first thought that the comet must have passed *behind* the sun. But this proved not to have been the case. The observers at the Cape had witnessed a genuine transit. Nor could non-visibility be explained by equality of lustre. For the gradations of light on the sun's disc are amply sufficient to bring out against the dusky background of the limb any object matching the brilliancy of the centre; while an object just equally luminous with the limb must inevitably show dark at the centre. The only practicable view, then, is that the bulk of the comet was of too filmy a texture, and its presumably solid nucleus too small, to intercept any notice-

[1] *Observatory*, vol. v. p. 355.

able part of the solar rays—a piece of information worth remembering.

On the following morning, the object of this unique observation showed (in Dr. Gill's words) "an astonishing brilliancy as it rose behind the mountains on the east of Table
Bay, and seemed in no way diminished in brightness when the
sun rose a few minutes afterward. It was only necessary to
shade the eye from direct sunlight with the hand at arm's
length, to see the comet with its brilliant white nucleus and
dense, white, sharply-bordered tail of quite half a degree in
length."[1]   All over the world, wherever the sky was clear
during that day, September 18, it was obvious to ordinary
vision.   Since 1843 nothing had been seen like it.   From Spain,
Italy, Algeria, Southern France, despatches came in announcing the extraordinary appearance.   At Cordoba, in South
America, the "blazing star near the sun" was the one topic
of discourse.[2]   Moreover, and this is altogether extraordinary,
the records of its daylight visibility to the naked eye extend
over three days.   At Reus, near Tarragona, it showed bright
enough to be seen through a passing cloud when only three
of the sun's diameters from his limb, just before its final rush
past perihelion on September 17 ; while at Carthagena in
Spain, on September 19, it was kept in view during two hours
before, and two hours after noon, and was similarly visible in
Algeria on the same day.[3]

But still more surprising than the appearance of the body
itself, were the nature and relations of the path it moved in.
The first rough elements computed for it by Mr. S. C. Chandler
of Harvard, and by Mr. White, assistant at the Melbourne
Observatory, showed at once a striking resemblance to those
of the twin comets of 1843 and 1880. This suggestive fact
became known in this country, September 27, through the
medium of a Dunecht Circular.   It was fully confirmed by
subsequent inquiries, for which ample opportunities were

---

[1] *Observatory*, vol. v. p. 354.   [2] Gould, *Astr. Nach.*, No. 2481.
[3] Flammarion, *Comptes Rendus*, t. xcv. p. 558.

luckily provided. The likeness was not, indeed, so absolutely perfect as in the previous case; it included some slight, though real differences; but it bore a strong and unmistakable stamp, broadly challenging explanation.

Two hypotheses only were really available. Either the comet of 1882 was an accelerated return of those of 1843 and 1880, or it was a fragment of an original mass to which they also had belonged. For the purposes of the first view the "resisting medium" was brought into full play; the opinion invoking it was, for some time, both prevalent and popular, and formed the basis, moreover, of something of a sensational panic. For a comet which, at a single passage through the sun's atmosphere, encountered sufficient resistance to shorten its period from thirty-seven, to two years and eight months, must, in the immediate future, be brought to rest on his surface; and the solar conflagration thence ensuing was represented in some quarters, with more license of imagination than countenance from science, as likely to be of catastrophic import to the inhabitants of our little planet.

But there was a test available in 1882 which it had not been possible to apply either in 1843 or in 1880. The two bodies visible in those years had been observed only after they had already passed perihelion;[1] the third member of the group, on the other hand, was accurately followed for a week before that event, as well as during many months after it. Mr. Finlay's and Dr. Elkin's observation of its disappearance at the sun's edge formed, besides, a peculiarly delicate test of its motion. The opportunity was thus afforded, by directly comparing the comet's velocity before and after its critical plunge through the solar surroundings, of ascertaining with some approach to certainty whether any considerable retardation had been experienced in the course of that plunge. The answer distinctly given was that *there had not*. The computed and observed places on both sides of the sun, fitted harmo-

---

[1] Captain Ray's sextant-observation of the comet of 1843 a few hours before perihelion, was too rough to be of use.

niously together. The effect, if any were produced, was too
small to be perceptible.

This result is, in itself, a memorable one. It seems to give
the *coup de grâce* to Encke's theory—somewhat discredited, in
addition, by Backlund's investigation—of a resisting medium
growing rapidly denser inwards. For the perihelion distance
of the comet of 1882, though somewhat greater than that of
its predecessors, was nevertheless extremely small. It passed
at less than 300,000 miles of the sun's surface. But the
ethereal substance long supposed to obstruct the movement of
Encke's comet, would there be nearly 2000 times denser than
at the perihelion of the smaller body, and must have exerted
a conspicuous retarding influence. That none such could be
detected seems to argue that no such medium exists.

Further evidence of a decisive kind was not wanting on the
question of identity. The "Great September Comet" of 1882
was in no hurry to withdraw itself from curious terrestrial
scrutiny. It was discerned with the naked eye at Cordoba
as late as March 7, 1884, and still showed in the field of the
great equatoreal on June 1 as an "excessively faint whiteness." [1]
It was then about 470 millions of miles from the earth—a
distance to which no other comet—save the exceptional one
of 1729—has been pursued.[2] Moreover, an arc of 340 out of
the entire 360 degrees of its circuit had been described under
the eyes of astronomers; so that its course came to be very
well known. It may then be taken as ascertained that its
movement is in a very eccentric ellipse, traversed in several
hundred years. The lowest estimate of period, founded on sufficiently extensive data, is of 652½ years (Morrison); the highest
deserving any confidence that by Kreutz of 843.[3] There is
reason to believe that this last is not very far from the truth.

Now this conclusion of a period to be counted by centuries,
must be taken to apply to all the three bodies so curiously
related by the nature of their movements. For to assert (as

---

[1] *Astr. Nach.*, No. 2538.   [2] *Nature*, vol. xxix. p. 135.
[3] *Astr. Nach.*, No. 2482.

many astronomers of repute still do) that the comets of 1843
and 1880 are one and the same body revolving regularly in
nearly thirty-seven years, is virtually to cut off all connection
between them and the comet of 1882. If the *length* of the
ellipses they respectively trace out be thus totally and widely
different, then the likeness between their other elements must
be purely superficial—a mere freak of circumstance—and
means nothing. But this no one has ever ventured to assert.
We have no alternative, then, but to regard all three as moving
in nearly the same orbit, with nearly the same period—that is,
as individually distinct, though members of a single system.
So that the visibility of none of them can again be looked for
until the twenty-sixth or twenty-seventh century, when they
will probably return successively to perihelion in the same
order, and presenting much the same appearances, as in the
nineteenth.

The idea of cometary systems was first suggested by Thomas
Clausen in 1831.[1] It was developed by the acute inquiries of
the late M. Hoek, director of the Utrecht Observatory, in 1865
and some following years.[2] He found that in quite a considerable number of cases, the paths of two or three comets had a
common point of intersection far out in space, indicating with
much likelihood a community of origin. This consisted,
according to his surmise, in the disruption of a parent mass
during its sweep round the star latest visited. Be this as it
may, the fact is undoubted that numerous comets fall into
groups, in which affinity of geometrical relations betrays a pre-
existent physical connection. Never before, however, had
geometrical affinity been so notorious as between the three
comets now under consideration ; and never before, in a comet
still, it might be said, in the prime of life, had physical peculiarities tending to account for that affinity been so obvious as
in the last-comer of the group.

Observation of a granular structure in cometary nuclei dates

[1] Gruithuisen's *Analekten*, Heft vii. p. 48.
[2] *Month. Not.*, vols. xxv., xxvi., xxviii.

far back into the seventeenth century, when Cysatus and Hevelius described the central parts of the comets of 1618 and 1652 respectively, as made up of a congeries of minute stars. Analogous symptoms of a loose state of aggregation have of late been not unfrequently detected in telescopic comets, besides the instances of actual division offered by those connected with the names of Biela and Liais. The forces concerned in producing these effects seem to have been peculiarly energetic in the great comet of 1882.

The segmentation of the nucleus was first noticed in the United States and at the Cape of Good Hope, September 30. It proceeded rapidly. At Kiel, on October 5 and 7, Professor Krüger perceived two centres of condensation. A definite and progressive separation into *three* masses was observed by Professor Holden, October 13 and 17.[1] A few days later, M. Tempel found the head to consist of *four* lucid aggregations, ranged nearly along the prolongation of the caudal axis;[2] and Mr. Common, January 27, 1883, saw *five* nuclei in a line " like pearls on a string."[3] This remarkable character was preserved to the last moment of the comet's distinct visibility.

There were, however, other curious proofs of a marked tendency in this body to disaggregation. On October 8, Schmidt discovered at Athens a nebulous object 4° south-west of the great comet, and travelling in the same direction. It remained visible for a few days, and, from Oppenheim's and Hind's calculations, there can be little doubt that it was really the offspring by fission of the body it accompanied.[4] This is rendered more probable by the unexampled spectacle offered, October 14, to Mr. E. E. Barnard of Nashville, Tennessee, of *six* or *eight* distinct cometary masses within 6° south by west of the comet's head, none of which reappeared on the next opportunity for a search.[5] A week later, however, one similar object was discerned by Mr. Brooks, of Phelps, N.Y., in the

---

[1] *Nature*, vol. xxvii. p. 246.   [2] *Astr. Nach.*, No. 2468.
[3] *Athenæum*, Feb. 3, 1883.   [4] *Astr. Nach.*, Nos. 2462, 2466.
[5] *Ibid.*, No. 2489.

opposite direction from the comet. Thus, space appeared to be strewn with the filmy débris of this extraordinary body all along the track of its retreat from the sun.

Its tail was only equalled (if it were equalled) in length by that of the comet of 1843. It extended in space to the vast distance of two hundred millions of miles from the head; but, so imperfectly were its proportions displayed to terrestrial observers, that it at no time covered an arc of the sky of more than 30°. This apparent extent was attained, during a few days previous to September 25, by a faint, thin, rigid streak, noticed only by a few observers—by Elkin at the Cape Observatory, Eddie at Grahamstown, and Cruls at Rio Janeiro. It diverged at a low angle from the denser curved train, and was produced, according to Bredichin,[1] by the action of a repulsive force twelve times as strong as the counter-pull of gravity. It belonged, that is, to type 1; while the great forked appendage, obvious to all eyes, corresponded to the lower rate of emission characteristic of type 2. This was remarkable for the perfect definiteness of its termination, for its strongly forked shape, and for its unusual permanence. Down to the end of January 1883, its length, according to Schmidt's observations, was still 93 million miles; and a week later it remained visible to the naked eye, without notable abridgment.'

Most singular of all was an anomalous extension of the appendage *towards* the sun. During the greater part of October and November, a luminous "tube" or "sheath" of prodigious dimensions seemed to surround the head, and project in a direction nearly opposite to that of the usual outpourings of attenuated matter. Its diameter was computed by Schmidt to be, October 15, no less than *four million miles,* and it was described by Cruls as a "truncated cone of nebulosity," stretching 3° or 4° sunwards.[2] There can be little doubt that this abnormal kind of efflux was a consequence of the tremendous physical disturbance suffered at perihelion;

[1] *Annales,* Moscow, t. ix. pt. ii. p. 52.
[2] *Comptes Rendus,* t. xcvii. p. 797.

and it is worth remembering that something analogous was observed in the comet of 1680 (Newton's), also noted for its excessively close approach to the sun. The only plausible hypothesis as to the mode of its production is that of an opposite state of electrification in the particles composing the ordinary and extraordinary appendages.

The spectrum of the great comet of 1882 was, in part, a repetition of that of its immediate predecessor, thus confirming the inference that the previously unexampled sodium-blaze was in both a direct result of the intense solar action to which they were exposed. But the D-line was, this time, not seen alone. At Dunecht, on September 17, Drs. Copeland and Lohse succeeded in identifying six brilliant rays in the green and yellow with as many prominent iron-lines;[1] a very significant addition to our knowledge of cometary constitution, and one which goes far to justify Bredichin's assumption of various kinds of matter issuing from the nucleus with velocities inversely as their atomic weights. All the lines equally showed a slight displacement, indicating recession from the earth at the rate of 37 to 46 miles a second. A similar observation made by M. Thollon at Nice on the following day, supplied a highly satisfactory test of the accuracy of the spectroscopic method of estimating movement in the line of sight. Before anything was as yet known of the comet's path or velocity, he announced, from the position of the double sodium line alone, that at three P.M. on September 18 it was increasing its distance from our planet by from 61 to 76 kilometres per second.[2] M. Bigourdan's subsequent calculations showed that its actual swiftness of recession was at that moment 73 kilometres.

Changes in the inverse order to those seen in the spectrum of Comet Wells, soon became apparent. In the earlier body, carbon-bands had died out with *approach* to perihelion, and had been replaced by sodium-emissions; in its successor, sodium-emissions became weakened and disappeared with

[1] *Copernicus*, vol. ii. p. 235.   [2] *Comptes Rendus*, t. xcvi. p. 371.

retreat from perihelion, and found their substitute in carbon-bands. Professor Riccò was, in fact, able to infer, from the sequence of prismatic phenomena, that the comet had already passed the sun; thus establishing a novel criterion for determining the position of a comet in its orbit by the varying quality of its radiations.

Recapitulating what has been learnt from the five conspicuous comets of 1880–82, we find that the leading facts acquired to science were these three. First, that groups of comets may be met with pursuing each other, after intervals of many years, in the same, or nearly the same track; so that identity of orbit can no longer be regarded as a sure test of individual identity. Secondly, that no appreciable resistance to motion is experienced by such bodies in traversing the sun's corona. Finally, that their chemical constitution is a highly complex one, and that they possess, in some cases at least, a metallic core resembling the meteoric masses which occasionally reach the earth from planetary space.

As to the origin of comets, there has been of late years much speculation, ingenious or inane, which, however, it were quite superfluous to review. Yet we are not wholly without the guidance of ascertained fact on the subject. Laplace assumed that the fundamental shape of comets' orbits, when unmodified by planetary perturbations, is that of a hyperbola—a circumstance which, if true, would imply their total disconnection from our system, save by fortuitous encounter. But Gauss and Schiaparelli separately proved, on the contrary, that these bodies move *naturally* in prodigiously long ellipses,[1] the hyperbolic form, in the extremely rare cases where it may exist, being a result of disturbance. This being so, it follows that their condition previous to being attracted by the sun was one of relative repose.[2] In other words, they shared the movement of translation through space of the solar system.

This significant conclusion had been indicated, on other

---

[1] Thury and Meyer, *Arch. des Sciences*, t. vi. (3d ser.), p. 187.
[2] W. Förster, *Pop. Mitth.*, 1879, p. 7.

grounds, as the upshot of researches undertaken independently
by Carrington [1] and Mohn [2] in 1860, with a view to ascertaining the anticipated existence of a relationship between the
general *lie* of the paths of comets, and the direction of the
sun's journey. It is tolerably obvious that, if they wander at
haphazard through the interstellar regions, a preponderance
of their apparitions should seem to arrive from the vicinity of
the constellation Hercules; that is to say, we should meet
considerably more comets than would overtake us. Just for
the very same reason that falling stars are more numerous
after than before midnight. Moreover, the comets met by
us should be apparently swifter-moving objects than those
coming up with us from behind; because, in the one case, our
own real movement would be added to, in the other, subtracted
from theirs.   But nothing of all this can be detected.   Comets
approach the sun indifferently from all quarters, and with
velocities quite independent of direction.

We conclude then, with Schiaparelli and Förster, that the
"cosmical current" which bears the solar system towards its
unknown goal, carries also with it nebulous masses of undefined
extent, and at an undefined remoteness, fragments detached
from which, continually entering the sphere of the sun's attraction, flit across our skies under the form of comets. These
are, however, almost certainly so far strangers to our system
that they had no part in the long processes of development by
which its present condition was attained.   They are, perhaps,
survivals of an earlier, and by us scarcely and dimly conceivable state of things, when the chaos from which sun and
planets were, by a supreme edict, to emerge, had not as yet
separately begun to be.

[1] *Mem. R. A. Soc.*, vol. xxix. p. 355.
[2] *Month. Not.*, vol. xxiii. p. 203.

CHAPTER XII.

STARS AND NEBULÆ.

THAT a science of stellar chemistry should not only have
become possible, but should already have made material
advances, is assuredly one of the most amazing features in the
swift progress of knowledge our age has witnessed. Custom
can never blunt the wonder with which we must regard the
achievement of compelling rays emanating from a source
devoid of sensible magnitude through immeasurable distance, to reveal, by its peculiarities, the composition of that
source. The discovery of revolving double stars assured us
that the great governing force of the planetary movements, and
of our own material existence, sways equally the courses of the
farthest suns in space; the application of prismatic analysis
certified to the presence in the stars of the familiar materials,
no less of the earth we tread, than of the bodies built up out
of its dust and circumambient vapours.

We have seen that, as early as 1823, Fraunhofer ascertained
the generic participation of stellar light in the peculiarity by
which sunlight, spread out by transmission through a prism,
shows numerous transverse rulings of interrupting darkness.
No sooner had Kirchhoff supplied the key to the hidden meaning of those ciphered characters, than it was eagerly turned to
the interpretation of the dim scrolls unfolded in the spectra of
the stars. Donati made at Florence, in 1860, the first efforts
in this direction; but with little result, owing to the imperfections of the instrumental means at his command. His
comparative failure, however, was a prelude to others' success.

Almost simultaneously, in 1862, the novel line of investigation was entered upon by Huggins and Miller near London, by Father Secchi at Rome, and by Lewis M. Rutherfurd in New York. Fraunhofer's device of using a cylindrical lens for the purpose of giving a second dimension to stellar spectra, was adopted by all, and was indeed indispensable. For a luminous point, such as a star appears, becomes, when viewed through a prism, a variegated line, which, until broadened into a band by the intervention of a cylindrical lens, is all but useless for purposes of research. This process of *rolling out* involves, it is true, much loss of light—a scanty and precious commodity, as coming from the stars; but the loss is an inevitable one. And so fully is it compensated by the great light-grasping power of modern telescopes, that important information can now be gained from the spectroscopic examination of stars far below the range of the unarmed eye.

The effective founders of stellar spectroscopy, then (since Rutherfurd shortly turned his efforts elsewhither), were Father Secchi, the eminent Jesuit astronomer of the Collegio Romano, where he died, February 26, 1878, and Dr. Huggins, with whom the late Professor W. A. Miller was associated. The work of each was happily directed so as to supplement that of the other. With less perfect appliances, the Roman astronomer sought to render his extensive rather than precise; at Upper Tulse Hill, searching accuracy over a narrower range was aimed at and attained. To Father Secchi is due the merit of having executed the first spectroscopic survey of the heavens. Above 4000 stars were in all passed in review by him, and classified according to the varying qualities of their light. His provisional establishment (1863–67) of four types of stellar spectra [1] has proved a genuine aid to knowledge through the facilities afforded by it for the arrangement and comparison of rapidly accumulating facts. Moreover, it is scarcely doubtful that

---

[1] *Report Brit. Ass.*, 1868, p. 166. Rutherfurd gave a rudimentary sketch of a classification of the kind in December 1862, but based on imperfect observations. See *Am. Jour. of Sc.*, vol. xxxv. p. 77.

these spectral distinctions correspond to differences in physical condition of a marked kind.

The first order comprises more than half the visible stars, and a still larger proportion of those eminently lustrous. Sirius, Vega, Regulus, Altair, are amongst its leading members. Their spectra are distinguished by the breadth and intensity of the four dark bars due to the absorption of hydrogen, and by the extreme faintness of the metallic lines, of which, nevertheless, hundreds are disclosed by careful examination. The light of these "Sirian" orbs is white or bluish; and it is found to be rich in ultra-violet rays.

Capella and Arcturus belong to the second, or solar type of stars, which is about one-sixth less numerously represented than the first. Their spectra are quite closely similar to that of sunlight, in being ruled throughout by innumerable fine dark lines; and they share its yellowish tinge.

The third class includes most red and variable stars (commonly synonymous), of which Betelgeux in the shoulder of Orion, and "Mira" in the Whale are noted examples. Their characteristic spectrum is of the "fluted" description. It shows like a strongly illuminated colonnade seen in perspective, the light falling from the red end towards the violet. This *kind* of absorption is produced by the vapours of metalloids or of compound substances.

To the fourth order of stars belongs also a colonnaded spectrum, but *reversed;* the light is thrown the other way. The individuals composing it are few, and apparently insignificant, the brightest of them not exceeding the fifth magnitude. They are commonly distinguished by a deep red tint, and gleam like rubies in the field of the telescope. Father Secchi, who detected the peculiarity of their analysed light, ascribed it to the presence of carbon in some form in their atmospheres; and this has been confirmed by the latest researches of H. C. Vogel,[1] now director of the Astro-physical Observatory at Potsdam. The hydro-carbon bands, in fact, seen bright in

[1] *Publicationen,* Potsdam, No. 14, 1884, p. 31.

comets, are dark in these singular objects—the only ones in the heavens (save, perhaps, a coronal streamer or a rare meteor) [1] which display a cometary analogy of the fundamental sort revealed by the spectroscope.

The members of all four orders are, however, emphatically *suns.* They possess, it would appear, photospheres radiating all kinds of light, and differ from each other (so far as we are able to judge) solely in the varying qualities of their absorptive atmospheres. The principle that the colours of stars depend, not on the intrinsic nature of their light, but on the kinds of vapours surrounding them, and stopping out certain portions of that light, was laid down by Huggins in 1864. [2] The phenomena of double stars seem to indicate a connection between the state of the investing atmospheres by the action of which their often brilliantly contrasted tints are produced, and their mutual physical relations. A remarkable tabular statement put forward by Professor Holden in June 1880 [3] made it, at any rate, clear that inequality of magnitude between the components of binary systems accompanies unlikeness in colour, and that stars more equally matched in one respect are pretty sure to be so in the other. Besides, blue and green stars of a decided tinge are *never* (so far as is certainly known) solitary ; they invariably form part of systems. So that association has undoubtedly a predominant influence upon colour.

Nevertheless, the crude notion thrown out by Zöllner in 1865, [4] that yellow and red stars are simply white stars in various stages of cooling, has obtained undeserved currency. D'Arrest, it is true, protested against it, but Vogel adopted it in 1874 as the "rational" basis of his classification. [5] This differs from Father Secchi's only in presenting his third and fourth types as

---

[1] Von Konkoly *once* derived from a slow-moving meteor a hydro-carbon spectrum. A. S. Herschel, *Nature*, vol. xxiv. p. 507.

[2] *Phil. Trans.*, vol. cliv. p. 429.

[3] *Am. Jour. of Sc.*, vol. xix. p. 467.   [4] *Photom. Unters.*, p. 243.

[5] *Astr. Nach.*, No. 2000.

subdivisions of the same order ; but the seductive, though possibly misleading idea of progressive development is added. Thus, the white Sirian stars are represented as the *youngest* because the hottest of the sidereal family ; those of the solar pattern as having already wasted much of their store by radiation, and being well advanced in middle life ; while the red stars with banded spectra figure as effete suns, hastening rapidly down the road to final extinction.[1]

Now the truth is, that we are just as ignorant of the relative, as of the absolute ages of the stars, the arguments employed on the point being, as it were, reversible. For instance, if there be any truth in the theory of nebular condensation, we should expect to find a forming sun surrounded by a dense and extensive atmosphere, not unlike that of a " hydro-carbon " star ; while the decay of luminous power would probably be attended by a falling-off in absorptive action, resulting in a feebly continuous spectrum. Stars of the latter description may exist ; but the absence of characterisation, no less than of intensity in their light renders them both a difficult, and an unattractive subject of study.

A spectroscopic star-catalogue (the first attempted) is now in course of preparation at Potsdam and Lund by Drs. Vogel and Dunér. It will include all stars down to magnitude $7\frac{1}{2}$ situated between the north pole of the heavens and one degree south of the equator. The first part, giving the results of observations upon the spectra of 4051 stars (12,000 were incidentally examined), was published in 1883,[2] and a further instalment will shortly follow. The provision of such a vast and accurate store of data for future reference is a duty, in Vogel's estimation, which the present generation owes to posterity, and may prove of inestimable importance to the progress of discovery.

---

[1] Mr. J. Birmingham, in the Introduction to his valuable Catalogue of Red Stars, comments upon this "singular conceit," and alleges various instances of change of colour in a direction the opposite of that which it supposes to be the inevitable result of time. *Trans. R. Irish Ac.*, vol. xxvi. pp. 251-253.   [2] *Publicationen*, No. 11, Potsdam, 1883.

A fairly complete answer to the question, What are the stars made of? was given by Dr. Huggins in 1864.[1] By laborious processes of comparison between stellar dark lines and the bright rays emitted by terrestrial substances, he made quite sure of his conclusions, though at much cost of time and pains. He assures us, indeed, that—taking into account restrictions by weather and position—the thorough investigation of a *single* star-spectrum would be the work of some years. Of two, however—those of Betelgeux and Aldebaran—he was able to furnish detailed and accurate drawings. The dusky flutings in the prismatic light of the first of these stars have not been identified with the absorption of any particular substance; but associated with them are dark lines telling of the presence of sodium, iron, calcium, magnesium, and bismuth. Hydrogen rays are also inconspicuously present. That an exalted temperature reigns, at least in the lower strata of the atmosphere, is certified by the vaporisation there of matter so refractory to heat as iron.[2]

Nine elements—those identified in Betelgeux, with the addition of tellurium, antimony, and mercury—were recognised as having stamped their signature on the spectrum of Aldebaran; while the existence in Sirius, and nearly all the other stars inspected, of hydrogen, sodium, iron, and magnesium was rendered certain or highly probable. This was admitted to be a bare gleaning of results; nor is there reason to suppose any of his congeners inferior to our sun in complexity of constitution.

The evidence given by the spectroscope of fluctuations in *quality* as well as in *quantity*, in the light of variable stars, suggests a rationale of the surprising appearances presented by them, which may eventually be expected to supersede all others. Speaking generally, stellar variability is an accompaniment of a ruddy tint and a banded spectrum. In other

---

[1] *Phil. Trans.*, vol. cliv. p. 413. Some preliminary results were embodied in a "note" communicated to the Royal Society, February 19, 1863 (*Proc. Roy. Soc.*, vol. xii. p. 444).   [2] *Ibid.*, p. 429, *note.*

words, it prevails in stars surrounded by powerfully absorptive atmospheres. Moreover, the strength of their absorption increases as the light diminishes; perhaps we might say, the light diminishes *because* it increases.

Heretofore the explanations of variability chiefly entertained had been these two: rotation on an axis, showing alternately a darker and a brighter side, and the interposition of a non-luminous body, revolving round a star periodically eclipsed by it. But in truth, the facts, for the most part, fitted very ill with either, and were, not unfrequently, in glaring disaccord with both. There are, however, a few exceptional cases to which the "eclipse" theory has been thought to be peculiarly applicable; and assuredly, if it fail in them, it will succeed nowhere else. The leading member of this small group is the star called Algol in the head of Medusa.

It stands apart in several respects from most other variables. In the first place, it is a white star, and shows a spectrum of the Sirian pattern—two circumstances highly favourable to stability in lustre. Further, the diminution of its light is by a strictly impartial process; no individual rays are attacked more than others; it remains unchanged in kind even when reduced to a sixth of its original amount. Finally, the accomplishment of its decline and revival, instead of being distributed, with more or less of irregularity, over its entire period, is restricted to a perfectly definite fractional part of it. During somewhat more than two days and a half it shines quite steadily as a star of the second magnitude; its fall to, and recovery from the fourth are hurried over in about seven hours.

This manner of procedure suggested to Goodricke, who discovered in 1782 the periodical variability of Algol, the interposition of a large satellite; and the explanation has been generally accepted. The conditions under which it must be available were, however, first seriously investigated by Professor Pickering in 1880.[1] He found that the appearances in ques-

---

[1] *Proc. Am. Ac. Sc.*, vol. xvi. p. 17; *Observatory*, vol. iv. p. 116. For a preliminary essay by T. S. Aldis in 1870, see *Phil. Mag.* vol. xxxix. p. 363.

tion could be quite satisfactorily accounted for by admitting the revolution round the star of an obscure body 0.764 of its own diameter, in a period of two days twenty hours forty-nine minutes. It needs, indeed, a mind trained to the docile adoption of views authoritatively recommended, to contemplate without some measure of incredulity a system in which a satellite of the same relative magnitude that 446 Jupiters would bear to our sun, circulates in a relative contiguity to its primary only a little less close than that of his inner satellite to Mars. But, as Professor Pickering remarks in a similar connection, "what could be more improbable than the phenomenon itself, were it not verified by observation?"[1]

The Algol class of variables includes only seven or eight members. If the hypothesis of an eclipsing body (for which a cloud of meteorites may be substituted) represent the truth in one case, it must be capable of adaptation to all. But the attempt to fit it to a remarkable star in the constellation Cepheus, discovered by M. Ceraski at Moscow, June 23, 1880, may be said to have broken down. Its phases are of the same rapid and well-defined description as those of Algol, and recur in the still shorter period of two and a half days. Its bluish white rays, however, turn ruddy at minimum, which implies, not mere stoppage, but selective absorption. Besides, the interposing satellite should (according to Pickering's calculations) be almost as large as its primary, so that the eclipse-rationale seems here burdened with intolerable difficulties.

The number of recognised variables of all classes already reaches some hundreds, and is continually increasing. Indeed, Dr. Gould is of opinion that most stars fluctuate slightly in brightness through surface-alternations similar to, but on a larger scale than those of the sun. The solar analogy might, perhaps, be pushed somewhat further. It may be found to contain a clue to much that is perplexing in stellar behaviour. Wolf pointed out in 1852 the striking resemblance in character between curves representing sun-spot frequency, and curves

[1] *Proc. Am. Ac.*, vol. xvi. p. 259.

representing the changing luminous intensity of many variable stars. There were the same steep ascent to maximum and more gradual decline to minimum, the same irregularities in heights and hollows, and, it may be added, the same tendency to a double maximum, and complexity of superposed periods. It is impossible to compare the two sets of phenomena thus graphically portrayed, without reaching the conclusion that they are of closely related origin—that our sun, in fact, is of the kindred of variable stars, though the family peculiarities have, for some reason, remained comparatively undeveloped.

Every kind and degree of variability is exemplified in the heavens. At the bottom of the scale are stars like the sun, of which the lustre is—tried by our instrumental means—sensibly steady. At the other extreme are ranged the astounding apparitions of " new," or " temporary " stars. Within the last score of years three of these stellar guests (as the Chinese call them) have presented themselves, and we meet with a fourth no farther back than April 27, 1848. But of the " new star " in Ophiuchus found by Mr. Hind on that night, little more could be learnt than of the brilliant objects of the same kind observed by Tycho and Kepler. The spectroscope had not then been invented. Let us hear what it had to tell of later arrivals.

Between thirty and fifteen minutes before midnight of May 12, 1866, Mr. John Birmingham of Millbrook, near Tuam, in Ireland, saw with astonishment a bright star of the second magnitude unfamiliarly situated in the constellation of the Northern Crown. Four hours earlier, Schmidt of Athens had been surveying the same part of the heavens, and was able to testify that it was not visibly there. That is to say, a few hours, or possibly a few minutes, sufficed to bring about a conflagration, the news of which may have occupied hundreds of years in travelling to us across space. The rays which were its messengers, admitted within the slit of Dr. Huggins's spectroscope, May 16, proved to be of a composition highly significant as to the nature of the catastrophe. The star—which had already declined below the third magnitude—showed what was described as a double spectrum. To the dusky flutings of Secchi's third type four brilliant rays were added.[1] The chief of these agreed in position with lines of hydrogen; so that the immediate cause of the outburst was plainly perceived to have been the eruption, or ignition, of vast masses of that subtle kind of matter, the universal importance of which throughout the cosmos is one of the most curious facts revealed by the spectroscope.

T Coronæ (as the new star was called) quickly lost its adventitious splendour. Nine days after its discovery it was again invisible to the naked eye. It is now a pale yellow, slightly variable star near the tenth magnitude, and finds a place as such in Argelander's charts. It was thus obscurely known before it made its sudden leap into notoriety.

The mantle of gaseous incandescence in which it was temporarily wrapt, is a recurrent, or even a permanent feature in some other stars. Two of these—β Lyræ, a white star variable (by a rare exception) in a period of twelve days and nearly twenty-two hours, and γ Cassiopeiæ—were noticed by Father Secchi at the outset of his spectroscopic inquiries. Both show *bright* lines of hydrogen and 'helium,' so that the peculiarity of their condition probably consists in the unusual extent, and intense ignition of their chromospheric surroundings. But this condition is subject to fluctuations. The brilliant rays indicative of it died out during nine years, 1874–83, and the first symptom of their reappearance was caught by M. Eugen von Gothard in a twinkling of the crimson C line in the spectrum of γ Cassiopeiæ, August 13, 1883.[2] Before the end of the month, the whole range was vividly apparent, and the Lyre variable followed suit in the course of the autumn. An ebb and flow of brightness in a period of seven days, has since been found by M. von Gothard to affect the hydrogen and

---

[1] *Proc. Roy. Soc.*, vol. xv. p. 146.
[2] *Astr. Nach.*, Nos. 2539, 2548, 2581.

helium spectrum of the latter star, and he suspects an analogous inconstancy in the emissions of its fellow.[1]

These two luminaries formed the nucleus of what is now generally regarded as a distinct stellar class. To it belong the extraordinary variable η Argûs, with γ in the same constellation, examined by Respighi in 1871 ; and it includes three small stars in the Swan, the peculiar character of which was discovered in 1867 by MM. Wolf and Rayet of the Paris Observatory.[2] Their light betrayed a mainly gaseous origin, separating into three bands, identical in each star, but corresponding to no known substance, and scarcely connected by an almost evanescent continuous spectrum. No sign of change has been detected in them. Three analogous objects have since been discovered by Professor Pickering, and five more were found by Dr. Copeland in 1883 in the course of an excursion exploratory of visual possibilities in the Andes.[3]

Now the question arises, have we here to do with *stars* in the ordinary sense at all? that is, with suns like our own, reduced by the immensity of their distance to sparkling points of light? We should reply in the negative were the above definition to be adopted ; but our readers will have already gathered that it requires much extension and qualification. How far we may yet be led to extend it, and how profoundly our ideas of what constitutes a " star " may eventually have to be modified, a recent noteworthy event has gone a great way to indicate.

On the 24th of November 1876, at Athens, Dr. Schmidt discovered a new star in the constellation Cygnus. It was then nearly of the third magnitude, and in its previous state must have been below the ninth, since Argelander had made no record of its existence. Its spectrum was examined December 2, by Cornu at Paris,[4] and a few days later by Vogel and Lohse at Potsdam.[5] It proved of a closely similar

[1] *Bull. Astr.*, t. ii. p. 149.   [2] *Comptes Rendus*, t. lxv. p. 292.
[3] *Copernicus*, vol. iii. p. 207.   [4] *Comptes Rendus*, t. lxxxiii. p. 1172.
[5] *Monatsb.*, Berlin, 1877, pp. 241, 826.

character to that of T Coronæ. A range of bright lines, including those of hydrogen, helium, and perhaps of the coronal
gas (1474), stood out from a continuous background strongly
"fluted" by absorption. It may be presumed that in reality
the gaseous substances, which, by their sudden incandescence
had produced the apparent conflagration, lay comparatively
near the surface of the star, while the screen of cooler materials
rhythmically intercepting large portions of its light, was situated
at a considerable elevation in its atmosphere.

The object, meanwhile, steadily faded. By the end of the
year it was of no more than seventh magnitude. After the
second week of March 1877, strengthening twilight combined
with the decline of its radiance to arrest further observation.
It was resumed, September 2, at Dunecht, with a strange
result. Practically the whole of its scanty light (it had then
sunk below the tenth magnitude) was perceived to be gathered
into a single bright line in the green, and that the most
characteristic line of gaseous nebulæ.[1] The star had, in fact,
so far as outward appearance was concerned, become transformed into a planetary nebula, many of which are so minute
as to be distinguishable from small stars only by the quality of
their radiations. The nebular phase, however, seems to have
been transient. In the course of 1880, Professor Pickering
found that Nova Cygni gave an ordinary stellar spectrum of
barely perceptible continuous light ;[2] and his observation was
negatively confirmed at Dunecht, February 1, 1881.

This enigmatical object has now dropt to (if not below) the
fourteenth magnitude, being thus out of reach of spectroscopic
scrutiny, save (possibly) with a few of the most powerful telescopes in the world. The lesson learnt from its changes appears to be no less than this : That no clear dividing-line can
be drawn between stars and nebulæ ; but that in what are
called "planetary nebulæ" on the one side, and in "gaseous
stars" (those giving a spectrum of bright lines) on the other,
we meet with transitional forms, serving to bridge the gap

[1] *Copernicus,* vol. ii. p. 101.   [2] *Annual Report,* 1880, p. 7.

between such vast and *highly finished* orbs—if we may be permitted the expression—as Sirius, and the inchoate, faintly-
lucent stuff which curdles round the trapezium of Orion.

We have been compelled somewhat to anticipate our narrative as regards inquiries into the nature of this latter kind of
object. The fluctuations of opinion on the point came to an
abrupt end with the application to them of the spectroscope.
On August 29, 1864, Dr. Huggins sifted through his prisms
the rays of a bright planetary nebula in Draco.[1] To his
infinite surprise, they proved to be mainly of one colour. In
other words, they avowed their origin from a mass of glowing
vapour. As to what *kind* of vapour it might be by which
Herschel's conjecture of a "shining fluid" variously diffused
throughout the cosmos, was thus unexpectedly verified, an
answer was also at hand. The conspicuous bright line of the
Draco nebula was found to belong very probably to nitrogen;
of its two fainter companions, one was unmistakably the F
line of hydrogen, while the other, in position intermediate
between the two, still remains unidentified. The extreme
faintness of nebular light was experimentally shown to be
reason sufficient for the solitariness in its spectrum of the lines
emanating respectively from nitrogen and hydrogen; the surviving nebular rays being precisely those which resist extinction longest.

By 1868, Dr. Huggins had satisfactorily examined the
spectra of about seventy nebulæ, of which one-third displayed a gaseous character.[2] In *all* of these (and the rule has
hitherto proved without exception) the nitrogen line appeared;
though in some cases—as "the Dumb-bell" nebula in Vul-
pecula—it appeared alone. On the other hand, a fourth line,
the dark blue of hydrogen—in addition to the normal three,
was subsequently detected in the light of the great Orion
nebula. But, fundamentally, the composition of all bodies of
this class may be assumed the same. The differences in their
radiations seem to be of intensity, not of kind. All planetary

[1] *Phil. Trans.*, vol. cliv. p. 437.   [2] *Ibid.*, vol. clviii. p. 540.

and annular nebulæ belong to it, as well as those termed "irregular" which frequent the region of the Milky Way. Thus the signs of resolvability noted at Parsonstown and Cambridge (U.S.) in Orion and the "Dumb-bell," were proved fallacious, so far, at least, as they had been taken to indicate a stellar constitution; though they may have quite faithfully corresponded to the existence in discrete masses of the glowing vapours elsewhere more equably diffused.

The well-known nebula in Andromeda, and the great spiral in Canes Venatici are amongst the more remarkable of those giving a continuous spectrum; and, as a general rule, the emissions of all such nebulæ as present the appearance of star-clusters grown misty through excessive distance, are of the same kind. It would, however, be eminently rash to conclude thence that they are really aggregations of sun-like bodies. The improbability of such an inference has been vastly enhanced by the recent outbreak of a new star apparently in the very heart of the Andromeda nebula. First seen by Mr. Isaac W. Ward, August 19, 1885, as an ordinary $7\frac{1}{2}$ magnitude star, it already shows a diminished lustre. It gives a continuous spectrum precisely similar to that of the nebula—that is, truncated in the red as if by absorptive action. Hence, the cause of its sudden development of light must have been a totally different one from that occasioning the flaming apparitions in Corona and Cygnus.

Among the ascertained analogies between the stellar and nebular systems is that of variability of light. On October 11, 1852, Mr. Hind discovered a small nebula in Taurus. Chacornac observed it at Marseilles in 1854, but was confounded four years later to find it vanished. D'Arrest missed it October 3, and re-detected it December 29, 1861. It was easily seen in 1865–66, but invisible in the most powerful instruments 1877–80.[1] This was the first undisputed instance of nebular variability. Brought to the notice of astronomers

[1] Chambers, *Descriptive Astronomy* (3d ed.), p. 543; Flammarion, *L'Univers Sidéral*, p. 818.

STARS AND NEBULÆ. 425

by D'Arrest in 1862,[1] it has since been confirmed by others of
the same nature. Two such have recently been adduced by
Winnecke ;[2] and Professor Holden, having co-ordinated in his
admirable " Monograph of the Nebula of Orion "[3] the results
of all the more prominent inquiries into the structure of that
marvellous object since 1758, reaches the conclusion, that
while the figure of its various parts has (with only one possible
exception) remained the same, their brightness has been and
is in a state of continual fluctuation. This accords precisely
with the conviction expressed by O. Struve in 1857,[4] and may
now be safely accepted an an ascertained fact.

More dubious is the case of the "trifid" nebula in Sagit-
tarius, investigated by Professor Holden in 1877.[5] That
change of some kind has occurred, is indeed established by a
comparison of his own and others' observations with those of
the two Herschels ; but he inclines to the view that motion,
with or without an accompanying variation in light, is here the
agent of change. What is certain is, that a remarkable triple
star which, during the years 1784–1833, was centrally situated
in a dark space between the three great *lobes* of the nebula,
has since become involved in one of them ; and since the
star gives no sign of sensible displacement, the movement—if
movement there should prove to be—must be thrown upon
the nebula.

A similar example was last year alleged by Mr. H. Sadler,[6]
but the evidence upon which it rests is disputed. The ascertainment of true proper motion in a nebula would be the
more interesting from its absolute novelty. Hitherto this
class of bodies have shown no sign of sharing the busy journey-
ings of the stars.[7] They have remained as seemingly fixed in
their places as if exempt from all relation with the multitudi-

[1] *Astr. Nach.*, No. 1366.
[2] *Month. Not.*, vol. xxxviii. p. 105 ; *Astr. Nach.*, No. 2293.
[3] *Wash. Obs.*, vol. xxv. App. 1.   [4] *Month. Not.*, vol. xvii. p. 230.
[5] *Am. Jour. of Sc.*, vol. xiv. p. 433.   [6] *Observatory*, vol. viii. p. 127.
[7] For some instances of supposed orbital movement in "double"
nebulæ, see Flammarion, *Comptes Rendus*, t. lxxxviii. p. 27.

nous hosts of the galactic world. This singular immobility might, on a casual view, be set down to the account of enor-mous distance, no single nebula having, so far, exhibited the faintest trace of parallactic displacement. But there is a method of estimating motion independent of distance, and to this also nebulæ have hitherto proved unresponsive.

The principle upon which "motion in the line of sight" can be detected and measured with the spectroscope, has already been explained.[1] It depends, as our readers will remember, upon the removal of certain lines, dark or bright (it matters not which), from their normal places by almost infinitesimal amounts. The whole spectrum of the moving object, in fact, is very slightly *shoved* hither or thither, according as it is travelling towards or from the eye; but, for convenience of measurement, one line is usually picked out from the rest, and attention concentrated upon it. The application of this method to the stars, however, is encompassed with difficulties. It needs a powerfully dispersive spectroscope to show line-displacements of the minute order in question; and powerful dispersion involves a strictly proportionate enfeeblement of light. This, where the supply is already to a deplorable extent niggardly, can ill be afforded; and it ensues that the operation of determining a star's approach or recession is, even apart from atmospheric obstacles, an excessively deli-cate one.

It was first successfully executed by Dr. Huggins early in 1868.[2] The brightest star in the heavens was selected as the most promising subject of experiment, and proved amenable. In the spectrum of Sirius, the F line was perceived to be just so much displaced towards the red as to indicate (the orbital motion of the earth being deducted) recession at the rate of twenty-nine miles a second. Of this an undetermined propor-tion was no doubt attributable to the advance through space of the solar system, for which Struve's estimate of four miles a

---

[1] See *ante*, p. 243.   [2] *Phil. Trans.*, vol. clviii. p. 529.

second was almost certainly too small. Still there remained a large surplus of Sirian proper motion. Its reality and direction were placed beyond doubt by Vogel and Lohse's observation, March 22, 1871, of a similar, but even more considerable displacement.[1] The inquiry was resumed by Dr. Huggins with improved apparatus in the following year, when the movements of thirty stars were approximately determined.[2] The retreat of Sirius was now diminished in estimated velocity to about twenty miles per second, and it was discovered to be shared, at rates varying from twelve to twenty-nine miles, by Betelgeux, Rigel, Castor, Regulus, and five of the principal stars in the Plough. Arcturus, on the contrary, gave signs of rapid approach (fifty-five miles a second), as well as Pollux, Vega, Deneb in the Swan, and the brightest of the Pointers.

The realisation of this method of investigating stellar motions has an importance far beyond that of which the idea is conveyed by the bare enumeration of its preliminary results. It may confidently be expected to play a leading part in the unravelment of the vast and complex relations which we can dimly detect as prevailing amongst the innumerable orbs of the sidereal world; for it supplements the means which we possess of measuring by direct observation movements transverse to the line of sight, and thus completes our knowledge of the courses and velocities of stars at ascertained distances, while supplying for all a valuable index to the amount of perspective foreshortening of apparent movement. Thus some, even if an imperfect, knowledge may at length be gained of the revolutions of the stars—of the systems they unite to form, of the paths they respectively pursue, and of the forces under the compulsion of which they travel.

Already, though the method can scarcely be said to have passed the tentative stage, a most curious fact has been brought to light. Since 1874, spectroscopic measures of the visual component of stellar motions have been made part of

---

[1] Schellen, *Die Spectralanalyse*, Bd. ii. p. 326 (ed. 1883).
[2] *Proc. Roy. Soc.*, vol. xx. p. 386.

the regular work at the Royal Observatory, Greenwich. The
results have proved, on the whole, strongly confirmatory of
Dr. Huggins's. But in the movement of Sirius a perplexing
change has taken place. In March 1876 it was estimated to
be adding to its distance from the earth by twenty-seven miles
each second. In 1877 a slackening was perceived; and
this progressively advanced, until, in 1882, the rate of recession was diminished to, or below, seven miles a second. A
reversal of direction was even anticipated, and shortly occurred.
The spectrum was markedly shifted towards the *blue* end,
November 16, 1883;[1] and a series of forty-five measures executed by Mr. Maunder on thirteen nights in 1884, gave to
the star a mean motion of *approach* of twenty-two miles a
second.[2] It does not appear that the known elliptic revolution
of Sirius round its companion will account for these vicissitudes,
although it is remarkable that they are suspected also to affect,
in some degree, the course of Procyon, a star similarly circumstanced to Sirius in its vicinity to a comparatively obscure
source of disturbance. The further development of these
significant changes will be of the highest interest.

None of the nebulæ hitherto examined show the slightest
trace of displacement in the line of sight.[3] And this conclusion,
unlike estimates of apparent movement across the sky, has
absolutely no connection with their greater or less remoteness.
So that we seem compelled to draw an inference which must
largely affect our ideas of the whole structure of the heavens ;
namely, that nebulæ, as a class, are very much slower-moving
bodies than stars.

The uses of photography in celestial investigations become
every year more manifold and more apparent. The earliest
chemical star-pictures were those of Castor and Vega, obtained

---

[1] *Month. Not.,* vol. xliv. p. 91.

[2] *Observatory,* vol. viii. p. 109.   Dr. Huggins in 1872 anticipated as a
possible consequence of its circulation in an orbit the occurrence of such
changes in the movement of Sirius as have actually been observed.   *Proc.
Roy. Soc.,* vol. xx. p. 387.

[3] Huggins, *Proc. Roy. Soc.,* vol. xxii. p. 251.

with the Cambridge refractor in 1845 by Whipple of Boston
under the direction of W. C. Bond. Double-star photography
was inaugurated under the same auspices in 1857, with an
impression of Mizar, the middle star in the handle of the
Plough, and its small companion Alcor, the old Arab test of
keen eyesight, but now a comparatively easy naked-eye object.
The matter next fell into the able hands of Rutherfurd, who
completed in 1864 a fine object-glass, corrected for the ultra-
violet rays, consequently useless for visual purposes. The
sacrifice was recompensed by conspicuous success. A set of
measurements from his photographs of nearly fifty stars in the
Pleiades, enabled Dr. Gould in 1866 to ascertain, by comparison with Bessel's places for the same stars, that during the
intervening quarter of a century no changes of importance
had occurred in their relative positions.[1] The construction of
photographic star-maps of real and permanent value was thus
demonstrated to be a possibility, and is rapidly being converted into a reality of the utmost moment to the future of
science. In carrying on the work of eclliptical charting, left
half completed by Chacornac, the MM. Henry encountered
sections of the Milky Way which defied the enumerating
efforts of eye and hand, and resolved in consequence to have
recourse to the camera. The perfect success of some preliminary trials made with an instrument constructed expressly
for the purpose, was announced to the Academy of Sciences
at Paris, May 11, 1885. By its means, stars down to the
sixteenth magnitude clearly record their presence and their
places ; and we are hence doubtless on the eve of seeing the co-
operative photographic survey of the heavens, recommended by
Dr. Gill, carried into execution. It will include the uncounted
host of separate stars, showing the significant character-
istics of their distribution ; will individualise the hundreds, or

---

[1] Gould on Celestial Photography, *Observatory*, vol. ii. p. 16. Professor
Pritchard communicated to the Roy. Astr. Soc., May 9, 1884, his detection
of some small movements *inter se* of members of the Pleiades group.
Observatory, vol. vii. p. 163.

even thousands, of components forming each of those strange systems apart, known to us as " star-clusters ;" will determine the configurations and apparent distances of the members of binary and multiple groups, with an enormous saving of labour, and with the elimination of vexatious personal peculiarities in error ; besides faithfully recording the forms and positions of those baffling crowds of nebulæ, the yearly discoveries of which are counted by the score ; thus providing in all branches of sidereal astronomy a sure criterion of future change.

In the use of photography as an engine of research into the physical condition of the stars, Dr. Huggins led the way. In March 1863 he obtained with his coadjutor, Dr. Miller, microscopic prints of the spectra of Sirius and Capella.[1] But they told nothing. No lines were visible in them. They were mere characterless streaks of light. He tried again in 1876, when the 18-inch speculum of the Royal Society had come into his possession, using prisms of Iceland spar, and lenses of quartz ; and this time with better success. A photograph of the spectrum of Vega showed seven strong lines.[2] Still he was not satisfied. He waited and worked for three years longer. At length, on December 18, 1879, he was able to communicate to the Royal Society[3] results answering to his expectations. The delicacy of eye and hand needed to attain them may be estimated from the single fact, that the image of a star had to be kept, by continual minute adjustments, exactly projected upon a slit $\frac{1}{350}$ of an inch in width during nearly an hour, in order to give it time to imprint the characters of its analysed light upon a gelatine plate raised to the highest pitch of sensitiveness.

The ultra-violet spectrum of the white stars—of which Vega was taken as the type—was by this means shown to be a very remarkable one. Twelve strong lines, arranged at intervals diminishing regularly upwards, intersected it. They belonged presumably to one substance ; and since the two least refrangible

[1] *Month. Not.*, vol. xxiii. p. 180.   [2] *Proc. Roy. Soc.*, vol. xxv. p. 446.
[3] *Phil. Trans.*, vol. clxxi. p. 669.

were known hydrogen rays, that substance could scarcely be any other than hydrogen This was rendered certain by direct photographs of the hydrogen-spectrum taken by H. W. Vogel at Berlin a few months earlier.[1] In them seven of the white-star series were visible; and the remaining five were absent only because the higher rays failed to get through the glass prism employed.

In yellow stars, such as Capella and Arcturus, the same rhythmical series was *partially* represented, but associated with a great number of other lines; their state, as regards ultra-violet absorption, thus approximating to that of the sun; while the redder stars betrayed so marked a deficiency in actinic rays, that from Betelgeux, with an exposure *forty times* that required for Sirius, only a faint spectral impression could be obtained, and from Aldebaran, in the strictly invisible region, almost none at all.

The same process was successfully applied to the Orion nebula, March 7, 1882.[2] Five lines in all stamped themselves upon the plate during forty-five minutes of exposure. Of these, four were the known visible rays, and the fifth seemed to agree with one of the hydrogen set displayed by Vega. Almost simultaneously, this notable feat in celestial photography was achieved by Dr. Draper at New York,[3] and with the additional result of obtaining from the nebulous "knots" preceding the trapezium, a continuous spectrum. This was thought to indicate an advance of central condensation— possibly even the beginning of the long birth-process of an orderly revolving system, reserved for the future habitation of rational beings. It may be so; the ways of creative power are dark. Yet we cannot help remarking that the presence of so many stars fully formed, yet seemingly wrapt up and involved in the prodigious masses of nebulosity filling that portion of the sky, appears in some degree to discount the expectation of stellar development from them.

[1] *Astr. Nach.*, No. 2301.   [2] *Proc. Roy. Soc.*, vol. xxxiii. p. 425.
[3] *Comptes Rendus*, t. xciv. p. 1243.

The first promising photograph of the Orion nebula itself was obtained by Draper, September 30, 1880.[1] The marked approach towards a still more perfectly satisfactory result shown by his plates of March 1881 and 1882, was unhappily cut short by his premature death. Meanwhile, M. Janssen was at work in the same field from 1881, with his accustomed success.[2]   But Mr. Ainslie Common left all competitors far behind with a splendid picture, taken January 30, 1883, by means of an exposure of thirty-seven minutes in the focus of his three-foot silvered glass-mirror.[3]   Photography may thereby be said to have definitively assumed the office of historiographer to the nebulæ; since this one impression embodies a mass of facts hardly to be compassed by months of labour with the pencil, and affords a record of shape and relative brightness in the various parts of the stupendous object it delineates, which must prove invaluable to the students of its future condition.

The sublime problem of the construction of the heavens has not been neglected amid the multiplicity of tasks imposed upon the cultivators of astronomy by its rapid development. But data of a far higher order of precision, and indefinitely greater in amount, than those at the disposal of Herschel or Struve, must be accumulated before any definite conclusions on the subject are possible.   The first organised effort towards realising this desideratum, was made by the German Astronomical Society in 1865, two years after its foundation at Heidelberg.   The scheme, as originally proposed, consisted in the *exact* determination of the places of about 100,000 stars, from the re-observation of which, say, in the year 1950, astronomers of two or three generations hence may gather a vast store of knowledge—directly of the apparent motions, indirectly of the mutual relations binding together the suns and systems of space.   Fourteen observatories in Europe and America joined in the work, which is now far advanced.

[1] *Wash. Obs.*, vol. xxv. App. i. p. 226.
[2] *Comptes Rendus*, t. xcii. p. 261.
[3] *Month. Not.*, vol. xliii. p. 255.

Its scope, however, has, since its inception, been widened so as
to include southern zones as far as the Tropic of Capricorn ;
and a preliminary survey of the new region on Argelander's
plan has just been made by Schönfeld at Bonn.

Through Dr. Gould's unceasing labours, during his fifteen
years' residence at Cordoba, a detailed acquaintance with
southern stars has at length been brought about.   His *Urano-
metria Argentina* (1879) enumerates the magnitudes of 8198
out of 10,649 stars visible to the naked eye under those transparent skies ; 73,160 down to 9½ magnitude are embraced in
his "zones ; " besides which, he has brought back with him to
Boston materials for a catalogue including 35,000 entries.
Valuable work of the same kind is being done at Virginia by
Professor O. Stone ; while the present Radcliffe observer's
"Cape Catalogue for 1880" affords an aid to the practical
astronomer south of the line, of which it would be difficult to
over-estimate the importance.   Moreover, the gigantic task
undertaken in 1860 by Dr. C. H. F. Peters, director of the
Litchfield Observatory, Clinton (N.Y.), and of which a large
instalment was finished in 1882, deserves honourable mention.
It is nothing less than to map all stars down to, and even below,
the fourteenth magnitude, situated within 30° degrees on either
side of the ecliptic, and so to afford "a sure basis for drawing
conclusions with respect to the changes going on in the starry
heavens."[1]

In the arduous matter of determining star-distances, too,
progress has been made.   Together, yet independently, Drs.
Gill, and Elkin carried out, at the Cape Observatory in 1882–83,
an investigation of remarkable accuracy into the parallaxes of
nine southern stars.   One of these was the famous α Centauri,
the distance of which from the earth was ascertained to be just
one-third greater than Henderson had made it.   The parallax
of Sirius, on the other hand, was doubled, or its distance
halved ; while Canopus was discovered to be quite immeasur-

---

[1] Gilbert, *Sidereal Messenger*, vol. i. p. 288.

ably remote—a circumstance which, considering that, amongst all the stellar multitude, it is outshone only by the radiant Dog-star, gives a stupendous idea of its real splendour and dimensions.

Dr. Ball, the Astronomer Royal for Ireland, has recently devoted much attention to inquiries of this kind. Besides approximately confirming Struve's parallax of half a second of arc for 61 Cygni, he discovered in 1881 that another very similar double star in the same constellation is situated at a sensibly equal distance from us;[1] and by a sweeping search for (so-called) "large" parallaxes disposed of certain baseless conjectures of comparative nearness to the earth, in the case of red and temporary stars.[2] Amongst other noteworthy results may be mentioned Otto Struve's detection of a parallax of half a second for Aldebaran, and Professor A. Hall's measures of 61 Cygni and Vega with the great Washington refractor, 1880–81.

Foremost among living observers of double stars ranks Mr. S. W. Burnham of Chicago. His discoveries in this line numbered one thousand (including some of the most difficult objects known) in May 1882, when he brought his regular astronomical work to a close.[3] The curious phenomenon of one star revolving round another in a period shorter than that in which Jupiter circulates round the sun, came to his notice in 1883.[4] The very close pair in question, discovered by Otto Struve in 1852, is known as δ Equulei, and the period probably assigned to it is of 10.8 years—by far the shortest attributable to any member of a stellar system.

Another fact of interest in this connection is that 61 Cygni at length gives signs of yielding up its secret. The seemingly parallel tracks followed by its components during a century and a quarter of observation, were found by Struve in 1875 to exhibit deviations countenancing the inference

[1] *Nature,* vol. xxvii. p. 210.   [2] *Ibid.,* vol. xxiv. p. 91.
[3] *Mem. R. A. Soc.,* vol. xlvii. p. 178.
[4] *Observatory,* vol. vii. p. 13.

of mutual revolution; for which, in 1880, a period of about
eleven hundred years was arrived at as a first approximation.[1]
From a fresh discussion three years later, Mr. N. M. Mann
of Rochester (N.Y.) concluded it 1159 years, giving (with a
parallax taken at 0.55″) a value for the combined mass of the
connected bodies only one-seventh the solar mass.[2]

Stellar photometry, initiated by the elder Herschel, has of
late years assumed the importance of a separate department
of astronomical research. More systematically than elsewhere
it has been cultivated at Harvard, under the direction of Professor Pickering. His photometric catalogue of 4260 stars,
constructed from ninety thousand observations of light-intensity
during the years 1879–82, constitutes one more of the precious
seeds of discovery laid in the ground by the present generation
of astronomers, for their successors to reap the fruits of.

Meanwhile, thought cannot be held aloof from the great
subject upon the future illustration of which so much patient
industry is being expended. Nor are partial glimpses denied
to us of relations fully discoverable perhaps only by the slow
efflux of time. Some important points in cosmical economy
have, indeed, become quite clear within the last thirty years,
and scarcely any longer admit of a difference of opinion.
One of these is that of the true status of nebulæ.

This was virtually settled by Sir J. Herschel's description in
1847 of the structure of the Magellanic clouds; but it was
not until Whewell in 1853, and Herbert Spencer in 1858,[3]
enforced the conclusions necessarily to be derived therefrom,
that the conception of the nebulæ as remote galaxies, which
Lord Rosse's resolution of many into stellar points had appeared to support, began to withdraw into the region of discarded and half-forgotten speculations. In the Nubeculæ,
as Whewell insisted,[4] " there co-exists, in a limited compass,

---

[1] *Mém. de l'Ac.*, St. Pétersbourg, t. xxvii. p. 16.
[2] *Sidereal Messenger*, vol. ii. p. 22.
[3] *Essays* (2d ser.), *The Nebular Hypothesis.*
[4] *On the Plurality of Worlds*, p. 214 (2d ed.)

and in indiscriminate position, stars, clusters of stars, nebulæ, regular and irregular, and nebulous streaks and patches. These, then, are different kinds of things in themselves, not merely different to us. There are such things as nebulæ side by side with stars and with clusters of stars. Nebulous matter resolvable occurs close to nebulous matter irresolvable."

This argument from co-existence in nearly the same region of space, was reiterated and reinforced, with others, by Mr. Spencer, and has more lately been urged with his accustomed force and freshness by Mr. Proctor. It is unanswerable. There is no maintaining nebulæ to be simply remote worlds of stars in the face of an agglomeration like the Nubecula Major, containing in its (certainly capacious) bosom *both* stars and nebulæ. Add the evidence of the spectroscope to the effect that a large proportion of these perplexing objects are gaseous, with the facts of their distribution telling of an intimate relation between the mode of their scattering and the lie of the Milky Way, and it becomes impossible to resist the conclusion that both nebular and stellar systems are parts of a single scheme.[1]

As to the stars themselves, the presumption of their approximate uniformity in size and brightness has been effectually dissipated. Differences of distance can no longer be invoked to account for dissimilarity in lustre. Minute orbs, altogether invisible without optical aid, are found to be indefinitely nearer to us than such radiant objects as Capella, Regulus, or Procyon. Moreover, intensity of light is perceived to be a very imperfect index to real magnitude. Brilliant suns are swayed from their courses by the attractive power of massive, yet imperfectly luminous companions, and are suspected of suffering eclipse from obscure interpositions. Besides, effective lustre is now known to depend no less upon the qualities of the investing atmosphere, than upon the extent and radiative power of the stellar surface. Red stars must be far larger in proportion to

---

[1] Proctor, *Month. Not.*, vol. xxix. p. 342.

the light diffused by them than white stars.[1] It is highly
probable that our sun would at least double its brightness
were the absorption suffered by its rays to be reduced to the
Sirian standard ; and, on the other hand, that it would lose half
its present efficiency as a light-source, if the atmosphere partially veiling its splendours were rendered as dense as that
of Aldebaran.

Thus, variety of all kinds is seen to abound in the heavens ;
and it must be admitted that the inevitable abolition of all
hypotheses as to the relative distances of the stars singularly
complicates the question of their allocation in space. Nevertheless, something has been learnt even on that point ; and the
tendency of modern research is, on the whole, strongly confirmatory of the views expressed by Herschel in 1802. He
then no longer regarded the Milky Way as the mere visual
effect of an enormously extended stratum of stars, but as an
actual aggregation, highly irregular in structure, made up of
stellar clouds and groups and nodosities. All the facts since
ascertained fit in with this conception ; and to them Mr.
Proctor has added, what we may almost call the discovery
that the stars forming the galactic stream are not only situated
more closely together, but are also really, as well as apparently,
of smaller dimensions than the lucid orbs studding our skies.
By the laborious process of isographically charting the whole
of Argelander's 324,000 stars, he made it clear, in 1871,[2] that
the brighter stars show, in their distribution, a detailed relationship to the complex branchings of the Milky Way, avoiding, to
a marked extent, its vacuities, and thronging its denser convolutions. It follows that they must be actually intermingled
with them. So that, for every triton sun there are doubtless
swarms of minnows—bodies not perhaps larger than our own
little planet, yet self-luminous and diffusive of beneficent influences according to the inscrutable design of the Creator.

The first step towards the unravelment of the tangled web of

---

[1] This remark is due to the late Mr. J. Birmingham.
[2] *Month. Not.*, vols. xxxi. p. 175 ; xxxii. p. 1.

stellar movements was taken when Herschel established the reality, and indicated the direction of the sun's journey. But the gradual shifting backwards of the whole of the celestial scenery amid which we advance, accounts for only a part of the observed displacements. The stars have motions of their own besides those reflected upon them from ours. All attempts, however, to grasp the general scheme of these motions, have hitherto failed. Yet they have not remained wholly fruitless. The community of slow movement in Taurus, upon which Mädler based his famous theory, has proved to be a fact, and one of very extended significance.

In 1870 Mr. Proctor undertook to chart down the directions and proportionate amounts of about 1600 proper motions, as determined by Messrs Stone and Main, with the result of bringing to light the remarkable phenomenon termed by him "star-drift."[1] Quite unmistakably, large groups of stars, otherwise apparently disconnected, were seen to be in progress together, in the same direction, and at the same rate, across the sky. A striking instance of this kind of unanimity is afforded by the five intermediate stars of the Plough. So clearly were they marked out from their companions in the same asterism, that Mr. Proctor ventured to invite the application of the spectroscope as a sure means of ratifying the distinction. And so indeed it proved. The five associated stars were discerned by Dr. Huggins in 1872[2] to be in rapid retreat from the earth, while the brightest of the Pointers, and the last star in the tail of the Great Bear, verified their surmised independence by displaying, the one a diametrically opposite, the other a widely different rate of motion.

Here then we have a system on a scale so vast that the imagination shrinks from the effort to conceive it. None of the stars forming it have any sensible parallax, so that they certainly surpass our sun many, perhaps thousands of times in dimensions and splendour. Moreover, the distances separating them one from the other must be enormous—to be reckoned

[1] *Proc. Roy. Soc.*, vol. xviii. p. 169.   [2] *Ibid.*, vol. xx. p. 392.

by billions of miles, or years of light-travel. Yet a special tie
unites them; they are subject to the stress of an identical
force, swaying their movements into harmonious accord; and
they doubtless shed one upon the other mutual influences
apart from which their function in the cosmos would be imperfectly fulfilled.

And this is by no means a solitary example. Particular
association, indeed—as was surmised by Michell six-score years
ago—appears to be the rule, rather than an exception in the
sidereal scheme. Stars are bound together by twos, by threes,
by dozens, by hundreds. Our own sun is perhaps not exempt
from this gregarious tendency. Dr. Gould conjectures that it
belongs to a group of about four hundred of the brightest
visible stars, forming a subordinate system within the confines
of the Milky Way.[1] Such another would be the Pleiades. The
laws and revolutions of such majestic communities lie, for the
present, far beyond the range of possible knowledge; centuries
may elapse before even a rudimentary acquaintance with them
begins to develop; while the economy of the higher order of
association, which we must reasonably believe that they unite
to compose, will possibly continue to stimulate and baffle
human curiosity to the end of time.

[1] *Month. Not.*, vol. xl. p. 249.

CHAPTER XIII.

METHODS OF RESEARCH.

COMPARING the methods now available for astronomical inquiries with those in use thirty years ago, we are at once struck
with the fact that they have multiplied. The telescope has
been supplemented by the spectroscope and the photographic
camera. Now this really involves a whole world of change.
It means that astronomy has left the place where she dwelt
apart in rapt union with mathematics, indifferent to all things
on earth save only to those mechanical improvements which
should aid her to penetrate further into the heavens, and has
descended into the forum of human knowledge, at once a
suppliant and a patron, alternately invoking help from, and
promising it to each of the sciences, and patiently waiting upon
the advance of all. The science of the heavenly bodies has,
in a word, become a branch of terrestrial physics, or rather a
higher kind of integration of all their results. It has, however,
this leading peculiarity, that the materials for the whole of its
inquiries are telescopically furnished. They are such as the
unarmed eye takes no, or a very imperfect cognisance of.

Spectroscopic and photographic apparatus are simply additions to the telescope. They do not supersede, or render it
of less importance. On the contrary, the efficacy of their action
depends primarily upon the optical qualities of the instrument
they are attached to. Hence the development, to their fullest
extent, of the powers of the telescope is of vital moment to
the progress of modern physical astronomy, while the older

mathematical astronomy could afford to remain comparatively indifferent to it.

The colossal Rosse reflector still marks, as to size, the *ne plus ultra* of performance in that line. No existing mirror comes nearer to it than that, four feet in diameter, sent out to Melbourne by the late Thomas Grubb of Dublin in 1870. This is mounted in the Cassegrainian manner; so that the observer looks straight through it towards the object viewed, of which he really sees a twice-reflected image. It is of excellent definition and rare convenience in management; but the dust-laden atmosphere of Melbourne is said to impede very seriously its usefulness.

It may be doubted whether so large a speculum will ever again be constructed. A new material for the mirrors of reflecting telescopes was introduced by Lèon Foucault in 1857,[1] which has already in a great measure superseded the use of a metallic alloy. This is glass upon which a thin film of silver has been deposited by a process known as Drayton's. It gives a peculiarly brilliant reflective surface, throwing back more light than a metallic mirror of the same area, in the proportion of about sixteen to nine. Liability to tarnish in part counteracts this great advantage. The largest instrument successfully turned out on this plan is Mr. Common's 36-inch reflector, finished in 1879. To its excellent qualities his triumphs in celestial photography are largely due.

It is, however, in the construction of refracting telescopes that the most conspicuous advances have recently been made. The Harvard College 15-inch achromatic was mounted and ready for work in June 1847. A similar instrument had already for some years been in its place at Pulkowa. It was long before the possibility of surpassing these masterpieces of German skill presented itself to any optician. For fifteen years it seemed as if a line had been drawn just there. It was first transgressed in America. A portrait-painter of Cambridgeport, Massachusetts, named Alvan Clark, had for some time

---

[1] *Comptes Rendus*, t. xliv. p. 339.

amused his leisure with grinding lenses, the singular excellence of which was discovered in England by Mr. Dawes in 1853.[1] Seven years passed, and then an order came from the University of Mississippi for an object-glass of the unexampled size of eighteen inches. An experimental glance through it to test its definition resulted, as we have seen, in the detection of the companion of Sirius, January 31, 1862. It never reached its destination in the South. War troubles supervened; and it was eventually sent to Chicago, where it has served Professor Hough in his investigations of Jupiter, and Mr. Burnham in his scrutiny of double stars.

The next step was an even longer one, and it was again taken by a self-taught optician, Thomas Cooke, the son of a shoemaker at Allerthorpe, in the East Riding of Yorkshire. Mr. Newall of Gateshead ordered from him in 1863 a 25-inch object-glass. It was finished early in 1868, but at the cost of shortening the life of its maker, who died October 19, 1869, before the giant refractor he had toiled at for five years, was completely mounted. Although believed to be still the finest telescope in England, its high qualities have been largely neutralised by an unfavourable situation.

Close upon its construction followed that of the Washington 26-inch, for which twenty thousand dollars were paid to Alvan Clark. Set to work in 1873, the most illustrious point in its career, so far, has been the discovery of the satellites of Mars. Once known to be there, these were, indeed, found to be perceptible with very moderate optical means (Mr. Wentworth Erck saw Deimos with a nine-inch Clark); but the first detection of such minute objects is a feat of a very different order from their subsequent observation.

For a little over eight years the Washington refractor held the primacy. It had to yield the place of honour in December 1880 to a giant achromatic, twenty-seven inches in aperture, built by Howard Grubb (son and successor of Thomas Grubb) for the Vienna Observatory. This, in its turn, has been sur-

[1] Newcomb, *Pop. Astr.*, p. 137.

passed by one of thirty inches sent by Alvan Clark to Pulkowa ; and an object-glass, fully *three feet* in diameter, is now in course of construction by the same firm for the Lick Observatory in California.   The difficulties, however, encountered in procuring discs of glass of the size and purity required for this last venture, seem to indicate that a term to progress in this direction is near at hand.   The flint was indeed cast with comparative ease in the workshops of M. Feil at Paris.   The flawless mass weighed 170 kilogrammes, was over 38 inches across, and cost 2000 pounds.   But with the crown part of the designed achromatic combination, things have gone less smoothly.   The production of a successful disc was preceded by *nineteen* failures, involving a delay of more than two years, and postponing the probable completion of the great telescope until the year 1887 or 1888.[1]

Nor is the difficulty in obtaining suitable material—almost overwhelming though it be—the only obstacle to increasing the size of refractors.   Colour-fringes also step in and bar the way, their complete, or approximately complete, correction demanding, in the case of such vast apertures as have recently been attempted, a focal length so exorbitant as to be practically, under the ordinary conditions of mounting, out of the question.   Besides, a refracting telescope loses one of its chief advantages over a reflector when its size is increased beyond a certain limit.   That advantage is the greater luminosity of the images given by it.   Considerably more light is transmitted through a glass lens than is reflected from an equal metallic surface.   But only so long as both are of moderate dimensions.   For the glass necessarily grows in thickness as its area augments, and consequently stops a larger percentage of the rays it refracts.   So that a point at length arrives—fixed by the late Dr. Robinson at a diameter a little short of three feet [2]—where the glass and the metal are, in this respect, on an equality ; while above it, the metal has the

[1] Holden, *Observatory*, vol. viii. p. 84.
[2] H. Grubb, *Trans. Roy. Dub. Soc.*, vol. i. (new ser.), p. 2.

advantage. And since silvered glass gives back considerably more light than speculum-metal, the stage of equalisation with lenses is reached proportionately sooner where this material is employed.

It will thus probably be long before the light-grasp of Mr. Common's three-foot mirror is surpassed by a refractor. But in the inquiries for which the great telescopes of modern times are more especially designed, light-grasp is everything. For the spectroscopic examination of stars, for the measurement of their motions in the line of sight, for the study of nebulæ, for stellar and nebular photography, the cry continually is, "More light." Apart from the exigencies of these, and a few other enticing branches of research, there would be little to be gained in adding to the power of optical apparatus. And there is much lost. The penalties of bigness are heavy. Perfect definition becomes, with increasing size, more and more difficult to attain; once attained, it becomes more and more difficult to keep. For the huge masses of material employed to form great object-glasses or specula, tend, with every movement, to become deformed by their own weight. Gravity exacts the further toll of unwieldiness. Each glance through a large instrument is highly paid for in time and trouble. Nor is the glance thus paid for often a satisfactory one. Atmospheric troubles intervene.

These are the worst plagues of all those that afflict the astronomer. No mechanical skill avails to neutralise or alleviate them. They augment, in a rapidly increasing ratio, with each addition to the aperture of the telescope, or of the magnifying powers applied to it. To them chiefly is due the growing discontent with the performance of the colossal instruments of modern times. It is admitted on all hands that, for the ordinary work of an observatory, an aperture of ten or twelve inches is the outside limit of usefulness. But it is also found, with disappointment, that even in the field of descriptive research, where it might be expected that luminosity and magnification would be all-important, results fall far short of anticipation.

Schiaparelli, with an eight-inch achromatic, obtained views of Mars such as were never vouchsafed to Harkness or Hall, though using the Washington 26-inch; and, according to Mr. Denning,[1] details of the Jovian surface are shown by an insignificant 4½-inch, which remain invisible with the majestic refractor of the Dearborn Observatory, Chicago.

Now this is due to no imperfections inherent in the instruments themselves; it is due to the conditions of our habitation on an air-wrapt globe. It is not only that much less than half the light incident upon the surface of the atmospheric ocean penetrates to the bottom of it. That loss might, in some measure, be repaired; but what no optical contrivance can get rid of, is the disturbance suffered by the rays that reach us. The twinkling of stars to the naked eye is but a faint symptom of their behaviour in the telescope; while the images of sun, moon, and planets "boil" at the edges, or are suffused and distorted by waves of agitation caused by the magnified surgings of the turbulent vapours we see through. The mischief, Dr. Robinson estimated in the case of reflectors, grows with the *cubes* of their diameters; and it is commonly found, in practice, that the "seeing" will be perfectly good with a small telescope, but altogether intolerable with a large one standing beside it. Under such skies as ours, in fact, there are not more than three or four nights in the year when an aperture of as much as eighteen inches can be used to real advantage; and Mr. Newall remarked in 1885 that during fifteen years he had known but *one* fine night [2]—fine, that is, in the sense of availability for observation with his great refractor.

Thus it seems clear that we have reached a turning-point in the history of telescopic improvement. Not alone have the material obstacles to any further increase of size become all but insuperable, but the conviction is forced upon us that, were instruments of greater power than any now possessed by astronomers actually in their hands, they must remain wholly useless save on one condition—that of an improved climate.

[1] *Observatory*, vol. viii. p. 79. [2] *Ibid.*, p. 80.

Ever since the Parsonstown telescope was built, it has been obvious that the limit of profitable augmentation of aperture had been reached, if not overpassed; and Lord Rosse himself was foremost to discern the need of pausing to look round the world for a clearer and stiller air than was to be found within the bounds of the United Kingdom.   With this express object Mr. Lassell transported his two-foot Newtonian to Malta in 1852, and mounted there, in 1860, a similar instrument of four-fold capacity, with which in the course of about two years 600 new nebulæ were discovered.   Professor Piazzi Smyth's experiences during a trip to the Peak of Teneriffe in 1856 in search of astronomical opportunities,[1] gave countenance to the most sanguine hopes of deliverance, at suitably elevated stations, from some of the oppressive conditions of low-level star-gazing; yet for a number of years nothing effectual was done for their realisation.   Now at last, however, mountain observatories are not only an admitted necessity, but an accomplished fact; and Newton's long forecast of a time when astronomers would be compelled, by the developed powers of their telescopes, to mount high above the "grosser clouds" in order to use them,[2] has been justified by the event.

Mr. James Lick, a millionaire of San Francisco, had already chosen when he died, October 1, 1876, a site for the new observatory, to the building and endowment of which he had devoted a part of his large fortune.   The establishment now only awaits the completion of the 36-inch refractor and its great sheltering dome, to be in a state of perfect efficiency. Indeed, its present instrumental outfit—including a twelve-inch Clark's achromatic—is one of high excellence.   The situation of the "Lick" Observatory is exceptional and splendid.   Planted on one of the three peaks of Mount Hamilton, a crowning summit of the Californian Coast Range, at an elevation of 4200 feet above the sea, in a climate scarce rivalled throughout the world, it commands views both celestial and terrestrial which the lover of nature and astronomy may

[1] *Phil. Trans.*, vol. cxlviii. p. 465.   [2] *Optice*, p. 107 (2d ed., 1719.)

alike rejoice in. Impediments to observation are there found to be most materially reduced. Professor Holden tells us that during six or seven months of the year an unbroken serenity prevails, and that half the remaining nights are clear.[1] . The power of continuous work thus afforded is of itself an inestimable advantage; and when combined with the high visual excellences testified to by Mr. Burnham's discovery, during a two months' trip to Mount Hamilton in the autumn of 1879, of forty-two new double stars with a six-inch achromatic, it gives hopes of a brilliant future for the Lick establishment.

A higher altitude than the comparatively modest one at which it is placed, would hardly prove suitable to a great permanent observatory; but considerably more elevated posts for temporary astronomical occupation are being provided, and will shortly be looked upon as indispensable. One such was fitted up near the summit of Mount Etna in 1882. The building is the highest in Europe, standing 9655 feet above the sea, and includes within its walls the "Casa Inglese," in which travellers were used to seek repose before attempting the final ascent of the cone. Splendid telescopic opportunities are indicated by Professor Langley's experimental observations, carried through under every disadvantage in the winter of 1879–80; and the Merz equatoreal of nearly fourteen inches aperture, provided for the Etnean establishment, may be expected, freed from the impeding mists and restless currents of the lower atmosphere, to prove of singular efficiency.

The Pic du Midi, too, is destined for astronomical occupation. A meteorological observatory was in 1881, thanks to the enterprise of General de Nansouty and M. Vaussenat, opened on its summit, at an altitude of 9600 feet; and the glowing account given by MM. Thollon and Trépied in 1883 [2] of the advantages offered by the dark translucency of its sky, determined Admiral Mouchez upon founding there a species of *succursale* to the Paris Observatory, whither despondent astronomers might repair within a few hours, in the sure hope

---

[1] *Observatory*, vol. viii. p. 85.   [2] *Comptes Rendus*, t. xcvii. p. 834.

of leaving their too-familiar weather-troubles behind, and of finding the heavens laid bare of all but the clearest and thinnest remnant of their atmospheric vesture. An eight-inch equatoreal has been appropriated to use on the Pic, but funds are not as yet forthcoming for the erection of a dome.

The diminution of "glare" at such elevated posts is all-important for solar inquiries; and if Dr. Huggins's ingenious devices for photographing the corona are not to remain a mere curiosity of science, but are to be turned to practical account for the increase of knowledge, it can only be by experiments liberated from the obliterating effects of confused reflections in dense air. For stellar and nebular photography, on the other hand, luminous and untroubled images are the chief requisite, and these can generally be secured by a judicious ascent. Indeed a store of materials may be collected during a few weeks' sojourn at a high altitude, for the due discussion and elucidation of which the whole year besides will hardly afford leisure. In the spectroscopy of the stars, Dr. Copeland's flying observations amongst the Andes show what can be done by climbing towards them. Peculiarities previously invisible become obvious; measurement is rendered easy; discoveries of curious interest crowd upon the enterprising observer. It may indeed be safely predicted that knowledge of the spectra of faint stars will never be made extensive and precise until ample means are available for studying them in the finer air of the mountains.

Vapours and air-currents, however, do not alone embarrass the use of giant telescopes. Mechanical difficulties also threaten to oppose an insuperable barrier to any further growth in size. But what seems to be an insuperable barrier often proves to be only a fresh starting-point; and signs are not wanting that it may be found so in this case. It is possible that the monumental domes and huge movable tubes of our present observatories will, in a few decades, be as much things of the past as Huygens's "aerial" telescopes. It is certain that the thin edge of the wedge of innovation has been driven into the old plan of equatoreal mounting.

M. Loewy, the present sub-director of the Paris Observatory, proposed to Delaunay in 1871 the erection of a telescope on a novel system. The design seemed feasible, and was adopted; but the death of Delaunay and the other untoward circumstances of the time interrupted its execution. Its resumption, after some years, was rendered possible by M. Bischoffsheim's gift of 25,000 francs for expenses, and the *Coudé* or "bent" equatoreal has been, since 1882, one of the leading instruments at the Paris establishment.

Its principle is briefly this. The telescope is, as it were, its own polar axis. The anterior part of the tube is supported at both ends, and is thus fixed in a direction pointing towards the pole, with only the power of twisting axially. The posterior section is joined on to it at right angles, and presents the object-glass accordingly to the celestial equator, in the plane of which it revolves. Stars in any other part of the heavens have their beams reflected upon the object-glass by means of a plane rotating mirror placed in front of it. The observer, meanwhile, is looking steadfastly down the bent tube towards the invisible *southern* pole. He would naturally see nothing whatever, were it not that a second plane mirror is fixed at the "elbow" of the instrument, so as to send the rays which have traversed the object-glass to his eye. He never needs to move from his place. He watches the stars seated in an arm-chair in a warm room, with as perfect convenience as if he were examining the seeds of a fungus with a microscope. Nor is this a mere gain of personal ease. The abolition of hardship includes a vast accession of power.[1]

Amongst other advantages of this method of construction are, first, that of added stability, the motion given to the ordinary equatoreal being transferred, in part, to an auxiliary mirror. Next, that of increased focal length. The fixed part of the tube can be made almost indefinitely long without in-

[1] Loewy, *Bull. Astr.*, t. i. p. 286; *Nature*, vol. xxix. p. 36.

convenience, and with enormous advantage to the optical qualities of a large instrument. Finally, the costly and unmanageable cupola is got rid of, a mere shed serving all purposes of protection required for the "Coudé."

The desirability of some such change as that which M. Loewy has realised, has been felt by others. Professor Pickering sketched in 1881 a plan for fixing large refractors in a permanently horizontal position, and reflecting into them, by means of a shifting mirror, the objects desired to be observed.[1] An instrument with "siderostatic" mounting by Mr. Howard Grubb has actually been in use at the Queen's College Observatory, Cork, since 1882; and in a paper read before the Royal Society, January 21, 1884, he proposed to carry out the principle on a more extended scale.[2] The chief honours, however, remain to the Paris inventor. None of the prognosticated causes of failure have proved effective. The loss of light from the double reflection is insignificant. The menaced deformation of images is, through the exquisite skill of the MM. Henry in producing plane mirrors of all but absolute perfection, quite imperceptible. The definition of the novel $10\frac{1}{2}$-inch equatoreal is admitted to be singularly good. Dr. Gill states that he had never measured a double star so easily as he did γ Leonis by its means.[3] Mr. Lockyer believes it to be "one of the instruments of the future;" and the principle of its construction has already been adopted by the directors of the Besançon and Algiers Observatories. At elevated stations especially, the abolition of the hitherto indispensable massive dome, obnoxious to all the winds of heaven, which there blow at times with exceeding violence, ought to be decisive in its favour; while its adaptation to reflectors [4] may be expected to turn the scale in favour of silvered glass mirrors as the great coming engines of physical research in astronomy.

Celestial photography is but forty years old; yet its earliest beginnings already seem centuries behind its present perfor-

---

[1] *Nature*, vol. xxiv. p. 389.
   Observatory, vol. vii. p. 167.
[2] *Ibid.*, vol. xxix. p. 470.
[4] Loewy, *Bull. Astr.*, t. i. p. 265.

mances. The details of its gradual, yet rapid improvement are of too technical a nature to find a place in these pages. Suffice it to say that the "dry-plate" process, with which such wonderful results have been obtained, appears to have been first made available by Dr. Huggins in photographing the spectrum of Vega in 1876, and was then successively adopted by Common, Draper, and Janssen. Nor should Captain Abney's remarkable extension of the powers of the camera be left unnoticed. He began his experiments on the chemical action of red and infra-red rays in 1874, and at length succeeded in obtaining a substance—the "blue" bromide of silver— highly sensitive to these slower vibrations of light. With its aid he explored a vast, unknown, and for ever invisible region of the solar spectrum, presenting to the Royal Society, December 5, 1879,[1] a detailed map of its infra-red portion (wave-lengths 7600 to 10,750), from which valuable inferences may yet be derived as to the condition of the various kinds of matter ignited in the solar atmosphere.

The chemical plate has two advantages over the human retina.[2] First, it is sensitive to rays which are utterly powerless to produce any visual effect; next, it can accumulate impressions almost indefinitely, while from the retina they fade after one-tenth part of a second, leaving it a continually renewed *tabula rasa.*

It is accordingly quite possible to photog aph objects so faint as to be altogether beyond the power of any telescope to reveal; and we may thus eventually learn whether a blank space in the sky truly represents the end of the stellar universe in that direction, or whether farther and farther worlds roll and shine beyond, veiled in the obscurity of immeasurable distance.

The means at the disposal of astronomers have not multi-plied faster than the tasks imposed upon them. Looking back to the year 1800, we cannot fail to be astonished at the change. The comparatively simple and serene science of the heavenly

[1] *Phil. Trans.*, vol. clxxi. p. 653.
[2] Janssen, *L'Astronomie*, t. ii. p. 121.

452 *HISTORY OF ASTRONOMY.*

bodies known to our predecessors, almost perfect so far as it went, incurious of what lay beyond its grasp, has developed into a body of manifold powers and parts, each with its separate mode and means of growth, full of strong vitality, but animated by a restless and unsatisfied spirit, haunted by the sense of problems unsolved, and tormented by conscious impotence to sound the immensities it perpetually confronts.

Knowledge might then be said to be bounded by the solar system ; but even the solar system presented itself under an aspect strangely different from that it now wears. It consisted of the sun, seven planets, and twice as many satellites, all circling harmoniously in obedience to an universal law, by the compensating action of which the indefinite stability of their mutual relations was secured. The occasional incursion of a comet, or the periodical presence of a single such wanderer chained down from escape to outer space by planetary attraction, availed nothing to impair the symmetry of the majestic spectacle.

Now, not alone the ascertained limits of the system have been widened by a thousand millions of miles, with the addition of one more giant planet and six satellites to the ancient classes of its members, but a complexity has been given to its constitution baffling description or thought. Two hundred and fifty circulating planetary bodies bridge the gap between Jupiter and Mars, the complete investigation of the movements of any one of which would overtask the energies of a lifetime. Meteorites, strangers apparently to the fundamental ordering of the solar household, swarm, nevertheless, by millions in every cranny of its space, returning at regular intervals like the comets so singularly associated with them, or sweeping across it with hyperbolic velocities, brought perhaps from some distant star. And each of these cosmical grains of dust has a theory far more complex than that of Jupiter ; it bears within it the secret of its origin, and fulfils a function in the universe. The sun itself is no longer a semi-fabulous, fire-girt globe, but the vast scene of the play of forces as yet imperfectly known to us, offering a boundless field for the most arduous and inspiring researches. Amongst the planets, the widest variety in physical habitudes is seen to prevail, and each is recognised as a world apart, inviting inquiries which, to be effective, must necessarily be special and detailed. Even our own moon threatens to break loose from the trammels of calculation, and commits " errors " which sap the very foundations of the lunar theory, and suggest the formidable necessity for its revision. Nay, the steadfast earth has forfeited the implicit confidence placed in it as a time-keeper, and questions relating to the stability of the earth's axis, and the constancy of the earth's rate of rotation, are amongst those which it behoves the future to answer. Everywhere there is multiformity and change, stimulating a curiosity which the rapid development of methods of research offers the possibility of at least partially gratifying.

Outside the solar system, the problems which demand a practical solution are all but infinite in number and extent. And these have all arisen and crowded upon our thoughts within less than a hundred years. For sidereal science became a recognised branch of astronomy only through Herschel's discovery of the revolutions of double stars in 1802. Yet already it may be, and has been called, "the astronomy of the future." So rapidly has the development of a keen and universal interest attended and stimulated the growth of power to investigate this sublime subject. What has been done is little—is scarcely a beginning; yet it is much in comparison with the total blank of a century past. And our knowledge will, we are easily persuaded, appear in turn the merest ignorance to those who come after us. Yet it is not to be despised, since by it we reach up groping fingers to touch the hem of the garment of the Most High.

INDEX.

462INDEX.

THE END.

Printed in the United States
By Bookmasters